AF389372

Selected Papers from the 8th International Conference of Biotechnologies, Present and Perspectives

Selected Papers from the 8th International Conference of Biotechnologies, Present and Perspectives

Editors

Mircea Oroian
Georgiana Gabriela Codină

MDPI • Basel • Beijing • Wuhan • Barcelona • Belgrade • Manchester • Tokyo • Cluj • Tianjin

Editors

Mircea Oroian
Ștefan cel Mare University
Romania

Georgiana Gabriela Codină
Ștefan cel Mare University
Romania

Editorial Office
MDPI
St. Alban-Anlage 66
4052 Basel, Switzerland

This is a reprint of articles from the Special Issue published online in the open access journal *Applied Sciences* (ISSN 2076-3417) (available at: https://www.mdpi.com/journal/applsci/special_issues/8th_Conference_Biotechnologies).

For citation purposes, cite each article independently as indicated on the article page online and as indicated below:

LastName, A.A.; LastName, B.B.; LastName, C.C. Article Title. *Journal Name* **Year**, *Volume Number*, Page Range.

ISBN 978-3-0365-4855-5 (Hbk)
ISBN 978-3-0365-4856-2 (PDF)

Contents

About the Editors

Mircea Oroian

Mircea Oroian (Prof. PhD Eng) joined the Stefan cel Mare University of Suceava, Faculty of Food Engineering, in 2009 where he teaches different courses focused on food rheology, food authentication and adulteration detection, as well as the development of techniques for food characterization. He holds a PhD in Material Science and he defended the habilitation thesis in 2015 in the field of food engineering. He has published more than 70 articles in ISI-quoted journals and has a h-index score of 25, according to Google Scholar.

Georgiana Gabriela Codină

Georgiana Gabriela Codină (Professor Habilitate, PhD Eng.) joined the Ștefan cel Mare University, Faculty of Food Engineering (Suceava, Romania), in 2005 where she teaches different courses in the fields of food science and technology. As of 2009, she holds a PhD in Industrial Engineering, and she became a PhD supervisor in food engineering in 2017 when she sustained her habilitation thesis. Dr. Codină has expertise in food rheology; food sensory analysis; bread making; the beer industry; food quality analysis through the use of different rheological, textural, and sensory methods; and the design of experiments and data analysis. Her research activities focus on improving the technology of food products and the quality of different foods, such as baked goods, beer, dairy products, etc. She was involved in more than 10 interdisciplinary research projects, has published more than 130 scientific papers, and is the author of more than 35 patents under evaluation from which 5 were published.

Editorial

Current Research in Food Safety and Biotechnology

Mircea Oroian and Georgiana Gabriela Codină *

Ștefan cel Mare, Faculty of Food Engineering, University of Suceava, 13 Universitatii Street, 720229 Suceava, Romania; m.oroian@fia.usv.ro
* Correspondence: codina@fia.usv.ro

Citation: Oroian, M.; Codină, G.G. Current Research in Food Safety and Biotechnology. *Appl. Sci.* **2022**, *12*, 6640. https://doi.org/10.3390/app12136640

Received: 15 June 2022
Accepted: 29 June 2022
Published: 30 June 2022

Publisher's Note: MDPI stays neutral with regard to jurisdictional claims in published maps and institutional affiliations.

Biotechnology is a field at the interface of biology and industry, being part of the applied sciences field. The term biotechnology encompasses quite a large field and includes the processing of raw materials with the obtainment of food products via enzymatic processes or with the help of microorganisms. Different food products, such as milk, beer, alcohol, starch, meat, fruit juices, sugar, bakery products, etc., may be obtained using different enzymatic processes or with the help of microorganisms. This Special Issue of the 8th International Conference of Biotechnologies, Present and Perspectives, organized by Ștefan cel Mare University, Faculty of Food Engineering, Suceava, Romania, presents the latest developments in the food industry field, with a focus on the biotechnological processes that take place in various branches of the food industry, which have major implications in determining the overall quality of food products.

Today, the role of the food industry is to process different raw materials of vegetable or animal provenience or by-products at a higher level and to increase their value through processing. The food industry ensures not only the preservation of raw materials but also increases the nutritional value of food products. The development of biotechnologies has led to the production of food products for which their maximum nutritional values can be ensured. The current orientation of the food industry is focused on minimum food processing, which implies simplified, economical processing that results in maintaining a nutritional value of food products that is as high as possible. The article written by Ghinea et al. [1] underlined the possibility of obtaining apple chips from three different low-cost apple cultivars through a drying process with good physico-chemical and sensory characteristics. This will lead to the possibility of their preservation and consumption for a long period of time. Taking into account the fact that apple fruits present high nutritional value, being rich in phenolic compounds, dietary fibers, vitamins, etc., their transformation in value-added products such as apple chips may help us to obtain health benefits from their consumption. The possibility of using a heating treatment in order to improve the nutritional value of two sorghum varieties (red and white) with different particle sizes was reported by Batariuc et al. [2]. According to them, a dry heat treatment applied to sorghum flour will improve its nutrition value, which may be helpful for food producers who want to use sorghum flour as a raw material at the industrial level. The application of a minimum food processing procedure with low costs may lead to improvements in the nutritional value of various raw materials which may be used to obtain different food products. According to Coțovanu and Mironeasa [3], the use of different particle sizes of amaranth flour obtained through milling and sieving may lead to different nutritional values of amaranth flour. This may be an important criterion when obtaining food products with amaranth addition with high nutritional values. From the technological perspective, the addition of amaranth with different particle sizes in bread making leads to optimal results; a 9.41% addition level of amaranth enables large changes to nutritional values, a level of 9.39% enables medium changes, and a level of 7.89% enables small changes. Due to amaranth's high nutritional value, amaranth seed compounds have been intensively studied in recent decades. In our Special Issue, amaranth seeds were also studied by Procopet and Oroian [4], who concluded that they are a valuable source of fatty acids, from which 16.54% were saturated acids and 83.45% were unsaturated acids. Additionally, amaranth seeds are rich in essential amino acids such as isoleucine, histidine, leucine, methionine, lysine, phenylalanine, tryptophan, threonine, and valine and polyphenols such as vanillic acid, rosmarinic acid, chlorogenic acid, p-coumaric acid, and caffeic acid.

The increases in the nutritional value of the food products obtained using different biotechnological processes through the addition of active components had been intensively studied in recent years. In this Special Issue, the effects of the use of hemp seed oil on dough and bread quality were analyzed by Ropciuc et al. [5]. Hemp seed oil contains high amounts of essential fatty acids and is a valuable ingredient due to its omega-6 and omega-3 fatty acid ratio of 3:1. It is used in bread making, and a 5–10% addition in wheat flour will lead to good bread quality from the textural and sensorial perspectives. Another improvement by using an ingredient that may increase the nutritional value of food products in which the main raw material is wheat flour, such as pasta, was reported by Ungureanu-Iuga and Mironeasa [6]. According to them, the use of grape peels, which are rich in polyphenols and fibers, in levels of up to 5% in wheat flour will lead to improvements in quality characteristics from the technological and nutritional perspectives. As an ingredient in bakery products, the analysis of nine samples of margarine was also carried out due to its nutritional and technological characteristics. The data reported by Păduret [7] showed that bakery margarine is a plastic material with a high fracturability, and from the nutritional perspective, palmitic acid is the dominant fatty acid within it.

The use of fibers (psyllium and pectin) and plant-based milks from almonds and hemps in order to obtain vegetable milk ice cream with no additive additions was discussed in this Special Issue by Leahu et al. [8]. They concluded that it is possible to obtain a vegetable ice cream with high nutritional value as an alternative to milk-based ice cream, which may be marketed to consumers who follow a plant-based diet or have an intolerance to milk products.

Trans-resveratrol accumulation in pruned vine shoots was also investigated by Crăciun and Gutt [9] in order to obtain this supplement for pharmacological and medical purposes. Additionally, increased amounts of macronutrients and micronutrients in different herbal raw materials such as summer savory, marjoram, and lemon balm through the use of biostimulants were reported in this Special Issue by Majkowska-Gadomska et al. [10].

Food safety is the most important criteria that food must meet. Food products must be useful to the human body and must not become a source of life-threatening illness. The consumption of crops contaminated with a certain level of mycotoxins that has been established by European Union legislation as harmful can lead to toxic effects. An overview analysis of the presence of mycotoxins such as deoxynivalenol, fumonisins, and zearalenone in maize and maize-derived products from eastern Romania was presented by Mihalcea and Amariei [11]. According to their study, the mycotoxins analyzed were within legal limits in the products they examined, which indicates that these products are safe for food production. Additionally, another aspect of food safety that was discussed in this Special Issue concerned water quality in bottled drinking water for babies. For this purpose, Ungureanu et al. [12] used 19 samples of bottled baby water to analyze 12 potentially toxic elements (Ba, Cu, Co, Zn, Ni, Mn, Li, Fe, Cd, Pb, Cr, and Sb) using ICP-MS. According to the data obtained, the levels of all of the elements analyzed, except iron, adhered to the legislation in force. Therefore, all of the samples analyzed can be safely consumed by children and babies.

The works published in this Special Issue presented various aspects of research relating to the main current trends regarding the obtainment of safe food with high preservation and nutritional values by improving the biotechnologies applied in the food industry. Thus, based on their own research and data from the literature, the authors presented food safety aspects and different issues regarding the use or obtainment of raw materials that lead to the creation of products with high nutritional value. Biotechnological aspects of obtaining food products such as bakery products, pasta, and ice cream were also described.

By presenting this complex theme and high-level approaches, the works from this Special Issue of the 8th International Conference of Biotechnologies, Present and Perspectives organized by Ştefan cel Mare University, Faculty of Food Engineering, Suceava, Romania, will be of use to those interested in the future trends regarding biotechnologies in the food industry, applied engineering sciences, food product quality, food safety, environment protection, and equipment in the food industry.

Funding: This work was funded by the Ministry of Research, Innovation and Digitalization within Program 1—Development of national research and development system, Subprogram 1.2—Institutional Performance—RDI excellence funding projects, under contract no. 10PFE/2021.

Conflicts of Interest: The authors declare no conflict of interest.

References

1. Ghinea, C.; Prisacaru, A.E.; Leahu, A. Physico-Chemical and Sensory Quality of Oven-Dried and Dehydrator-Dried Apples of the Starkrimson, Golden Delicious and Florina Cultivars. *Appl. Sci.* **2022**, *12*, 2350. [CrossRef]
2. Batariuc, A.; Ungureanu-Iuga, M.; Mironeasa, S. Effects of Dry Heat Treatment and Milling on Sorghum Chemical Composition, Functional and Molecular Characteristics. *Appl. Sci.* **2021**, *11*, 11881. [CrossRef]
3. Coţovanu, I.; Mironeasa, S. Features of Bread Made from Different Amaranth Flour Fractions Partially Substituting Wheat Flour. *Appl. Sci.* **2022**, *12*, 897. [CrossRef]
4. Procopet, O.; Oroian, M. Amaranth Seed Polyphenol, Fatty Acid and Amino Acid Profile. *Appl. Sci.* **2022**, *12*, 2181. [CrossRef]
5. Ropciuc, S.; Apostol, L.C.; Damian, C.; Prisacaru, A.E. Effect of Hemp Seed Oil Addition on the Rheological Properties of Dough and Bread. *Appl. Sci.* **2022**, *12*, 2764. [CrossRef]
6. Ungureanu-Iuga, M.; Mironeasa, S. Advance on the Capitalization of Grape Peels By-Product in Common Wheat Pasta. *Appl. Sci.* **2021**, *11*, 11129. [CrossRef]
7. Pǎdureţ, S. The Quantification of Fatty Acids, Color, and Textural Properties of Locally Produced Bakery Margarine. *Appl. Sci.* **2022**, *12*, 1731. [CrossRef]
8. Leahu, A.; Ropciuc, S.; Ghinea, C. Plant-Based Milks: Alternatives to the Manufacture and Characterization of Ice Cream. *Appl. Sci.* **2022**, *12*, 1754. [CrossRef]
9. Crăciun, A.L.; Gutt, G. Study on Kinetics of Trans-Resveratrol, Total Phenolic Content, and Antioxidant Activity Increase in Vine Waste during Post-Pruning Storage. *Appl. Sci.* **2022**, *12*, 1450. [CrossRef]
10. Majkowska-Gadomska, J.; Jadwisieńczak, K.; Francke, A.; Kaliniewicz, Z. Effect of Biostimulants on the Yield and Quality of Selected Herbs. *Appl. Sci.* **2022**, *12*, 1500. [CrossRef]
11. Mihalcea, A.; Amariei, S. Study on Contamination with Some Mycotoxins in Maize and Maize-Derived Foods. *Appl. Sci.* **2022**, *12*, 2579. [CrossRef]
12. Ungureanu, E.L.; Mustatea, G.; Popa, M.E. Assessment of Potentially Toxic Elements and Associated Health Risk in Bottled Drinking Water for Babies. *Appl. Sci.* **2022**, *12*, 1914. [CrossRef]

Article

Physico-Chemical and Sensory Quality of Oven-Dried and Dehydrator-Dried Apples of the Starkrimson, Golden Delicious and Florina Cultivars

Cristina Ghinea *, Ancuta Elena Prisacaru and Ana Leahu

Faculty of Food Engineering, Stefan cel Mare University of Suceava, 720229 Suceava, Romania;
ancuta.prisacaru@fia.usv.ro (A.E.P.); analeahu@fia.usv.ro (A.L.)
* Correspondence: cristina.ghinea@fia.usv.ro

Abstract: Apple fruits are high in phenolic compounds, sugar and dietary fiber content and are rich in malic acid and vitamins, with a significant impact on the organoleptic quality and its health-promoting properties. They can be turned out in value-added product such as apple chips due to the low cost of raw material. The aim of the study was to obtain apple chips, fat-free, healthy, traditionally dried and without added sugar, which can be easily obtained and capitalized economically, as well as the evaluation of their physico-chemical and sensory qualities. The apple chips were produced from three apple cultivars ('Starkrimson', 'Golden Delicious' and 'Florina') by drying the apple fruits in an oven and a dehydrator at 65 °C. To inactivate the browning enzymes, the apple slices were immersed in a solution of lemon salt (4%) for 7 min before drying. Apple chips were sensory-evaluated and relevant parameters were analyzed at defined intervals during storage at room temperature up to 21 days. The water activity (a_w) values of apple chip samples dried in the oven ranged from 0.544 to 0.650, while for the samples dried in the dehydrator, a_w values were between 0.374 and 0.426. During the storage, the pH of apple chips varied very little, while titratable acidity increased for all samples. Compared with fresh apple slices, it was observed that the total soluble solids (TSS) content of all dried apple chip samples decreased. Color parameters and browning and whitening indexes differed depending on the apple cultivars and dryer type used.

Keywords: apple chips; drying; physico-chemical characteristics; sensory analysis

Citation: Ghinea, C.; Prisacaru, A.E.; Leahu, A. Physico-Chemical and Sensory Quality of Oven-Dried and Dehydrator-Dried Apples of the Starkrimson, Golden Delicious and Florina Cultivars. *Appl. Sci.* **2022**, *12*, 2350. https://doi.org/10.3390/app12052350

Academic Editors: Luca Mazzoni, Mircea Oroian and Georgiana Gabriela Codină

Received: 1 February 2022
Accepted: 22 February 2022
Published: 24 February 2022

Publisher's Note: MDPI stays neutral with regard to jurisdictional claims in published maps and institutional affiliations.

1. Introduction

Apple (*Malus domestica*) is an important fruit produced in large quantities worldwide, ranked third after bananas and watermelons according to FAO [1], while in the European Union (EU), apples are ranked first, followed by oranges [2]. Fresh apple consumption in the EU is approximately 15 kg per capita and is expected to increase by 1 kg by 2030, while consumption of processed apples is around 8 kg per capita [3]. Apples are popular due to their availability throughout the year in various forms: fresh and dried fruits, juice, cider, puree. It is well known that the consumption of apples has various benefits for human health [4–6]. Current demand for natural products involves the continued development of small producers who process apples in various forms, apple chips or apple cider in food markets, restaurants and bars. Koutsos et al. [7] affirmed that apple consumption positively affects lipid metabolism, vascular function, inflammation and weight management. Apple snacks can increase the amount of consumed fruits. In the food industry, drying is a widely used method of producing apple snacks [8,9]. Apple fruits are not only preserved by drying, but the dried products obtained can be stored and transported at a relatively low cost [10,11]. Dried fruits can be consumed at breakfast and as snacks or incorporated in prepackaged cereals and baked goods [12–14]. Carughi et al. [15] suggested that dried fruits contribute to good health as well as fresh fruits, and they ensure satiety and have beneficial influence on the glycemic index and blood pressure.

Kowalska et al. [11] stated that consumers are interested in food products rich in dietary fiber and macro- and micronutrients. Traditional dried fruits (with no added sugar) are appreciated for their sweetness, higher fiber content and long-term stability. They have a nutritional profile similar to the original food. In addition, it is considered that increasing the consumption of natural sweeteners can have a positive effect on body weight and metabolism [16]. Oven-drying and microwave-drying techniques are the most used for fruit drying. It is considered that drying with hot air is the most commonly utilized drying method [17]. There are various pre-treatment methods applied; among them are sulfite treatment, osmotic dehydration, steam blanching, ultrasonic treatment, etc. [18–20], which may influence the dry products' properties. Color and visual appearance are evaluated first by the consumers, followed by taste and aroma [21]. The color and sensory attributes change during fruit drying [22,23]. The dried apple color can change due to the drying method: the apple may darken, and the color may change from green to red or from yellow to blue [12]. Marzec et al. [24] suggested that polyphenol oxidase and non-enzymatic reactions can be responsible for product darkening. It is considered that the water activity of the fruit decreases during the dehydration process, thus influencing the enzymatic activity and possible alteration [25]. The water activity during apple drying is reduced to a level that ensures microbiological safety [12,26].

The apple fruits used in this study were purchased from the North East of the country, known for its centuries-old tradition of growing fruit trees, especially apples. Between 1978 and 1982, the state system organized the first huge orchard in Curtesti (10 km from Botosani) on a scientific basis. The orchard had over 400 hectares all cultivated with apples. Even though only 60 hectares are still cultivated with apples today, the tradition is continued, and the orchard has been rejuvenated. The fruit-growing area from North East of the country is revived following the investments of the last years, is known as a major apple production nationwide and has many small farms that use family and local labor [4]. Current demand for natural products involves the continued development of small producers who process apples in various forms, apple chips or apple cider in food markets, restaurants and bars. To support these small producers, dried apple chips were produced from readily available raw materials, from abundant cultivars of apples existing in this area of the country, and evaluated in terms of physico-chemical and sensory properties. These products can be easily obtained and economically capitalized. The easiest way to obtain apple chips is by drying them in the oven, maintaining the temperature between 60 and 70 °C, without the need for special equipment. Drying in the oven is faster compared to drying in the sun or in a food dehydrator [27]. For small producers, who operate in the food markets, restaurants and bars, it can be quite useful in preserving the fruits and for their economic aspects, especially if the fresh fruits remain unsold.

The aim of this study was to obtain a healthy and appreciated product, dried apple chips, fat-free, with no sugar added, from available and abundant low-cost raw materials, namely three apple cultivars 'Starkrimson', 'Golden Delicious' and 'Florina'.

In this study, drying of apple fruits was performed by using two types of dryers (oven and dehydrator). Physical and chemical quality attributes, including moisture, water activity, active acidity, titratable acidity, total soluble solids and total sugar contents, electrical conductivity, color of fresh apple slices and traditional dried apple chips, were evaluated. In addition, sensory and statistical evaluation was performed. Physico-chemical and sensory quality of dried apple chip samples were also determined at 7, 14 and 21 days during storage at room temperature. Finally, statistical correlations between physico-chemical parameters were obtained.

2. Materials and Methods

2.1. Sample Preparation and Drying

Apples ('Starkrimson' (S), 'Golden Delicious' (G) and 'Florina' (F) cultivars, Figure 1) were procured from a local apple orchard of Botosani, Romania. Fresh apples were selected, washed and sliced longitudinally, following their vertical axis, into slices 4 mm and 3–7 cm thick using a kitchen knife (after removing the apple core) and then dipped in a lemon salt solution (4%) for 7 min to inactivate browning enzymes, according to the methods described by Nyangena et al. [28] and ElHana [29] with minor modifications. The dipping was performed under normal conditions (light and room temperature). The apple slices were removed from the solution and washed with distilled water. After that, the adhering water was removed at the end of immersion by using paper towel. Samples were then loaded in perforated trays and subjected to drying, up to constant weight. Drying was performed by using two types of dryers: oven (Indesit on gas) and dehydrator (Gorenje, electric). The dehydrator controls and keeps the internal temperature constant, while the oven tends to fluctuate the temperature. The oven door was kept slightly open (1 to 2 inches) during drying, and a thermometer was used to check the oven temperature [27]. The drying temperature was 65 °C; all samples were dried to constant weight (Figure 1). There are numerous studies on apple chip drying at temperatures between 40 and 90 °C (convective drying) [30] but also between 35 and 85 °C (microwave, infrared or vacuum freeze-drying) [31]. According to Önal et al. [32], drying apples at 65 °C seems to be a good process to produce healthy, high-quality snacks for both consumers and the food industry.

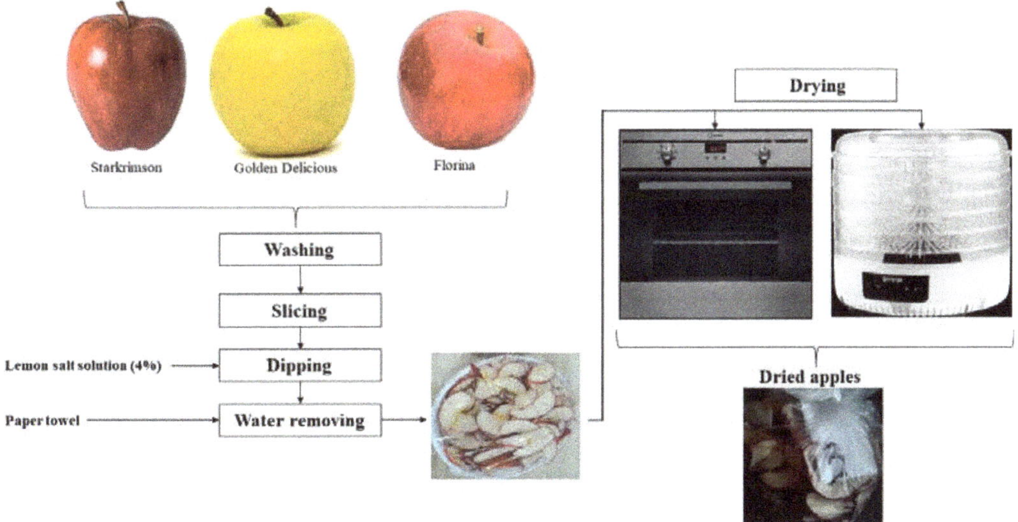

Figure 1. Steps included in sample preparation and drying.

Dried samples were cooled down and packed in plastic bags (HDPE) and stored at room temperature. Approximately 194 g of dried apple chips (172 g 'Golden Delicious' cultivar, 201 g 'Florina' cultivar and 209 g 'Starkrimson' cultivar) can be obtained from 1 kg of apples dehydrated in the oven, while from 1 kg of apples dried in dehydrator, an average of about 167 g of dried apple chips (158 g 'Starkrimson' cultivar, 169 g 'Florina' cultivar and 175 g 'Golden Delicious' cultivar) can be obtained. After removing the apple core, between 12 and 16% of the apple content can be lost, depending on the cultivar of apple (10.36–12.91% 'Starkrimson' cultivar, 15.44–16.33% 'Golden Delicious' cultivar, 11.79–19.21% 'Florina' cultivar).

2.2. Physico-Chemical Analysis

Apple water extract for determination of active acidity, titratable acidity, total soluble solids contents and electrical conductivity was prepared as follows: 50 mL of distilled water was added to 20 g of fresh/dried apples measured with a precision of 0.01 g. The content was heated on a water bath brought to boil for 30 min. After boiling, the extract was cooled, and the content was brought to 250 mL with distilled water and then filtered.

Moisture (*M*) was determined by oven-drying method; thus 5 g of apple slices from each sample was weighed and dried at 105 °C until constant weight [20,33].

Water activity (*aw*) was determined by using a water activity meter AquaLab Lite (Decagon, WA, USA). The determinations were performed at 25 °C, and all samples were measured in triplicates. In a special cuvette was introduced approximately 3 g of the sample; after that, the cuvette was placed in *aw* chamber, and the water activity was measured and displayed on the water activity meter screen [33].

Active acidity was determined by *pH* measurements at ambient temperature using laboratory *pH* meter Fisher Scientific ACCUMET Bio Set AE150 (Fisher, Waltham, MA, USA). For each sample, a 10 mL volume of apple extract was measured, the *pH* electrode was inserted, and after stabilization, the *pH* value was displayed on the *pH* meter screen [34].

Titratable acidity (*TA*) was obtained by titrating 50 mL of apple water extract with 0.1 N NaOH solution, in the presence of 1% phenolphthalein solution. The *TA* values were calculated according to Sadler and Murphy [35], and the results are expressed as malic acid (%) [36]. In apple, malic acid is the predominant acid (80–90% of the organic acids) that contributes to sourness of fruits [35,37].

Total soluble solids contents (*TSS*) were measured by using a digital hand-held refractometer standardized with distilled water. Two drops of fresh/dried apple extract were placed on the refractometer lens and measured. The obtained values were corrected to the equivalent reading at 20 °C [38,39] and reported as 'degrees Brix' (°Brix) which is equivalent to percentage (%) [40].

Total sugar contents in fresh and dried apple samples were determined by applying Fehling's method. Into a 300 mL Erlenmeyer flask, was placed 20 mL of the defecated extract, then 10 mL of Fehling I solution and 10 mL of Fehling II solution. The vessel was heated on an asbestos sieve, adjusting the flame for boiling, and boiled for exactly 2 min. The flask was cooled rapidly under running water, and then 20 mL of KI solution and 15 mL H_2SO_4 were added. Released iodine was titrated with 0.1 N $Na_2S_2O_3$ in the presence of starch (1%) as an indicator. Results were expressed in %.

Sugar/acid ratio was calculated as the total sugars content divided by titratable acidity.

Electrical conductivity (*EC*) of samples was measured with a laboratory conductometer VioLab COND 51+ Set (ROTH, Germany) and was expressed in µS/cm. The conductometer electrode, after calibration, was inserted into 10 mL of apple extract, and after stabilization, the electrical conductivity value was displayed on the conductometer screen. Three readings were performed for each sample analyzed.

Color analysis was performed by using Konika Minolta CR 400 colorimeter (Konica Minolta, Tokyo, Japan). The samples were introduced in a 20 mm cuvette and measured in a white spectrum. In order to establish the results, three readings were performed. The measurements were displayed in *L** (light–dark spectrum, from 0 = black to 100 = white), *a** (green–red spectrum; −a = green, +a = red), *b** (blue–yellow spectrum; −b = blue, +b = yellow) and Δ*E* (total change in the color of samples) values [9,39,41].

Browning index (*BI*) was calculated with Equation (1) and represents the brown color purity, due to enzyme activity, in the apple [17,33], while **whitening index** (*WI*), which corresponds to apple discoloration and lesser browning, was calculated according to Equation (2) [42,43].

$$BI = \frac{100}{0.17}\left(\frac{a^* + 1.75L^*}{5.645L^* + a^* - 3.012b^*} - 0.31\right) \qquad (1)$$

$$WI = 100 - \sqrt{(100 - L^*)^2 + a^{*2} + b^{*2}} \qquad (2)$$

2.3. Sensory Evaluation

Sensory analysis can be performed in order to establish if the obtained apple chips are different and what the magnitude of these differences is. In this case, according to Ackbarali and Mahara [44], 8 to 12 panelists or 6 to 10 trained panelists are required to assess intensity for sensory attributes, and analysis of variance assessment should be applied to determine whether the means are statistically different. Velickova et al. [33] indicated 15 evaluators with experience in descriptive evaluation of crunchy products that were used to evaluate apple chips. Cliff et al. [45] indicated that 20 students and staff performed a sensory evaluation of fresh-cut apple slices, while in Chauhan et al.'s [10] study, the sensory attributes of osmo-dried apples were evaluated by a trained panel of 10 staff members of their laboratory. In addition, a panel of 15 people (after training) evaluated the color, texture, aroma, sweetness and acidity of the dried apples obtained by Wang et al. [20], while dried red apple cubes produced by Kidoń and Grabowska [26] were evaluated only by a 5-member panel with formal classroom training in sensory evaluation, and a 5-member panel evaluated color, texture and taste of dried apples gamma-irradiated in Hussain et al.'s [46] study.

In the current study, the sensory evaluation was performed in laboratory conditions by a panel of 11 engineers from food industry, with experience and familiarity with sensory assessment method. Each of the panelists evaluated the appearance, color, aroma, taste and overall acceptability of the dried apple chip samples. The scoring was based on the nine points of the standard hedonic scale, where 9 denoted 'like extremely', 5 'neither like nor dislike', and 1 indicated 'dislike extremely' [9,20,21,47]. The sensory evaluation experiment was repeated three times.

2.4. Statistical Analysis

All experiments were performed in triplicates, and the results were expressed as mean values $\pm$ standard mean error. The procedure for performing the experiments in triplicate was simply adopted by tradition. Singer et al. [48] investigated the use of m-plicates and demonstrated that m-plicates do not significantly improve the accuracy of the estimates. Triplicate measurement for determination of chemical composition of apple fruits from eight apple cultivars ('Delicious', 'Golden Delicious', 'Ralls', 'Fuji', 'QinGuan', 'Jonagold', 'Granny Smith' and 'Orin') was performed by Wu et al. [49], while Preti and Tarola [50] adopted the same procedure in triplicate to determine antioxidant capacity and polyphenolic and major mineral composition of 14 ancient apple cultivars in Italy. In addition, Ergün [51] determined the biochemical contents of peel and pulp of fresh, oven-dried and sun-dried apple samples from the following apple cultivars: 'Amasya', 'Braeburn', 'Golden Delicious', 'Granny Smith' and 'Starking', in triplicate.

An analysis of variance (ANOVA) with a confidence interval of 95% ($p < 0.05$), with Tukey test, was used for comparison of physico-chemical and sensory evaluation results. Pearson correlations between measured and calculated parameters were determined. The statistical analysis was performed by using statistical software Minitab, version 17. According to Minitab [52]: *'the null hypothesis states that the parameter means are all equal'*.

3. Results and Discussion

3.1. Moisture and Water Activity

Moisture content in the fresh apple samples was in the range of 83.97 $\pm$ 0.08% ('Starkrimson' cultivar) to 86.27 $\pm$ 0.31% ('Florina' cultivar). In the case of fresh samples from the 'Golden Delicious' cultivar, the moisture content was determined as 86.03 $\pm$ 0.06%. These values are close to those reported in other papers. For example, Iordănescu et al. [53] obtained the following values for apple moisture content: 83.04 $\pm$ 0.87% 'Starkrimson' cultivar, 86.12 $\pm$ 1.35% 'Florina' cultivar and 84.61 $\pm$ 0.58% 'Golden Delicious' cultivar.

After drying in the oven, the following moisture values were obtained for the apple samples: 6.63 $\pm$ 0.01% ('Florina' cultivar), 6.83 $\pm$ 0.03% ('Golden Delicious' cultivar) and 7.08 $\pm$ 0.04% ('Starkrimson' cultivar). The moisture content of apple chip samples dried in the dehydrator ranges between 5.06 ± 0.03% ('Florina' cultivar) and 6.29 ± 0.03% ('Golden Delicious' cultivar). For samples of the 'Starkrimson' apple cultivar dried in a dehydrator, an average moisture of 5.55 ± 0.05% was obtained. These values are close to those reported by Velickova et al. [33] who obtained, by conventional drying, apple chips with moisture values between 5.17 ± 0.23 and 6.04 ± 0.31%, from fresh 'Idared' apples with 88.7 ± 3.9% water content.

Water activity shows how closely the water in an apple slice is bound. If aw values are less than 0.8, it means that the food has low water activity, and the most enzymatic reactions are slowing down. It is considered that the target value of dried products is $aw = 0.6$, which is the general level limit for the growth of molds, yeasts and bacteria [13,54–56]. In Figure 2 are illustrated the aw values obtained for fresh and dried apple chip samples. The aw of dried apple chip samples was investigated also at 7, 14 and 21 days during storage at room temperature. The highest value of aw (0.910) for the fresh apple sample was obtained for samples from the 'Golden Delicious' cultivar, followed by samples from the 'Starkrimson' cultivar with a mean value of 0.906. The aw mean value obtained for samples from the 'Florina' cultivar is slightly lower than those obtained for the samples of the other two cultivars. The water activity values for the oven-dried apples range from 0.544 to 0.650 with the highest value obtained for samples from the 'Starkrimson' cultivar and the lowest for the samples from the 'Florina' cultivar. In the case of apple chips dried in the dehydrator, it was observed that the highest mean value (0.426) was obtained for samples from the 'Golden Delicious' cultivar, while the lowest value (0.374) was determined for samples from the 'Starkrimson' cultivar. According to the results obtained, the water activity is below 0.6 for almost all samples of dried apple chips, with one exception. The obtained values are close to those reported by Klewicki et al. [55] for osmo-convectively-dried 'Idared' apple samples ($aw = 0.542$) and Kahraman et al. [17] (2021) who dried 'Gala' apples using hot air drying ($aw = 0.345$) and non-thermal ultrasound contact drying ($aw = 0.386$).

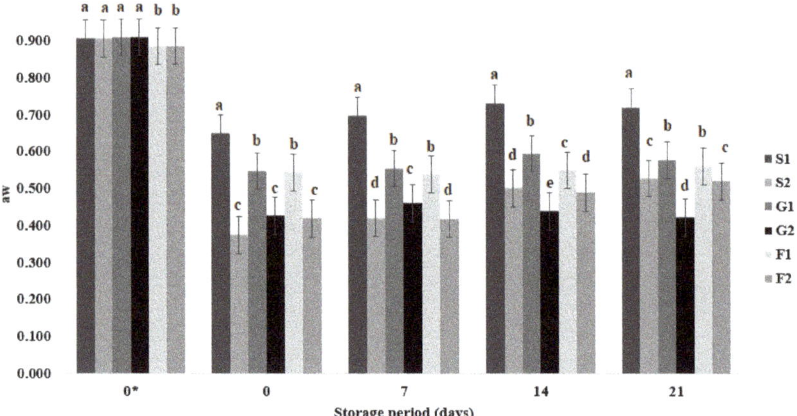

Figure 2. Water activity (aw) evaluation of fresh and dried apple chips during storage under ambient conditions (S1—apple chips 'Starkrimson', oven; S2—apple chips 'Starkrimson', dehydrator; G1—apple chips 'Golden Delicious', oven; G2—apple chips 'Golden Delicious', dehydrator; F1—apple chips 'Florina', oven; F2—apple chips 'Florina', dehydrator). Notes: 0*—fresh apple samples, 0—dried apple samples. Different lowercase letters (a–e) show significant differences between the groups ($p < 0.05$).

It can also be observed that the samples dried in the dehydrator have lower water activity values than those dried in the oven. Over time, up to 21 days, the aw values of dried apples vary, and it was observed that aw increases by about 41% and 10% in the case of 'Starkrimson' cultivar samples dried in a dehydrator and oven, respectively. The *aw* value of 'Florina' cultivar samples dried in a dehydrator increases by 24% until the twenty-first day, while for the samples dried in the oven, the *aw* value increase was only 3%. In the case of 'Golden Delicious' cultivar samples, the *aw* values increase only for the samples dried in the oven by 5%, while for the samples obtained in the dehydrator, an insignificant variation of *aw* values with time was observed.

3.2. pH, Titratable Acidity and Total Solids Content

In Figure 3, the *pH* values obtained for the samples of fresh and dried apples are illustrated. It can be seen that the samples of fresh apple slices have average values between 4.02 and 4.19, while the *pH* values of dried apple chips decreased compared to the fresh samples and ranged from 3.85 to 4.02. Patras [57] obtained, for the same apple cultivars, *pH* values between 3.68 and 4.20, while Iordănescu et al. [53] reported *pH* values between 5.34 and 5.60 for fresh samples. According to Kahraman et al. [17] and Owusu et al. [58], decreases in *pH* values of dried apple samples may be associated with the dissociation of organic acids with temperature. The *pH* of dried apple chip samples during storage varies very little compared to the *pH* obtained for the freshly dried samples, and on the 21st day of storage, the mean *pH* values were between 3.61 (S2) and 3.80 (F2).

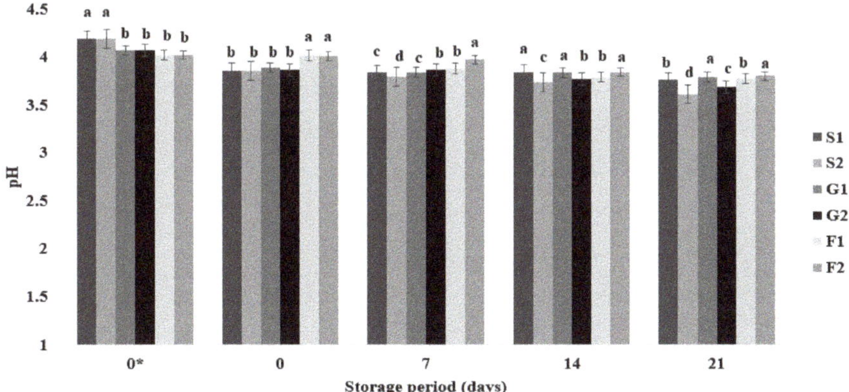

Figure 3. *pH* evaluation of fresh and dried apple chips during storage under ambient conditions (S1—apple chips 'Starkrimson', oven; S2—apple chips 'Starkrimson', dehydrator; G1—apple chips 'Golden Delicious', oven; G2—apple chips 'Golden Delicious', dehydrator; F1—apple chips 'Florina', oven; F2—apple chips 'Florina', dehydrator). Notes: 0*—fresh apple samples, 0—dried apple samples. Different lowercase letters (a–d) show significant differences between the groups ($p < 0.05$).

Perception of fruit taste is given by the apples' acidity. Moreover, it is considered that sour taste depends on organic acid contents such as malic or citric acids, while soluble solids content is responsible for the sweet taste [59]. According to Sadler and Murphy [35], the typical malic acid content in apple fruit is between 0.27 and 1.02%. In this study, the values obtained for the *TA* of fresh apple samples from 'Golden Delicious' and 'Florina' cultivars fall between these values. The higher content of malic acid was registered in apple samples from the 'Golden Delicious' cultivar (0.33 [a] ± 0.03% acid malic) followed by apple samples from the 'Florina' cultivar (0.29 [b] ± 0.01% malic acid) and 'Starkrimson' cultivar (0.17 [c] ± 0.03% acid malic). *TA* values are close to those obtained by Nour et al. [59]: 0.30% for 'Golden Delicious', 0.20% for 'Florina' and 0.10% for 'Starkrimson'. Butkeviciute et al. [6] reported a titratable acidity for other apple cultivars that varied between 0.24 and 0.61%. Titratable acidity values of dried apple chip samples are presented in Table 1. It can be observed that all dried apple chip samples have higher values of *TA* than fresh samples, regardless of cultivar or type of drying. The *TA* values of dried apple chips increased compared to the *TA* of fresh apple samples due to the organic acids in apples that become more concentrated as the temperature increases [58]. During storage, *TA* values increase for all samples.

Table 1. Titratable acidity of dried apple samples.

Sample		Titratable Acidity (*TA*), % Malic Acid	Sample		Titratable Acidity (*TA*), % Malic Acid	Sample		Titratable Acidity (*TA*), % Malic Acid	Sample		Titratable Acidity (*TA*), % Malic Acid
Dried apple chip samples	S1	0.72 [c] ± 0.11	Dried apple chip samples (7th day)	S1	0.72 [cd] ± 0.02	Dried apple chip samples (14th day)	S1	0.73 [c] ± 0.05	Dried apple chip samples (21st day)	S1	0.80 [b] ± 0.05
	S2	0.95 [a] ± 0.15		S2	0.95 [a] ± 0.09		S2	0.97 [a] ± 0.01		S2	1.04 [a] ± 0.05
	G1	0.63 [d] ± 0.01		G1	0.65 [e] ± 0.05		G1	0.67 [d] ± 0.01		G1	0.75 [b] ± 0.05
	G2	0.71 [c] ± 0.03		G2	0.74 [c] ± 0.05		G2	0.80 [b] ± 0.02		G2	0.81 [b] ± 0.01
	F1	0.87 [b] ± 0.05		F1	0.91 [b] ± 0.01		F1	0.95 [a] ± 0.00		F1	1.04 [a] ± 0.06
	F2	0.71 [c] ± 0.15		F2	0.71 [d] ± 0.05		F2	0.72 [c] ± 0.05		F2	0.75 [b] ± 0.01

Means that do not share a letter are significantly different ($p \leq 0.05$).

The total solids content of fresh apple slices varies from 15.00 °Brix ('Golden Delicious' cultivar samples) to 16.46 °Brix ('Starkrimson' cultivar samples). These values are close to those obtained by Patras [57]: 15.01 and 15.3 °Brix for 'Golden Delicious' and 'Starkrimson' cultivars available on the market in 2018. In the case of 'Florina' cultivar fresh samples, a mean value of 15.40 °Brix for *TSS* (Figure 4) was obtained, a value close to that obtained by Nour et al. [59] (15.00 °Brix) for apples of the same cultivar.

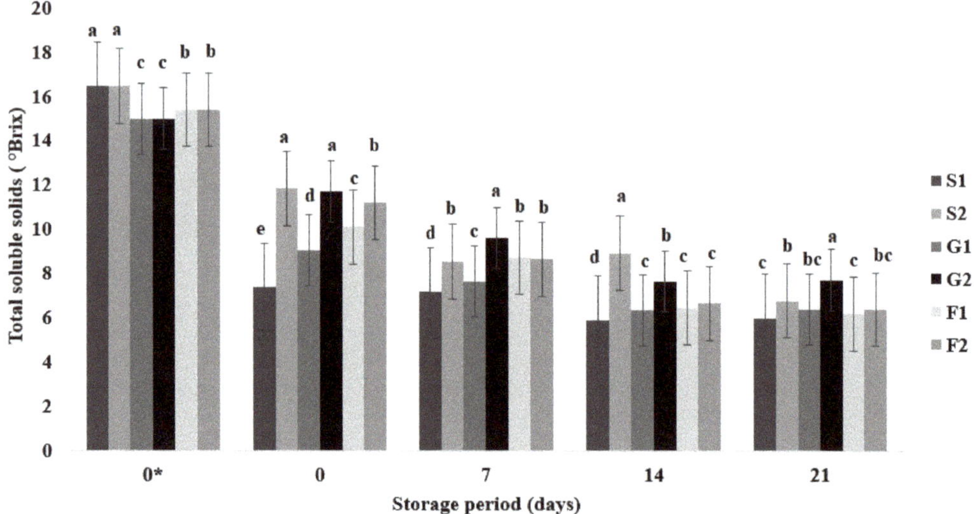

Figure 4. Total soluble solids (*TSS*) evaluation of fresh and dried apple chips during storage under ambient conditions. Notes: 0*—fresh apple samples, 0—dried apple samples. Different lowercase letters (a–e) show significant differences between the groups ($p < 0.05$).

Iordănescu et al. [53] reported *TSS* values ranging between 13.0 and 15.3 °Brix for 'Starkrimson' apples, between 10.6 and 14.2 °Brix for apples from the 'Florina' cultivar and 12.2–15.8 °Brix for 'Golden Delicious' apples. The total solids content for three apple cultivars determined by Egea et al. [60] varied from 13.78 to 17.61%, while Butkeviciute et al. [6] reported that the soluble solids content of ten different apple cultivars ranged between 12.40 and 14.00%.

After the drying of apple samples, the *TSS* values decreased for all types of apple cultivars (Figure 4). *TSS* values of dried apple samples decrease compared to values of fresh apple samples due to reactions that occur when the temperature rises [58]. It was observed that the *TSS* value decreased a lot (55%) for 'Starkrimson' cultivar samples dried in the oven, while the smallest decrease (22%) was recorded for 'Golden Delicious' cultivar samples dried with a dehydrator. According to Figure 4, it can be seen that the *TSS* values vary during the storage process, and all decreased on the 21st day compared to the *TSS* values obtained from the dried apple chip samples immediately after drying.

The flavor and quality of fruit products are given by acid and sugar content, and the sugar/acid ratio can better predict the flavor impact [35]. Apple fruits with sugar/acid ratios below 20 are characterized by a strong sour taste and are only suitable for processing; if the ratios are higher than this value, the fruits are perceived as sweet and suitable for consumption [61]. In the case of fresh samples investigated in this study, sugar/acid ratios were: 55.88 for the 'Starkrimson' cultivar, 46.06 for the 'Golden Delicious' cultivar and 39.31 for the 'Florina' cultivar. The values obtained are close to the values obtained in other papers. For example, Li et al. [61] obtained for the 'Starkrimson' cultivar sugar/acid ratios of between 28.07 and 59.73, while for the 'Golden Delicious' cultivar, the sugar/acid ratios were between 22.39 and 44.32 depending on the growth sites. Patras [57] reported a sugar/acid ratio of 41.36 for the 'Florina' cultivar. In this study, the sugar/acid ratio decreased in dried apple chip samples as follows: G1 (36.19) > S1 (33.33) > G2 (29.57) > F2 (26.90) > F1 (24.94) > S2 (24.31).

3.3. Electrical Conductivity

Electrical conductivity is a function of food components and a complex function of temperature and other physical properties; thus the components of salt, acids and moisture are very effective in increasing the conductivity of electricity, while total solids content, sugar, fats, lipids and alcohols decrease it [62,63]. It was demonstrated that with the increase of temperature, the electrical conductivity increases due to an increase in mobility of the tissue constituents as a result of higher temperature [63]. In addition, the nature of the product itself influences the conductivity of the food materials. According to [63], the interest in the *EC* of food is increasing greatly as advances in food processing technology, food quality assessment and food preservation increase. The mean values obtained for the electrical conductivity of apple samples (fresh and dried) are presented in Figure 5. Results show that fresh apple samples have *EC* values between 626 ('Golden Delicious' cultivar) and 727 ('Florina' cultivar) µS/cm, while apple samples dried in the oven have *EC* values ranging from 548 ('Florina' cultivar) to 1854 ('Starkrimson' cultivar) µS/cm. From Figure 5, it is possible to observe the increase of the *EC* values for all the samples of apple chips dried in the dehydrator; thus values between 1880 ('Florina' cultivar) and 2020 ('Golden Delicious' cultivar) µS/cm were obtained. Over time, a decrease in almost all the values recorded for the electrical conductivity of the dried apple samples can be observed (Figure 5). The *EC* values of dried 'Starkrimson' cultivar apple samples decrease with time with 42% in the case of samples dried in the oven and 45% for the samples dried in the dehydrator. In the case of the other apple cultivars, it was observed that the *EC* values increased with time for apple chip samples dried in the oven and decreased for the samples obtained by drying in the dehydrator.

3.4. Color Parameters, Browning and Whitening Index

Color is one of the principal quality factors that highly influence food acceptance [64]. It was observed that processed fruits are more acceptable by consumers if their colors are close to unprocessed fruits [65]. Results of color parameters measured for fresh and dried apple chips are presented in Table 2. The degree of freshness of the fruit is described by the lightness index (*L**) [6]. The *L** values of dried samples are higher than the *L** values of fresh samples, which means that the dried fruits are lighter in color than the fresh ones, possibly due to the lemon salt solution used in the preparation of the samples [66]. According to

Butkeviciute et al. [6], the yellow color of the apple fruit is given by the carotenoids, while the presence of anthocyanins leads to the red color of the fruit. All fresh apple samples had *a** values situated in the negative region more toward green, while *a** values measured for dried apple samples were situated in the positive region which means more toward red, in accord with [23]. The yellowness (*b**) of fresh apple samples was between 18.36 ± 0.45 and 27.30 ± 1.64. The *b** values increased for S1, S2 and F2 and decreased for G1 and F2 dried apple samples compared with the *b** values of fresh apple samples.

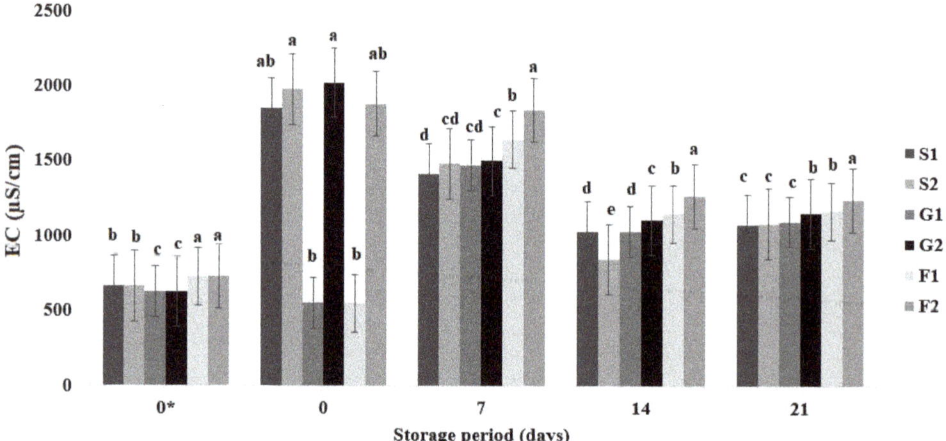

Figure 5. Electrical conductivity (*EC*) evaluation of fresh and dried apple chips during storage under ambient conditions (S1—apple chips 'Starkrimson', oven; S2—apple chips 'Starkrimson', dehydrator; G1—apple chips 'Golden Delicious', oven; G2—apple chips 'Golden Delicious', dehydrator; F1—apple chips 'Florina', oven; F2—apple chips 'Florina', dehydrator). Notes: 0*—fresh apple samples, 0—dried apple samples. Different lowercase letters (a–e) show significant differences between the groups ($p < 0.05$).

Table 2. Color parameters of fresh and dried apple chips during storage under ambient conditions.

	Sample	L*	a*	b*	ΔE	BI	WI
Fresh apple samples	S0	51.93 [a] ± 1.13	−1.42 [a] ± 0.14	18.36 [b] ± 0.45	-	40.27 [b] ± 1.23	48.52 [a] ± 1.14
	G0	64.28 [a] ± 4.86	−2.29 [a] ± 0.55	27.30 [a] ± 1.64	-	50.63 [a] ± 0.94	54.85 [a] ± 2.94
	F0	57.7 [a] ± 7.03	−3.47 [b] ± 0.33	19.83 [b] ± 1.51	-	36.29 [b] ± 1.09	53.02 [a] ± 5.73
Dried apple chip samples	S1	57.69 [e] ± 2.42	10.62 [a] ± 0.93	20.56 [c] ± 0.40	40.17 [a] ± 2.32	57.02 [a] ± 2.28	51.76 [d] ± 2.28
	S2	67.39 [c] ± 0.93	5.09 [c] ± 0.19	24.41 [b] ± 0.62	32.81 [c] ± 1.08	49.63 [b] ± 1.27	58.94 [c] ± 1.09
	G1	71.83 [b] ± 0.43	3.02 [d] ± 0.33	24.77 [b] ± 0.81	29.47 [de] ± 0.40	44.42 [b] ± 0.67	62.36 [ab] ± 0.38
	G2	76.10 [a] ± 1.05	2.09 [d] ± 0.37	27.56 [a] ± 0.47	28.76 [e] ± 0.26	45.86 [b] ± 0.13	63.44 [a] ± 0.36
	F1	65.85 [cd] ± 0.94	12.08 [a] ± 0.32	17.59 [d] ± 0.29	32.49 [cd] ± 0.93	44.05 [b] ± 0.84	59.73 [bc] ± 0.95
	F2	63.65 [d] ± 1.01	7.10 [b] ± 0.80	25.53 [b] ± 0.29	36.78 [b] ± 1.08	58.42 [a] ± 1.54	55.01 [d] ± 1.06
Dried apple chip samples (7th day)	S1	47.39 [c] ± 1.17	9.31 [a] ± 0.65	18.39 [d] ± 0.76	48.41 [a] ± 0.95	62.73 [ab] ± 1.51	43.48 [c] ± 0.97
	S2	71.51 [a] ± 3.01	3.98 [b] ± 1.12	25.44 [ab] ± 1.42	30.32 [c] ± 3.27	47.39 [cd] ± 3.93	61.58 [a] ± 3.29
	G1	61.63 [b] ± 1.01	9.14 [a] ± 0.91	25.96 [ab] ± 0.31	39.09 [b] ± 0.78	64.44 [a] ± 0.93	52.77 [b] ± 0.77
	G2	70.31 [a] ± 0.40	3.73 [b] ± 0.22	26.94 [a] ± 0.73	32.22 [c] ± 0.64	51.09 [cd] ± 1.10	59.72 [a] ± 0.63
	F1	58.09 [b] ± 1.26	8.21 [a] ± 0.03	21.13 [c] ± 0.48	39.43 [b] ± 0.90	54.84 [bc] ± 0.20	52.34 [b] ± 0.90
	F2	73.57 [a] ± 2.86	4.05 [b] ± 0.78	24.52 [b] ± 0.99	28.33 [c] ± 1.37	43.76 [d] ± 0.42	63.65 [a] ± 1.53

Table 2. *Cont.*

	Sample	L^*	a^*	b^*	ΔE	BI	WI
Dried apple chip samples (14th day)	S1	52.22 [d] ± 2.20	11.16 [a] ± 0.55	22.43 [c] ± 1.34	45.87 [a] ± 1.53	70.72 [a] ± 0.28	46.01 [d] ± 1.49
	S2	75.96 [a] ± 0.13	3.23 [d] ± 0.30	27.11 [a] ± 0.17	28.59 [c] ± 0.19	46.24 [c] ± 0.33	63.62 [b] ± 0.18
	G1	69.91 [b] ± 0.27	4.47 [c] ± 0.41	25.72 [ab] ± 0.45	31.69 [b] ± 0.37	49.58 [b] ± 0.80	60.16 [c] ± 0.34
	G2	77.17 [a] ± 0.80	0.64 [e] ± 0.15	24.28 [b] ± 0.29	25.37 [d] ± 0.67	37.46 [d] ± 0.60	66.66 [a] ± 0.71
	F1	64.49 [c] ± 1.13	8.09 [b] ± 0.17	20.52 [d] ± 0.34	33.65 [b] ± 0.75	46.93 [c] ± 0.17	58.19 [c] ± 0.77
	F2	63.78 [c] ± 0.87	2.77 [d] ± 0.20	19.90 [d] ± 0.11	32.98 [b] ± 0.84	39.82 [d] ± 0.66	58.57 [c] ± 0.83
Dried apple chip samples (21st day)	S1	52.26 [e] ± 1.03	9.84 [b] ± 0.75	20.02 [e] ± 0.42	44.57 [a] ± 0.98	61.34 [a] ± 1.42	47.29 [e] ± 0.95
	S2	69.63 [b] ± 1.09	5.97 [c] ± 0.10	26.31 [a] ± 0.60	32.58 [d] ± 0.54	52.78 [b] ± 0.48	59.36 [b] ± 0.59
	G1	64.76 [c] ± 0.78	6.19 [c] ± 0.24	23.70 [bc] ± 0.19	34.67 [c] ± 0.58	51.72 [b] ± 0.40	57.07 [c] ± 0.58
	G2	72.39 [a] ± 0.08	3.31 [d] ± 0.03	24.75 [b] ± 0.03	29.09 [e] ± 0.07	44.27 [c] ± 0.07	62.77 [a] ± 0.07
	F1	58.65 [d] ± 0.57	12.42 [a] ± 0.53	21.44 [d] ± 0.25	40.23 [b] ± 0.69	60.30 [a] ± 0.96	51.77 [d] ± 0.68
	F2	68.56 [b] ± 1.03	3.75 [d] ± 0.51	23.30 [c] ± 0.72	31.03 [d] ± 0.69	44.69 [c] ± 1.06	60.67 [b] ± 0.70

Values are mean ± standard deviation ($n = 3$). Means that do not share a letter are significantly different ($p \leq 0.05$).

The color difference between fresh and dry samples is greater when the browning index values are higher [17]. Temperature and the duration of exposure influence the evolution of color parameters [67]. Cruz et al. [23] stated that the change in color during drying is associated with the development of browning reactions, attributed to the activity of the enzyme polyphenol oxidase. Generally, the browning increases with temperature [58] (observed for S1, S2, F1 and F2 samples); thus the decrease in the browning index was unexpected for samples G1 and G2. The browning index (BI) at 0 days of storage was between 44.05 ± 0.84 (F1) and 58.42 ± 1.54 (F2), while the whitening index (WI) varied between 51.76 ± 2.28 and 63.44 ± 0.36. WI values of S1 and S2 and F1 and F2 are statistically significantly different. WI values differ depending on the drying method for S and F cultivars, while for cultivar G, WI values are not so influenced by the drying method. During storage, BI values for S1 increased and then decreased and for S2 decreased and then increased, while for WI values, the reverse was observed. In the case of G1 and G2 samples, BI values fluctuated and finally increased on the 21st day; for F1 and F2 samples, the BI values also increased on the last day of the observations compared to the values on the first day. The differences between the color parameter values are statistically significantly different. It was observed that p-values are less than 0.05, the null hypothesis is rejected, and not all group means are equal [52]. In addition, according to results obtained by applying Tukey's method, the means that do not share a letter are significantly different (Table 2). The model fits our data, since R-sq explains 90.74% of the variation in the response in the case of the BI values of dried samples on the first day of drying, 89.48% on the 7th day of storage, 99.52% on the 14th day of storage and 96.75% on the 21st day of storage. R-sq explains 96.84% of the variation in the response in the case of WI values of dried samples on the first day of drying, 97.25% on the 7th day of storage, 98.83% on the 14th day of storage and 99.03% on the 21st day of storage.

3.5. Sensory Evaluation

For the statistical evaluation of the results, $\alpha = 0.05$ was set. Results show that p-value = $0.000 < \alpha$, and it can be considered that the mean of the results is statistically significant. By applying Tukey's pairwise comparisons, considering the scores obtained for the appearance, color, aroma, taste and overall acceptability of the dried apple samples, it was determined that the highest mean was obtained by sample F2 (8.63 [a] ± 0.18), followed by S2 (8.52 [ab] ± 0.11) that shares a letter with F2 but also with G1 (8.29 [b] ± 0.22), which means that they are not significantly different. In addition, the statistical evaluation of the scoring obtained for sensory assessment showed that sample S1 with a mean of 7.89 [c] ± 0.22 is not statistically different from the mean obtained for G2 (7.89 [c] ± 0.04). It can be observed that all panelists highly appreciated ('like very much' to 'like extremely') apple chips 'Florina' dried in a dehydrator (sample F2), while the last sample F1 (7.38 [d] ± 0.11) was liked moderately. It seems that the apple chips from 'Florina' and 'Starkrimson' cultivars are more liked if they are obtained with a dehydrator, while apple chips obtained from the 'Golden Delicious' cultivar are more appreciated if they are obtained by dehydration in an oven (Figure 6).

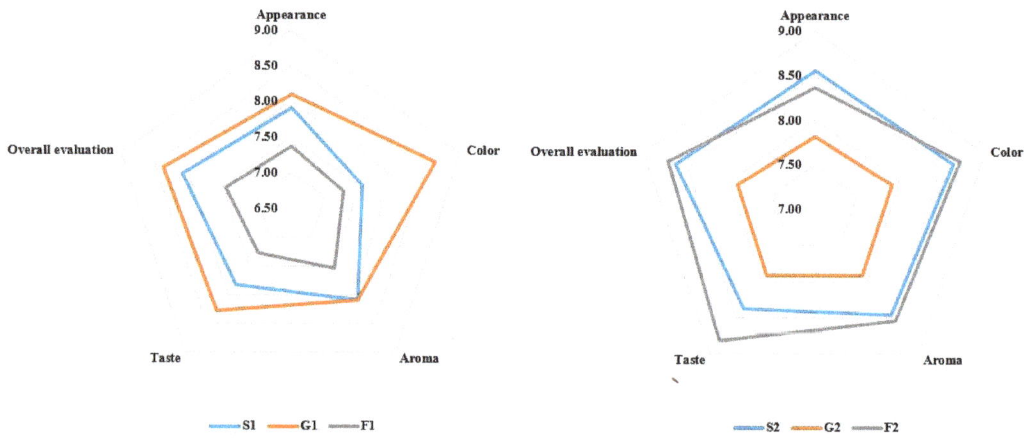

Figure 6. Sensory evaluation of apple chip samples (S1—apple chips 'Starkrimson', oven; S2—apple chips 'Starkrimson', dehydrator; G1—apple chips 'Golden Delicious', oven; G2—apple chips 'Golden Delicious', dehydrator; F1—apple chips 'Florina', oven; F2—apple chips 'Florina', dehydrator), on the first day.

3.6. Correlations of Measured Data

Positive and negative correlations between physico-chemical parameters measured or calculated for apple chips dried in the oven (Table 3) and apple chips dried in the dehydrator (Table 4) were determined.

Table 3. Correlations between physico-chemical parameters measured or calculated for apple chips dried in the oven (Pearson correlation).

	M	*aw*	*pH*	*TA*	*TSS*	*EC*	*L**	*a**	*b**	*ΔE*	*BI*
aw	**0.997** *										
pH	−0.722	−0.668									
TA	−0.246	−0.173	0.849								
TSS	**−0.959**	**−0.935**	0.888	0.511							
EC	**0.995**	**0.999** *	−0.646	−0.144	**−0.924**						
*L**	−0.858	−0.894	0.263	−0.287	0.676	**−0.906**					
*a**	0.264	0.336	0.477	0.870	0.020	0.363	−0.723				
*b**	0.007	−0.068	−0.697	**−0.971**	−0.290	−0.097	0.508	**−0.963**			
ΔE	**0.927**	**0.953**	−0.410	0.134	−0.783	**0.961**	**−0.988**	0.606	−0.368		
BI	**0.997** *	**0.999** *	−0.664	−0.168	**−0.933**	**0.999** *	−0.896	0.341	−0.073	**0.954**	
WI	**−0.941**	**−0.963**	0.444	−0.097	0.806	**−0.971**	**0.981**	−0.576	0.333	**−0.999** *	**−0.965**

* Statistically significant at $p < 0.05$. Values in bold show very high positive/negative correlation (according to Mukaka [68]).

Table 4. Correlations between physico-chemical parameters measured or calculated for apple chips dried in dehydrator (Pearson correlation).

	M	aw	pH	TA	TSS	EC	L^*	a^*	b^*	ΔE	BI
aw	0.473										
pH	−0.571	0.453									
TA	−0.373	**−0.994**	−0.549								
TSS	0.485	−0.541	**−0.995**	0.631							
EC	0.819	−0.119	**−0.939**	0.228	0.899						
L^*	**0.988**	0.331	−0.692	−0.225	0.615	0.898					
a^*	**−0.964**	−0.222	0.769	0.113	−0.700	**−0.942**	**−0.994**				
b^*	0.879	0.836	−0.110	−0.771	0.008	0.445	0.794	−0.72			
ΔE	**−0.930**	−0.116	0.833	0.006	−0.772	**−0.972**	**−0.976**	**0.994**	−0.641		
BI	−0.820	0.116	**0.938**	−0.225	−0.898	**−0.999** *	−0.899	**0.943**	−0.447	**0.973**	
WI	**0.942**	0.149	−0.814	−0.039	0.751	0.964	**0.982**	**−0.997** *	0.666	**−0.999** *	−0.965

* Statistically significant at $p < 0.05$. Values in bold show very high positive/negative correlation (according to Mukaka [68]).

Strongly positive correlations were observed between M and aw (0.997) or BI (0.997), and strongly negative correlations were noticed between M and TSS (−0.959) or WI (−0.941) in the case of apple samples dried in the oven (Table 3). From Table 4, it can be observed that there are strongly positive correlations between M and L^* (0.988) or WI (0.941) and strongly negative correlations between M and a^* (−0.964) or ΔE (−0.930) when apples are dried in the dehydrator. In the case of apple chip samples dried in the oven, strongly positive relationships were found between aw and EC or BI (0.999), between pH and TSS (0.888) or TA (0.849), between TA and a^* (0.870), between TSS and WI (0.806) and between EC and BI (0.999) (Table 3). In addition, strongly negative relationships were observed between aw and WI, pH and BI, TA and b^*, TSS and BI and EC and WI. Strongly positive correlations between parameters, when apples were dried in the dehydrator, were registered between aw and b^*, pH and BI, TSS and EC and EC and WI, while strongly negative relationships were observed between aw and TA, pH and TSS, TA and b^*, TSS and BI and EC and BI (Table 4). Many negative or positive significant correlations were observed between color parameters [42].

4. Conclusions

Apple quality is influenced by color, sugar content and acid composition. A ready-to-eat product, apple chips were produced from three different apple cultivars by drying. Apple chips preserved the good color, and it can be concluded, after analyzing the color parameters, that samples F1, G1 and S1 are closer to the color of fresh apple samples, followed by samples S2, G2 and F2. The drying of the apple chip samples in the dehydrator leads to a decrease in the values of water activity, which varied between 50 and 60% compared to the aw of the fresh samples, while for the samples obtained in the oven, the aw values decreased and varied between 28 and 30%. During storage, samples of apple chips dried in the oven are less influenced in terms of water activity, and the increase in aw values is smaller compared to the values obtained for samples from the dehydrator, except for the samples of the 'Golden Delicious' cultivar. Instead, it was observed that during apple chips storage, aw does not exceed the value 0.6, except for a single sample S1. During storage, TA values increased more for apple samples obtained in the oven (10.00–16.35%) than for those obtained in the dehydrator (5.33–12.35%), while TSS values decreased and varied from 34.11 (G2) to 42.87% (F2) (apple chips obtained in the dehydrator) and from 18.79 (S1) to 38.72% (F1) for apple chips obtained in the oven. After analyzing all the results obtained regarding the physico-chemical parameters and the sensory analysis, it can be concluded that the apple chips obtained by drying in a dehydrator on the first day are the most appreciated in the following order: S2, F2 and G2, followed by F1, G1 and S1; after twenty-one days, the order changes in the following way: F2 > G2 > G1 > S2 > F1 > S1.

Overall, this study concludes that the apple cultivar must also be considered in order to obtain apple chips, not just the dehydration method.

Author Contributions: Conceptualization, C.G. and A.E.P.; methodology, C.G. and A.E.P.; software, C.G.; validation, C.G., A.E.P. and A.L.; formal analysis, C.G. and A.E.P.; investigation, C.G. and A.E.P.; resources, C.G. and A.E.P.; writing—original draft preparation, C.G.; writing—review and editing, C.G., A.E.P. and A.L.; supervision, A.L.; funding acquisition, C.G. All authors have read and agreed to the published version of the manuscript.

Funding: This research was funded by 'Stefan cel Mare' University of Suceava and by Ministry of Research, Innovation and Digitalization within Program 1—Development of national research and development system, Subprogram 1.2—Institutional Performance—RDI excellence funding projects, under contract no. 10PFE/2021.

Institutional Review Board Statement: Not applicable.

Informed Consent Statement: Not applicable.

Data Availability Statement: Not applicable.

Acknowledgments: This research was supported by 'Stefan cel Mare' University of Suceava. This work was funded by Ministry of Research, Innovation and Digitalization within Program 1—Development of national research and development system, Subprogram 1.2—Institutional Performance—RDI excellence funding projects, under contract no. 10PFE/2021.

Conflicts of Interest: The authors declare no conflict of interest.

References

1. FAO. *World Food and Agriculture—Statistical Yearbook 2021*; FAO: Rome, Italy, 2021; p. 12.
2. FruitLogistica. *European Statistics Handbook*; Messe Berlin GmbH: Berlin, Germany, 2021.
3. EC. *EU Agricultural Outlook for Markets, Income and Environment, 2020–2030*; European Commission, DG Agriculture and Rural Development: Brussels, Belgium, 2020.
4. Leahu, A.; Ghinea, C.; Oroian, M.A. Osmotic dehydration of apple and pear slices: Color and chemical characteristics. *Ovidius Univ. Ann. Chem.* **2020**, *31*, 73–79. [CrossRef]
5. Średnicka-Tober, D.; Barański, M.; Kazimierczak, R.; Ponder, A.; Kopczyńska, K.; Hallmann, E. Selected Antioxidants in Organic vs. *Conventionally Grown Apple Fruits. Appl. Sci.* **2020**, *10*, 2997.
6. Butkeviciute, A.; Viskelis, J.; Viskelis, P.; Liaudanskas, M.; Janulis, V. Changes in the Biochemical Composition and Physicochemical Properties of Apples Stored in Controlled Atmosphere Conditions. *Appl. Sci.* **2021**, *11*, 6215. [CrossRef]
7. Koutsos, A.; Tuohy, K.M.; Lovegrove, J.A. Apples and cardiovascular health—Is the gut microbiota a core consideration? *Nutrients* **2015**, *7*, 3959–3998. [CrossRef]
8. Jakubczyk, E.; Kamińska-Dwórznicka, A.; Ostrowska-Ligeza, E.; Górska, A.; Wirkowska-Wojdyła, M.; Mańko-Jurkowska, D.; Górska, A.; Bryś, J. Application of Different Compositions of Apple Puree Gels and Drying Methods to Fabricate Snacks of Modified Structure, Storage Stability and Hygroscopicity. *Appl. Sci.* **2021**, *11*, 10286. [CrossRef]
9. Zhu, J.; Liu, Y.; Zhu, C.; Wei, M. Effects of different drying methods on the physical properties and sensory characteristics of apple chip snacks. *LWT-Food Sci. Technol.* **2022**, *154*, 112829. [CrossRef]
10. Chauhan, O.P.; Singh, A.; Singh, A.; Raju, P.S.; Bawa, A.S. Effects of Osmotic Agents on Colour, Textural, Structural, Thermal, and Sensory Properties of Apple Slices. *Int. J. Food Prop.* **2011**, *14*, 1037–1048. [CrossRef]
11. Kowalska, H.; Marzec, A.; Kowalska, J.; Samborska, K.; Tywonek, M.; Lenart, A. Development of apple chips technology. *Heat Mass Transf.* **2018**, *54*, 3573–3586. [CrossRef]
12. Figiel, A. Dehydration of apples by a combination of convective and vacuum-microwave drying. *Pol. J. Food Nutr. Sci.* **2007**, *57*, 131–135.
13. Royen, M.J.; Noori, A.W.; Haydary, J. Experimental Study and Mathematical Modeling of Convective Thin-Layer Drying of Apple Slices. *Processes* **2020**, *8*, 1562. [CrossRef]
14. Sullivan, V.K.; Na, M.; Proctor, D.N.; Kris-Etherton, P.M.; Petersen, K.S. Consumption of Dried Fruits Is Associated with Greater Intakes of Underconsumed Nutrients, Higher Total Energy Intakes, and Better Diet Quality in US Adults: A Cross-Sectional Analysis of the National Health and Nutrition Examination Survey, 2007–2016. *J. Acad. Nutr. Diet.* **2021**, *121*, 1258–1272. [CrossRef] [PubMed]
15. Carughi, A.; Feeney, M.J.; Kris-Etherton, P.; Fulgoni, V., 3rd; Kendall, C.W.; Bulló, M.; Webb, D. Pairing nuts and dried fruit for cardiometabolic health. *Nutr. J.* **2016**, *15*, 23. [CrossRef] [PubMed]
16. Sadler, M.J.; Gibson, S.; Whelan, K.; Ha, M.A.; Lovegrove, J.; Higgs, J. Dried fruit and public health—What does the evidence tell us? *Int. J. Food Sci. Nutr.* **2019**, *70*, 675–687. [CrossRef] [PubMed]

17. Kahraman, O.; Malvandi, A.; Vargas, L.; Feng, H. Drying characteristics and quality attributes of apple slices dried by a non-thermal ultrasonic contact drying method. *Ultrason. Sonochem.* **2021**, *73*, 105510. [CrossRef] [PubMed]
18. Nirmaan, A.M.C.; Rohitha Prasantha, B.D.; Peiris, B.L. Comparison of microwave drying and oven-drying techniques for moisture determination of three paddy (*Oryza sativa* L.) varieties. *Chem. Biol. Technol. Agric.* **2020**, *7*, 1. [CrossRef]
19. Zhu, R.; Jiang, S.; Li, D.; Law, C.L.; Han, Y.; Tao, Y.; Kiani, H.; Liu, D. Dehydration of apple slices by sequential drying pretreatments and airborne ultrasound-assisted air drying: Study on mass transfer, profiles of phenolics and organic acids and PPO activity. *Innov. Food Sci. Emerg. Technol.* **2022**, *75*, 102871. [CrossRef]
20. Wang, Y.; Zhao, H.; Deng, H.; Song, X.; Zhang, W.; Wu, S.; Wang, J. Influence of Pretreatments on Microwave Vacuum Drying Kinetics, Physicochemical Properties and Sensory Quality of Apple Slices. *Pol. J. Food Nutr. Sci.* **2019**, *69*, 297–306. [CrossRef]
21. Barrett, D.M.; Beaulieu, J.C.; Shewfelt, R. Color, flavor, texture, and nutritional quality of fresh-cut fruits and vegetables: Desirable levels, instrumental and sensory measurement, and the effects of processing. *Crit. Rev. Food Sci. Nutr.* **2010**, *50*, 369–389. [CrossRef]
22. Ayustaningwarno, F.; Dekker, M.; Fogliano, V.; Verkerk, R. Effect of Vacuum Frying on Quality Attributes of Fruits. *Food Eng. Rev.* **2018**, *10*, 154–164. [CrossRef]
23. Cruz, A.C.; Guiné, R.P.F.; Gonçalves, J.C. Drying Kinetics and Product Quality for Convective Drying of Apples (cvs. Golden Delicious and Granny Smith). *Int. J. Fruit Sci.* **2015**, *15*, 54–78. [CrossRef]
24. Marzec, A.; Kowalska, H.; Kowalska, J.; Domian, E.; Lenart, A. Influence of Pear Variety and Drying Methods on the Quality of Dried Fruit. *Molecules* **2020**, *25*, 5146. [CrossRef] [PubMed]
25. Prawiranto, K.; Defraeye, T.; Derome, D.; Verboven, P.; Nicolai, B.; Carmeliet, J. New insights into the apple fruit dehydration process at the cellular scale by 3D continuum modelling. *J. Food Eng.* **2018**, *239*, 52–63. [CrossRef]
26. Kidoń, M.; Grabowska, J. Bioactive compounds, antioxidant activity, and sensory qualities of red-fleshed apples dried by different methods. *LWT* **2021**, *136*, 110302. [CrossRef]
27. Singh, A.P.; Sahni, D.D.; Dubey, A. Experimental study of thermal effect of drying, heating, blowing machine with different stage. *Int. J. Res.-Granthaalayah* **2016**, *4*, 104–109. [CrossRef]
28. Nyangena, I.; Owino, W.; Ambuko, J.; Imathiu, S. Effect of selected pretreatments prior to drying on physical quality attributes of dried mango chips. *J. Food. Sci. Technol.* **2019**, *56*, 3854–3863. [CrossRef]
29. ElHana, A.N.H. Microwave drying of apple. *Misr J. Agric. Eng.* **2008**, *25*, 980–1003.
30. Fahim, U.; Kang, M. Impact of air flow rate on drying of apples and performance assessment of parabolic trough solar collector. *Appl. Therm. Eng.* **2017**, *127*, 275–280.
31. Seiiedlou, S.; Ghasemzadeh, H.R.; Hamdami, N.; Talati, F.; Moghaddam, M. Convective drying of apple: Mathematical modeling and determination of some quality parameters. *Int. J. Agric. Biol.* **2010**, *12*, 171–178.
32. Önal, B.; Adiletta, G.; Crescitelli, A.; Di Matteo, M.; Russo, P. Optimization of hot air drying temperature combined with pre-treatment to improve physico-chemical and nutritional quality of 'Annurca' apple. *Food Bioprod. Process.* **2019**, *115*, 87–99. [CrossRef]
33. Velickova, E.; Winkelhausen, E.; Kuzmanova, S. Physical and sensory properties of ready to eat apple chips produced by osmo-convective drying. *J. Food Sci. Technol.* **2014**, *51*, 3691–3701. [CrossRef]
34. Rydzak, L.; Kobus, Z.; Nadulski, R.; Wilczyński, K.; Pecyna, A.; Santoro, F.; Sagan, A.; Starek-Wójcicka, A.; Krzywicka, M. Analysis of Selected Physicochemical Properties of Commercial Apple Juices. *Processes* **2020**, *8*, 1457. [CrossRef]
35. Sadler, G.D.; Murphy, P. *pH* and Titratable Acidity. In *Food Anal.*, 4th ed.; Nielsen, S.S., Ed.; Springer: New York, NY, USA, 2010; pp. 219–238.
36. Karakasova, L.; Stefanovska, E.; Babanovska-Milenkovska, F.; Stamatovska, V.; Durmishi, N.; Culeva, B. Comparing the quality properties of fresh and dried apple fruit—Varieties Pinova and Red Delicious. *J. Food Agric. Environ.* **2019**, *73*, 36–45.
37. Khan, S.A.; Beekwilder, J.; Schaart, J.G.; Mumm, R.; Soriano, J.M.; Jacobsen, E.; Schouten, H.J. Differences in acidity of apples are probably mainly caused by a malic acid transporter gene on LG16. *Tree Genet. Genomes* **2013**, *9*, 475–487. [CrossRef]
38. Leahu, A.; Oroian, M.; Ropciuc, S. Total phenolics of fresh and frozen minor berries and their antioxidant properties. *Food Environ. Saf.* **2014**, *13*, 87–93.
39. Prisacaru, A.E.; Ghinea, C.; Apostol, L.C.; Ropciuc, S.; Ursachi, F. Physicochemical Characteristics of Vinegar from Banana Peels and Commercial Vinegars before and after in Vitro Digestion. *Processes* **2021**, *9*, 1193. [CrossRef]
40. Magwaza, L.S.; Opara, U.L. Analytical methods for determination of sugars and sweetness of horticultural products—A review. *Sci. Hortic.* **2015**, *184*, 179–192. [CrossRef]
41. Ghinea, C.; Leahu, A.; Priscarau, A.E.; Cojocaru, M.; Ladariu, V. Physico-chemical and sensory analyzes of muffins obtained with almond flour and coconut oil. In Proceedings of the 19th International Multidisciplinary Scientific GeoConference SGEM 2019, Vienna, Austria, 9–11 December 2019; Volume 19, pp. 165–172.
42. Alibas, I.; Yilmaz, A. Microwave and convective drying kinetics and thermal properties of orange slices and effect of drying on some phytochemical parameters. *J. Therm. Anal. Calorim.* **2021**. [CrossRef]
43. Cuccurullo, G.; Giordano, L.; Metallo, A.; Cinquanta, L. Drying rate control in microwave assisted processing of sliced apples. *Biosyst. Eng.* **2018**, *170*, 24–30. [CrossRef]

44. Ackbarali, D.; Mahara, R. Sensory Evaluation as a Tool in Determining Acceptability of Innovative Products Developed by Undergraduate Students in Food Science and Technology at The University of Trinidad and Tobago. *J. Curric. Teach.* **2014**, *3*, 10–27.
45. Cliff, M.A.; Toivonen, P.M.; Forney, C.F.; Lu, C. Quality of fresh-cut apple slices stored in solid and micro-perforated film packages having contrasting O_2 headspace atmospheres. *Postharvest Biol. Technol.* **2010**, *58*, 254–261. [CrossRef]
46. Hussain, P.R.; Wani, I.A.; Rather, S.A.; Suradkar, P.; Ayob, O. Effect of post-processing radiation treatment on physico-chemical, microbiological and sensory quality of dried apple chips during storage. *Radiat. Phys. Chem.* **2021**, *182*, 109367. [CrossRef]
47. Mohammed, S.; Edna, M.; Siraj, K. The effect of traditional and improved solar drying methods on the sensory quality and nutritional composition of fruits: A case of mangoes and pineapples. *Heliyon* **2020**, *6*, e04163. [CrossRef]
48. Singer, J.M.; Pedroso-de-Lima, A.C.; Tanaka, N.I.; González-López, V.A. To triplicate or not to triplicate? *Chemom. Intell. Lab. Syst.* **2007**, *86*, 82–85. [CrossRef]
49. Wu, J.; Gao, H.; Zhao, L.; Liao, X.; Chen, F.; Wang, Z.; Hu, X. Chemical compositional characterization of some apple cultivars. *Food Chem.* **2007**, *103*, 88–93. [CrossRef]
50. Preti, R.; Tarola, A.M. Study of polyphenols, antioxidant capacity and minerals for the valorisation of ancient apple cultivars from Northeast Italy. *Eur. Food Res. Technol.* **2021**, *247*, 273–283. [CrossRef]
51. Ergün, Z. Determination of Biochemical Contents of Fresh, Oven-Dried, and Sun-Dried Peels and Pulps of Five Apple Cultivars (Amasya, Braeburn, Golden Delicious, Granny Smith, and Starking). *J. Food Qual.* **2021**, *2021*, 9916694 [CrossRef]
52. Minitab. Interpret the Key Results for One-Way ANOVA. 2022. Available online: https://support.minitab.com/en-us/minitab-express/1/help-and-how-to/modeling-statistics/anova/how-to/one-way-anova/interpret-the-results/key-results/ (accessed on 1 February 2022).
53. Iordănescu, O.A.; Iuga, C.; Hădărugă, N.; Becherescu, A.; Drăgunescu, A.; Băla, M.; Scedei, D. Research on fruit quality of several apple cultivars from different locations in the area of Western Romania. *Sci. Pap. Ser. B Hortic.* **2021**, *LXV*, 1129–1138.
54. Shi, J.; Pan, Z.; McHugh, T.H.; Wood, D.; Hirschberg, E.; Olson, D. Drying and quality characteristics of fresh and sugar-infused blueberries dried with infrared radiation heating. *LWT-Food Sci. Technol.* **2008**, *41*, 1962–1972. [CrossRef]
55. Klewicki, R.; Konopacka, D.; Uczciwek, M.; Irzyniec, Z.; Piasecka, E.; Bonazzi, C. Sorption isotherms for osmo-convectively-dried and osmo-freeze-dried apple, sour cherry, and blackcurrant (ISAFRUIT Special Issue). *J. Hortic. Sci. Biotechnol.* **2009**, *84*, 75–79. [CrossRef]
56. Antal, T. Comparative study of three drying methods: Freeze, hot air assisted freeze and infrared-assisted freeze modes. *Agron. Res.* **2015**, *13*, 863–878.
57. Patras, A. Quality parameters of apple fruits marketed in Iaşi. *Sci. Pap. Ser. B Hortic.* **2018**, *61*, 281–286.
58. Owusu, J.; Ma, H.; Wang, Z.; Amissah, A. Effect of Drying Methods on Physicochemical Properties of Pretreated Tomato (lycopersicon esculentum mill.). *Slices. Croat. J. Food Tehnol. Biotech. Nutr.* **2012**, *7*, 106–111.
59. Nour, V.; Trandafir, I.; Ionica, M.E. Compositional characteristics of fruits of several apple (Malus domestica Borkh.) cultivars. *Not. Bot. Horti. Agrobot.* **2010**, *38*, 228–233.
60. Egea, M.B.; Ferreira da Silva, R.S.S.; Yamashita, F.; Borsato, D. Optimizing Dehydration of Apples Malus Domestica with Fructo-Oligosaccharide Incorporation. *Braz. Arch. Biol. Technol.* **2012**, *55*, 751–762. [CrossRef]
61. Li, Y.; Sun, H.; Li, J.; Qin, S.; Yang, W.; Ma, X.; Qiao, X.; Yang, B. Effects of Genetic Background and Altitude on Sugars, Malic Acid and Ascorbic Acid in Fruits of Wild and Cultivated Apples (*Malus* sp.). *Foods* **2021**, *10*, 2950. [CrossRef]
62. Varghese, K.S.; Pandey, M.C.; Radhakrishna, K.; Bawa, A.S. Technology, applications and modelling of ohmic heating: A review. *J. Food Sci. Technol.* **2014**, *51*, 2304–2317. [CrossRef]
63. Banti, M. Review on Electrical Conductivity in Food, the Case in Fruits and Vegetables. *World J. Food Sci. Technol.* **2020**, *4*, 80–89. [CrossRef]
64. Wrolstad, R.E.; Smith, D.E. Color Analysis. In *Food Anal.*, 4th ed.; Nielsen, S.S., Ed.; Springer: New York, NY, USA, 2010; pp. 573–586.
65. Coklar, H.; Akbulut, M.; Kilinc, S.; Yildirim, A.; Alhassan, I. Effect of Freeze, Oven and Microwave Pretreated Oven Drying on Color, Browning Index, Phenolic Compounds and Antioxidant Activity of Hawthorn (Crataegus orientalis) Fruit. *Not. Bot. Horti. Agrobot.* **2018**, *46*, 449–456. [CrossRef]
66. Moon, K.M.; Kwon, E.-B.; Lee, B.; Kim, C.Y. Recent Trends in Controlling the Enzymatic Browning of Fruit and Vegetable Products. *Molecules* **2020**, *25*, 2754. [CrossRef]
67. Li, L.; Li, X.; Wang, A.; Jiang, Y.; Ban, Z. Effect of heat treatment on physiochemical, colour, antioxidant and microstructural characteristics of apples during storage. *Int. J. Food Sci. Technol.* **2013**, *48*, 727–734. [CrossRef]
68. Mukaka, M.M. Statistics corner: A guide to appropriate use of correlation coefficient in medical research. *Malawi Med. J.* **2012**, *24*, 69–71. [PubMed]

applied sciences

MDPI

Article

Effects of Dry Heat Treatment and Milling on Sorghum Chemical Composition, Functional and Molecular Characteristics

Ana Batariuc [1], Mădălina Ungureanu-Iuga [1,2,*] and Silvia Mironeasa [1,*]

[1] Faculty of Food Engineering, "Ştefan cel Mare" University of Suceava, 13 University Street, 720229 Suceava, Romania; ana.batariuc@usm.ro
[2] Integrated Center for Research, Development and Innovation in Advanced Materials, Nanotechnologies, and Distributed Systems for Fabrication and Control (MANSiD), "Ştefan cel Mare" University of Suceava, 13th University Street, 720229 Suceava, Romania
* Correspondence: madalina.iuga@usm.ro (M.U.-I.); silviam@fia.usv.ro (S.M.)

Abstract: This study aimed to highlight the effects of grains dry heat treatment, flour particle size and variety on sorghum flour nutritional, functional, and molecular characteristics. The results obtained showed that dry heat treatment led to fat, fiber and water absorption capacity increase, while the moisture, protein, ash, water retention capacity, solubility index, foaming capacity, and FT-IR absorption bands characteristic to phytic acids decreased with temperature applied raised. Particle size reduction determined lower protein, solubility index, and emulsifying activity and higher fat content, oil absorption capacity, swelling power, and foaming capacity. White sorghum flour fractions presented lower protein content, except when they were treated at 140 °C, lower carbohydrates and fibers and higher fat content compared to those made of red sorghum. Moderate significant correlations ($p < 0.05$) were observed between some of the functional properties and proximate composition of flours. Thus, both dry heat treatment conditions and particle size exerted significant influences of sorghum flour chemical and functional properties. These results showed the importance of particle size and dry heat treatment on sorghum flours functionality, being helpful for further optimizations and choices for bakery products use.

Keywords: sorghum flour; roasting; proximate composition; functional properties; particle size; FT-IR spectra

Citation: Batariuc, A.; Ungureanu-Iuga, M.; Mironeasa, S. Effects of Dry Heat Treatment and Milling on Sorghum Chemical Composition, Functional and Molecular Characteristics. *Appl. Sci.* **2021**, *11*, 11881. https://doi.org/10.3390/app112411881

Academic Editor: Antonio Valero

Received: 5 November 2021
Accepted: 12 December 2021
Published: 14 December 2021

Publisher's Note: MDPI stays neutral with regard to jurisdictional claims in published maps and institutional affiliations.

1. Introduction

Sorghum (*Sorghum bicolor* (L.) Moench) is the fifth most cultivated cereal in the world, after corn, wheat, rice and barley [1], being a staple food crop in the semi-arid tropical regions of Africa, Asia, and Latin America [2]. Originally from East Africa (Sudan and Ethiopia), it has been cultivated all over the world due to its agronomic advantages, being resistant to drought, pests, and diseases, and flexible in planting time [3].

Sorghum is a small spheroid grain of approximately 3 mm in diameter that includes the germ and the endosperm inside the grain and the outer layers which give the red, brown, white, or black color [4]. Sorghum nutritional profile contains carbohydrates (54.6–77.2%), dietary fiber (4.5–26.3%), proteins (4.7–19.0%), and lipids (1.6–5.0%) [5–7]. Sorghum grains are also a good source of minerals (phosphorus, zinc, iron, calcium, magnesium, and potassium), B vitamins, vitamin E, β-carotene, and bioactive compounds (polyphenols and anthocyanins) [5,8]. Their high nutritional value varies depending on variety, cultivation region, and pericarp color [9]. Sorghum's outstanding nutritional properties make it a promising functional ingredient that offers the opportunity to produce foods with high levels of dietary fibers and antioxidants, and also with various natural colors [10,11] due to the specific pigment of each variety.

Different food products such as breads [12,13], tortillas [14,15], snacks [16], pasta [17] and noodles [18] were produced using sorghum flour as functional ingredient. In addition, due to its content in resistant starch and glucans, sorghum can be used as a prebiotic food ingredient [19]. Being deficient in gluten protein, sorghum grain can supply the needs of people with gluten intolerance [1,20], while also being suitable for patients with diabetes and cardiovascular diseases [21].

During roasting, a variety of changes can inevitably occur and it is thus necessary to control the roasting time and temperature to obtain optimal characteristics. Roasted cereal grains may influence milling yield since they are generally characterized by decreased kernel hardness, due to the increased internal porosity of the endosperm [22]. Thermal changes induced by roasting impact kernel hardness by making it softer due to the loss of endosperm structure [23]. The roasting method and conditions under which processing is performed impact the product properties [24]. As Schoeman et al. [24] states, roasting could potentially serve as a pre-processing step to enable the use of less energy for milling, to produce value-added products, or to extend the shelf life of products. Sorghum flour sieving to different particle sizes might be a necessary processing step, taking into account to improve the quality of certain food. Dayakar Rao et al. [25] affirmed that biscuits produced from sorghum flours of 180 and 251 μm particle sizes had a better consumer acceptance than those produced from flours with smaller particle sizes (below 180 μm). In contrast, 60% sorghum extraction flour used to produce gluten-free bread showed a higher specific volume and softer texture than gluten-free bread from whole-grain sorghum flour [26]. On the other hand, transferring results research from lab-scale mills to an industrial scale milling system represents a challenge. Recently, Rumler et al. [27] investigated the effectiveness of sorghum milling when using two different milling systems and showed their impact on the chemical and physical properties of flour fractions and whole sorghum flours obtained.

Flour particle size and composition affect the functionality of the sorghum flour such as water absorption capacity, water solubility index, swelling power, pasting properties, and product quality [26,28,29]. The particle size is frequently associated with the surface area available for enzymatic action [29]. The sorghum cultivar and climatic conditions impact grain hardness [30]. Moreover, the sorghum cultivar determines the amylose/amylopectin ratio, which influences the extent of gelatinization and retrogradation, higher amylose content promoting retrogradation [18], A decrease in starch gelatinization properties was found due to the presence of kafirins, storage proteins with high hydrophobicity [31]. In addition, these proteins, stabilized by disulfide bonds, determined and a decrease in starch and protein digestibility due to the tight starch–protein matrix formed [32]. These undesirable aspects can be diminished by using thermal treatment such as roasting. Additionally, this processing method can decrease the antinutritional components found in raw sorghum grains (e.g., tannins, phytic acid, and protein cross-linker) [33,34] which reduces the feed efficiency [8], improving thus their nutritional value and products. In comparison with other grains, sorghum develops a pleasant taste and crispiness after roasting, a treatment that extends the shelf-life and safety of products by lowering the water activity [34]. Several studies have been undertaken to investigate the effects of roasting as a pre-milling treatment of sorghum grains. Ranganathan et al. [35] found that roasting of sorghum grains increased the water absorption capacity, offering beneficial effects for the preparation of instant mixes, porridge, and soup. Roasted sorghum flour exhibited better functional, pasting, and antioxidant properties when microwave processing was applied [36].

The aim of the present study was to investigate the effects of dry heat treatment of two sorghum varieties (red and white) on the functional, chemical composition, and molecular characteristics of flours with different particle sizes. For this purpose, the proximate composition of flours in terms of proteins, fat, carbohydrates, moisture, and ash were determined along with the molecular structures and functional properties in terms of hydration capacity, water absorption capacity, oil absorption capacity, solubility index, water retention capacity, swelling power, bulk density, emulsifying and foaming properties.

2. Materials and Methods

2.1. Sorghum Treatment

White sorghum grains (ES Albanus hybrid) were purchased from the Secuieni Agricultural Development Research Station (Neamț, România) and red sorghum grains (ES Alize hybrid) were purchased from a farmer (Suceava, România).

Dry heat treatment of sorghum grains at different temperatures (121 °C—T1 and 140 °C—T2) was done for 15 min in a Binder ED53L convection oven (Binder, Tuttlingen, Germany). Untreated sorghum grains were considered as control samples.

Integral sorghum flours (I) were obtained by grains milling with a laboratory grinder (Grain Mill, KitchenAid, Model 106 5KGM, Benton Harbor, MI, USA). In order to obtain sorghum flour at three different particle sizes, large (L > 300 μm), medium (200 μm < M < 250 μm), small fractions (S < 200 μm), the integral flour was sieved in a Retsch Vibratory Sieve Shaker AS 200 basic (Haan, Germany).

2.2. Proximate Composition

The nutritional composition (moisture, protein, fat, ash) of sorghum flours were analyzed using ICC methods: moisture (101/1), fat (104/1), protein (105/2), and ash (105/1). Total dietary fiber was determined by using Megazyme kit (K-TDFR-200a 04/17), according to the AACC 32-05.01 method. The carbohydrate content was calculated by difference, by applying the equation [37] (Equation (1)):

$$\text{Carbohydrates } (\%) = 100 - (\text{protein} + \text{fat} + \text{ash} + \text{fiber} + \text{moisture}) \tag{1}$$

The energetic value (kcal/100 g) of the samples was also calculated by multiplying nutrients values by their corresponding conversion coefficients [37] (Equation (2)):

$$\text{Energy } (\text{kcal}/100 \text{ g}) = (4 \times \text{protein}) + (9 \times \text{fat}) + (4 \times \text{carbohydrates}) + (2 \times \text{fiber}) \tag{2}$$

2.3. Functional Properties of Sorghum Flours

2.3.1. Hydration Capacity

The hydration capacity was determined in duplicate, according to the method described by Bordei [38]. For this purpose, 5 g of flour were weighed into a 50 mL tube and 30 mL of tap water was added. After mixing with a rod for 30 s at 10 min intervals for 1 h, the rod was washed over the tube with 10 mL of water and the suspension was centrifuged for 20 min at 2300 rpm. After removal of the supernatant, the sample was kept at 50 °C for 25 min and weighed after cooling. The hydration capacity was calculated with Equation (3):

$$\text{Hydration capacity } (\%) = \frac{(m_2 - m_o) - m_1}{m_1} \times 100 \tag{3}$$

where m_o—the weight of the tube, m_1—the weight of sample taken into analysis, m_2—the weight of sample which absorbed water.

2.3.2. Water Absorption Capacity

The water absorption capacity was determined in duplicate, according to the method described by Oladiran and Emmambux [39] with modifications. An amount of 2.5 g of the sample was placed in a centrifuge tube with 30 mL of distilled water. The sample was kept in a water bath with continuous stirring at 30 °C for 30 min and after centrifugation at 3500 rpm for 15 min the supernatant was removed and the residue was weighed. The results were calculated by using Equation (4):

$$\text{Water absorption capacity } (\%) = \frac{m_1}{m_o} \times 100 \tag{4}$$

where m_o—the weight of sample taken into analysis, m_1—the weight of the sample after supernatant removal.

2.3.3. Oil Absorption Capacity

Oil absorption capacity was determined in duplicate according to the method described by Elkhalifa and Bernhardt [40] with modifications. An amount of 3 g of sample was placed in a centrifuge tube with 30 mL of sunflower oil. The sample was stirred for 1 min, every 10 min for 30 min. After centrifugation at 3000 rpm for 15 min, the supernatant was decanted and the tubes were allowed to drain for 5 min, then the residue was weighed. The results were calculated by using Equation (5):

$$\text{Oil absorption capacity } (\%) = \frac{m_1}{m_o} \times 100 \tag{5}$$

where m_o—the weight of sample taken into analysis, m_1—the weight of the sample after supernatant removal.

2.3.4. Determination of Solubility Index

The method for solubility index determination was adapted according to that presented by Oladiran and Emmambux [39]. The supernatant from the water absorption capacity determination was dried in metal capsules at 100 °C to constant mass and weighed after cooling in the desiccator. The amount of soluble solids expressed as a percentage was defined as the solubility index.

2.3.5. Water Retention Capacity

Water retention capacity was determined in duplicate according to the method described by Zhu et al. [41]. For this purpose, 1 g of sample was placed in a centrifuge tube with 30 mL of distilled water. After 18 h of resting, the sample was centrifuged at 3000 rpm for 20 min and the supernatant was removed. The sample was dried for 2 h at 105 °C in a convection oven. The results expressed as an average of two determinations were calculated with Equation (6):

$$\text{Water retention capacity } (\text{g/g}) = \frac{m_1 - m_2}{m_2} \times 100 \tag{6}$$

where m_1—the weight of the sample before drying, m_2—the weight of the sample after drying.

2.3.6. Swelling Power

The swelling power was determined according to the method described by Elkhalifa and Bernhardt [40] with modifications: 0.5 g of sample were mixed with 15 mL of distilled water in a weighed centrifuge tube, which was heated in a water bath at 90 °C for 30 min and mixed well to prevent lumps. After cooling to room temperature for 15 min, the sample was centrifuged at 3000 rpm for 25 min. The supernatant was carefully removed and the swollen flour sediment was weighed.

2.3.7. Emulsion Activity and Stability

The emulsifying properties were determined in duplicate according to the method presented by Elkhalifa and Bernhard [40]: 2 g of sample was mixed with 20 mL of distilled water cooled at 4 °C and 20 mL of sunflower oil, in a centrifuge tube. The sample was stirred for 20 min and centrifuged at 4000 rpm for 10 min, then the height of the emulsion layer formed was observed. The emulsion activity was calculated with Equation (7):

$$\text{Emulsifying activity } (\%) = \frac{\text{height of emulsion layer}}{\text{height of whole layer}} \times 100 \tag{7}$$

For emulsion stability evaluation, the emulsion formed was heated in a water bath at 80 °C for 30 min, followed by cooling to room temperature for 20 min. The tube was centrifuged again at 4000 rpm for 10 min. The height of the emulsified layer was measured and the stability of the emulsion was calculated with Equation (8):

$$\text{Emulsion stability (\%)} = \frac{\text{height of the emulsion layer after heating}}{\text{height of whole layer}} \times 100 \qquad (8)$$

2.3.8. Bulk Density

For bulk density measurement, 10 g of flour were placed in a 25 mL graduated cylinder. The cylinder was lightly beaten ten times to even out the flour and the final volume of the flour was measured and expressed as g/cm^3 [40].

2.3.9. Foaming Capacity and Stability

Foaming capacity and stability were determined according to the methods described by Elkhalifa and Bernhardt [40]. In a 500 mL beaker, 2 g of sample were transferred with 100 mL of distilled water, and the suspension was mixed with an electric blender, at room temperature, for 1 min. The contents were immediately transferred to a 250 mL graduated cylinder and the volume of foam was measured. The foaming capacity was expressed using the following formula (Equation (9)):

$$\text{Foaming capacity (\%)} = \frac{\text{volume after whipping} - \text{volume before whipping}}{\text{volume before whipping}} \times 100 \qquad (9)$$

Foam stability was determined by monitoring the fall in the volume of the foam as a function of time after every 10 min for 1 h and expressed using Equation (10):

$$\text{Foam stability (\%)} = \frac{\text{Foam volume after set of time}}{\text{Initial foam volume}} \times 100 \qquad (10)$$

2.4. FT-IR Spectra Collection and Interpretation

FT-IR spectra of sorghum flours were acquired in the range of 650 to 4000 cm^{-1}, from a Thermo Scientific Nicolet iS20 (Waltham, MA, USA) device equipped with and ATR module, at a resolution of 8 cm^{-1} and with 64 scans. The spectra were analyzed with OMNIC software and the carbohydrates, protein, lipid, and polyphenols structures were identified according to previous studies [42–44].

2.5. Statistical Analysis

Statistically significant differences at 95% confidence level were evaluated by means of three-way ANOVA with Tukey's test, by using XLSTAT for Excel 2021 version (Addinsoft, New York, NY, USA). Principal component analysis (PCA) based on Pearson correlations was employed to evaluate the relationships between the sorghum flour characteristics and to underline similarities or dissimilarities between them.

3. Results

3.1. Proximate Composition

Dry heat treatment, particle size, and sorghum variety significantly influenced ($p < 0.05$) sorghum flour proximate composition (Table 1). The protein content decreased with particle size decrease, while the increase of treatment temperature led to lower values, red sorghum fractions being richer in protein compared to white variety. Sorghum flour fat content registered proportional increases with particle size reduction, higher amounts being observed in white sorghum compared to the red one. Dry heat treatment produced an increase in fat content as the temperature was higher, depending on the particle size. The ash content showed significant raise as the particle size was lower, slightly higher values being obtained for white sorghum, while dry heat treatment induced a slight decrease of this parameter. Sorghum grain treatment caused flour moisture decrease, while in the case of particle sizes irregular trends were observed. Particle size reduction led to higher carbohydrates content compared to integral flour, except for L particle size, sorghum grain treatment temperature increase determining slightly higher values compared to the control, while the red variety showed higher flour carbohydrates contents than the white one. Sorghum flour fiber content increased until M particle size and then decreased for S samples, grains treatment causing significant increases compared to the control. Red sorghum showed higher fibers content compared to the white one.

Table 1. Effect of dry heat treatment on the proximate composition of sorghum flour fractions.

Treatment	Protein (%)	Fat (%)	Ash (%)	Moisture (%)	Total Dietary Fiber (%)	Carbohydrates (%)	Energetic Value (kcal/100 g)
CW_I	10.35 ± 0.08 [ef]	3.07 ± 0.03 [ijk]	1.16 ± 0.03 [j]	11.50 ± 0.01 [c]	8.35 ± 0.24 [cde]	65.56 ± 0.29 [i]	348.01 ± 0.44 [hi]
CW_L	11.84 ± 0.06 [a]	3.12 ± 0.01 [ijk]	0.76 ± 0.01 [l]	11.29 ± 0.01 [d]	5.38 ± 0.51 [k]	67.59 ± 0.51 [efg]	356.59 ± 1.12 [ef]
CW_M	8.84 ± 0.01 [mn]	3.15 ± 0.03 [ij]	1.15 ± 0.01 [j]	11.20 ± 0.03 [e]	5.81 ± 0.02 [jk]	69.86 ± 0.02 [abc]	354.74 ± 0.14 [fg]
CW_S	10.62 ± 0.19 [d]	3.20 ± 0.03 [hi]	2.26 ± 0.03 [a]	10.98 ± 0.01 [g]	5.30 ± 0.34 [k]	67.64 ± 0.43 [efg]	352.45 ± 0.82 [g]
T1W_I	10.28 ± 0.07 [fg]	3.45 ± 0.04 [f]	1.17 ± 0.01 [j]	8.19 ± 0.04 [P]	6.18 ± 0.31 [ijk]	70.73 ± 0.27 [a]	367.45 ± 0.65 [b]
T1W_L	11.14 ± 0.04 [c]	1.62 ± 0.06 [n]	0.30 ± 0.01 [n]	9.53 ± 0.01 [i]	6.49 ± 0.58 [hijk]	70.93 ± 0.62 [a]	355.74 ± 1.08 [f]
T1W_M	10.44 ± 0.05 [def]	4.64 ± 0.15 [c]	1.54 ± 0.03 [n]	9.52 ± 0.01 [i]	6.59 ± 0.10 [ghijk]	67.26 ± 0.09 [fgh]	365.78 ± 0.59 [bc]
T1W_S	8.68 ± 0.03 [no]	5.14 ± 0.06 [b]	2.15 ± 0.01 [b]	10.01 ± 0.01 [h]	6.48 ± 0.48 [hijk]	67.53 ± 0.51 [efg]	364.08 ± 0.87 [c]
T2W_I	9.48 ± 0.06 [jk]	3.01 ± 0.02 [k]	1.37 ± 0.01 [i]	8.74 ± 0.01 [n]	7.14 ± 0.60 [efghi]	70.25 ± 0.57 [ab]	360.35 ± 1.29 [d]
T2W_L	10.16 ± 0.03 [gh]	2.49 ± 0.04 [m]	0.80 ± 0.07 [l]	8.73 ± 0.01 [n]	6.81 ± 0.48 [fghij]	71.00 ± 0.46 [a]	360.69 ± 1.05 [d]
T2W_M	10.08 ± 0.03 [hi]	4.30 ± 0.03 [d]	1.53 ± 0.03 [h]	8.96 ± 0.01 [k]	7.57 ± 0.19 [efgh]	67.55 ± 0.17 [efg]	364.40 ± 0.37 [c]
T2W_S	8.58 ± 0.02 [o]	6.32 ± 0.03 [a]	2.03 ± 0.02 [c]	8.91 ± 0.01 [k]	5.42 ± 0.16 [k]	68.73 ± 0.18 [cde]	377.00 ± 0.28 [a]
CR_I	11.40 ± 0.05 [b]	2.84 ± 0.02 [l]	1.34 ± 0.00 [i]	11.08 ± 0.03 [f]	7.82 ± 0.28 [defg]	65.51 ± 0.28 [i]	348.89 ± 0.54 [h]
CR_L	11.94 ± 0.18 [a]	2.77 ± 0.04 [l]	0.88 ± 0.04 [k]	11.15 ± 0.06 [e]	9.94 ± 0.39 [ab]	63.32 ± 0.54 [j]	345.84 ± 0.85 [i]
CR_M	9.66 ± 0.18 [j]	3.10 ± 0.03 [ijk]	1.75 ± 0.03 [f]	11.80 ± 0.03 [b]	7.69 ± 0.39 [defgh]	66.00 ± 0.36 [hi]	345.91 ± 0.80 [i]
CR_S	10.58 ± 0.13 [d]	3.28 ± 0.04 [gh]	2.30 ± 0.03 [a]	12.01 ± 0.05 [a]	6.49 ± 0.31 [hijk]	65.33 ± 0.31 [i]	346.16 ± 0.60 [i]
T1R_I	10.53 ± 0.05 [de]	2.73 ± 0.16 [l]	1.39 ± 0.04 [i]	8.02 ± 0.03 [q]	9.41 ± 0.08 [bc]	67.92 ± 0.06 [defg]	357.21 ± 0.86 [ef]
T1R_L	11.48 ± 0.03 [b]	1.16 ± 0.04 [o]	0.66 ± 0.01 [m]	8.80 ± 0.02 [lm]	10.75 ± 0.95 [a]	67.14 ± 0.98 [gh]	346.43 ± 1.85 [hi]
T1R_M	9.29 ± 0.03 [l]	3.40 ± 0.03 [fg]	1.52 ± 0.01 [h]	8.75 ± 0.01 [mn]	8.38 ± 0.40 [cde]	68.65 ± 0.39 [cde]	359.13 ± 0.88 [de]
T1R_S	8.51 ± 0.03 [o]	4.04 ± 0.06 [e]	1.91 ± 0.03 [e]	8.94 ± 0.01 [k]	8.02 ± 0.01 [def]	68.58 ± 0.10 [cdef]	360.75 ± 0.07 [d]
T2R_I	9.35 ± 0.00 [kl]	3.04 ± 0.04 [jk]	1.54 ± 0.01 [h]	8.67 ± 0.01 [o]	7.64 ± 0.37 [defgh]	69.76 ± 0.37 [abc]	359.07 ± 0.66 [de]
T2R_L	9.93 ± 0.05 [i]	1.62 ± 0.03 [n]	0.93 ± 0.01 [k]	8.82 ± 0.02 [l]	8.11 ± 0.68 [cdef]	70.58 ± 0.67 [a]	352.88 ± 1.30 [g]
T2R_M	9.21 ± 0.05 [l]	3.40 ± 0.03 [fg]	1.64 ± 0.01 [g]	9.07 ± 0.01 [j]	7.64 ± 0.26 [defgh]	69.03 ± 0.29 [bcd]	358.90 ± 0.48 [de]
T2R_S	8.93 ± 0.03 [m]	4.13 ± 0.03 [e]	1.97 ± 0.01 [d]	8.96 ± 0.01 [k]	8.95 ± 0.51 [bcd]	67.05 ± 0.52 [gh]	359.02 ± 0.94 [de]

CW—Control white sorghum, CR—control red sorghum, T1—dry heat treatment at 121 °C, T2—dry heat treatment at 140 °C, L/M/S—particle sizes, I—integral. Each experiment was carried out in duplicate and data were reported as means ± standard deviation (SD). Means in the same column with different letters are significantly different ($p < 0.05$).

Dry heat treatment caused an increase in energetic value compared to the control and with temperature raise, the values for white and red varieties being close to each other. An increasing trend for the energetic values was observed in treated samples with particle size reduction, while for CW the opposite trend occurred.

3.2. Functional Properties

Sorghum grains dry heat treatment, milling and variety showed significant variations ($p < 0.05$) in flours functional properties. The water absorption capacity of flours generally decreased as the particle size was lower, except for CR, CW, and T1W which showed the highest values for the M fraction (Table 2). Sorghum grains dry heat treatment led to an increased water absorption capacity. Flours oil absorption capacity raise was proportional with particle size reduction, while red sorghum variety showed slightly higher values compared to the white one, and dry heat treatment produced irregular small changes depending on the particle size. Water retention capacity registered the lowest values in the case of M particle size, except for the T2R sample where it was the greatest, close values being observed between the two varieties, sorghum treatment determining a decrease of this parameter with temperature increase. Flours hydration capacity decreased with particle size reduction, except for treated white sorghum with M particle size and red sorghum with L particle size samples.

Table 2. Effect of dry heat treatment on the water absorption, oil absorption capacity, water retention capacity, hydration capacity, swelling, and solubility index of sorghum fractions.

Treatment	Water Absorption Capacity (%)	Oil Absorption Capacity (%)	Water Retention Capacity (g/g)	Hydration Capacity (%)	Swelling Power (g/g)	Solubility Index (%)
CW_I	208.80 ± 1.13 defghi	162.05 ± 0.25 g	1.29 ± 0.04 ab	92.41 ± 0.55 fgh	3.35 ± 0.01 kl	0.10 ± 0.00 ab
CW_L	202.00 ± 1.41 hij	152.36 ± 0.76 ij	1.31 ± 0.11 ab	98.11 ± 1.26 e	3.51 ± 0.01 h	0.04 ± 0.00 fgh
CW_M	208.60 ± 0.28 defghi	166.71 ± 0.54 e	1.06 ± 0.03 defg	97.80 ± 1.98 e	3.98 ± 0.01 d	0.12 ± 0.01 a
CW_S	207.40 ± 0.14 efghij	171.33 ± 0.61 bc	1.32 ± 0.02 ab	91.30 ± 2.12 ghi	4.31 ± 0.01 a	0.08 ± 0.00 bc
T1W_I	201.00 ± 0.54 hij	163.45 ± 0.00 f	1.30 ± 0.11 bcd	94.60 ± 0.30 f	3.38 ± 0.07 jk	0.08 ± 0.00 bc
T1W_L	209.10 ± 1.27 defgh	151.95 ± 0.41 jk	1.21 ± 0.03 abcd	94.58 ± 0.01 f	3.28 ± 0.00 m	0.02 ± 0.00 h
T1W_M	216.50 ± 0.71 bcd	176.99 ± 0.47 a	0.84 ± 0.01 h	99.60 ± 0.57 cde	3.82 ± 0.00 e	0.03 ± 0.01 gh
T1W_S	200.50 ± 0.71 ij	170.99 ± 0.15 bc	0.85 ± 0.05 h	89.00 ± 0.28 i	4.02 ± 0.00 cd	0.04 ± 0.00 fgh
T2W_I	199.00 ± 1.98 jk	166.38 ± 0.08 e	1.05 ± 0.01 defg	99.40 ± 1.13 cde	4.04 ± 0.00 c	0.03 ± 0.00 gh
T2W_L	221.60 ± 1.41 cde	151.01 ± 0.49 k	0.89 ± 0.13 fgh	101.29 ± 0.43 bcd	3.42 ± 0.00 ij	0.02 ± 0.00 h
T2W_M	219.40 ± 3.11 abc	166.53 ± 0.93 de	0.77 ± 0.00 h	102.29 ± 0.42 b	4.21 ± 0.00 b	0.03 ± 0.00 gh
T2W_S	209.20 ± 2.26 cdef	170.90 ± 0.61 bc	0.81 ± 0.04 h	89.91 ± 0.30 hi	4.35 ± 0.00 a	0.05 ± 0.03 efg
CR_I	208.78 ± 1.72 defghi	163.97 ± 0.61 f	1.27 ± 0.14 abc	93.30 ± 0.42 fg	3.32 ± 0.00 lm	0.11 ± 0.00 a
CR_L	199.50 ± 0.70 jk	160.11 ± 0.83 h	1.39 ± 0.01 a	98.74 ± 0.88 de	3.32 ± 0.00 lm	0.08 ± 0.00 bc
CR_M	209.55 ± 0.78 defg	171.66 ± 0.60 b	1.09 ± 0.01 cde	97.50 ± 0.99 e	3.63 ± 0.00 g	0.05 ± 0.00 efg
CR_S	209.30 ± 0.14 defgh	170.00 ± 0.82 c	1.30 ± 0.02 ab	91.65 ± 1.48 gh	3.70 ± 0.01 f	0.07 ± 0.00 cde
T1R_I	201.99 ± 1.68 ghij	168.05 ± 0.08 d	1.21 ± 0.11 abcd	93.04 ± 3.08 fg	3.43 ± 0.00 i	0.08 ± 0.01 bcd
T1R_L	226.80 ± 0.57 a	160.00 ± 0.48 h	0.87 ± 0.08 gh	107.71 ± 1.00 a	3.37 ± 0.00 k	0.05 ± 0.05 defg
T1R_M	210.08 ± 0.96 defg	168.16 ± 0.71 d	0.90 ± 0.03 efgh	101.50 ± 0.42 bc	3.60 ± 0.01 g	0.03 ± 0.00 gh
T1R_S	191.10 ± 0.99 k	170.99 ± 0.61 bc	0.88 ± 0.01 fgh	91.21 ± 0.83 ghi	3.68 ± 0.01 f	0.04 ± 0.00 fgh
T2R_I	203.39 ± 3.12 fghij	163.50 ± 0.70 f	1.07 ± 0.02 def	105.50 ± 1.27 a	3.77 ± 0.08 e	0.06 ± 0.02 cdef
T2R_L	224.76 ± 5.71 ab	153.65 ± 0.66 i	0.87 ± 0.06 gh	107.60 ± 0.28 a	3.32 ± 0.00 lm	0.02 ± 0.00 h
T2R_M	210.08 ± 0.57 abc	167.27 ± 0.87 de	1.31 ± 0.26 ab	107.10 ± 0.15 a	3.42 ± 0.00 ij	0.02 ± 0.01 h
T2R_S	210.08 ± 1.41 cde	170.09 ± 0.94 c	0.89 ± 0.01 fgh	93.01 ± 0.59 fg	3.79 ± 0.00 e	0.05 ± 0.01 efg

CW—Control white sorghum, CR—control red sorghum, T1—dry heat treatment at 121 °C, T2—dry heat treatment at 140 °C, L/M/S—particle sizes, I—integral. Each experiment was carried out in duplicate and data were reported as means ± standard deviation (SD). Means in the same column with different letters are significantly different ($p < 0.05$).

Dry heat treatment induced slight increases in flour hydration capacity in almost all cases, while red sorghum flours presented higher values compared to the white variety. A proportional raise of swelling power was obtained with particle size decrease, while white sorghum flours showed higher values compared to the red ones. Solubility index decreased with particle size reduction for almost all the tested samples (except for CW_M), and dry heat treatment led to the slight decrease of these parameter's values. Flour bulk density was influenced irregularly by particle size, the lowest values being observed for M particle sizes (Table 3). A similar irregular trend was determined by dry heat treatment, while the values between the two sorghum varieties were close.

The emulsifying activity recorded significant decreases ($p < 0.05$) with particle size reduction, except for L for both red and white varieties, while dry heat treatment induced irregular changes. Emulsion stability recorded higher values for L and S compared to M particle size for both sorghum varieties. The increase in temperature determined the increase of emulsion stability in all fractions of white sorghum, excepting T2W_S, while for the red variety the values followed an irregular trend.

Table 3. Effect of dry heat treatment on the bulk density and emulsifying properties of sorghum flour fractions.

Treatment	Bulk Density (g/cm^3)	Emulsifying Properties		Foaming Stability (%)					
		Emulsifying Activity (%)	Emulsion Stability (%)	10 min	20 min	30 min	40 min	50 min	60 min
CW_I	0.71 ± 0.01 [ij]	47.50 ± 0.72 [fg]	60.50 ± 0.72 [fg]	115.00 ± 0.00 [ab]	110.00 ± 0.00 [bc]	110.00 ± 0.00 [bc]	105.00 ± 0.00 [c]	105.00 ± 0.00 [c]	105.00 ± 0.00 [bc]
CW_L	0.83 ± 0.00 [d]	56.50 ± 0.72 [a]	66.50 ± 0.72 [a]	110.00 ± 0.00 [bc]	110.00 ± 0.00 [bc]	110.00 ± 0.00 [bc]	110.00 ± 0.00 [bc]	110.00 ± 0.00 [bc]	110.00 ± 0.00 [ab]
CW_M	0.70 ± 0.00 [jk]	45.50 ± 0.72 [ghi]	54.50 ± 0.72 [j]	120.00 ± 0.00 [a]	120.00 ± 0.00 [a]	120.00 ± 0.00 [a]	115.00 ± 0.00 [ab]	115.00 ± 0.00 [ab]	115.00 ± 0.00 [a]
CW_S	0.72 ± 0.00 [ghi]	41.50 ± 0.72 [jk]	62.50 ± 0.72 [de]	115.00 ± 0.00 [ab]	115.00 ± 0.00 [ab]	115.00 ± 0.00 [ab]	115.00 ± 0.00 [ab]	105.00 ± 0.00 [c]	105.00 ± 0.00 [bc]
T1W_I	0.70 ± 0.00 [jk]	50.00 ± 1.41 [de]	61.50 ± 0.72 [ef]	115.00 ± 0.00 [ab]	110.00 ± 0.00 [bc]	110.00 ± 0.00 [bc]	110.00 ± 0.00 [bc]	110.00 ± 0.00 [bc]	110.00 ± 0.00 [ab]
T1W_L	0.88 ± 0.00 [b]	48.50 ± 0.72 [ef]	59.50 ± 0.72 [g]	-	-	-	-	-	-
T1W_M	0.62 ± 0.00 [m]	41.50 ± 0.72 [jk]	61.50 ± 0.72 [ef]	115.00 ± 0.00 [ab]	110.00 ± 0.00 [bc]	110.00 ± 0.00 [bc]	110.00 ± 0.00 [bc]	110.00 ± 0.00 [bc]	110.00 ± 0.00 [ab]
T1W_S	0.73 ± 0.00 [gh]	41.50 ± 0.72 [jk]	62.50 ± 0.72 [de]	120.00 ± 0.00 [a]	120.00 ± 0.00 [a]	120.00 ± 0.00 [a]	120.00 ± 0.00 [a]	115.00 ± 0.00 [ab]	115.00 ± 0.00 [a]
T2W_I	0.64 ± 0.01 [l]	47.50 ± 0.72 [fg]	63.50 ± 0.72 [cd]	110.00 ± 0.00 [bc]	110.00 ± 0.00 [bc]	110.00 ± 0.00 [bc]	110.00 ± 0.00 [bc]	110.00 ± 0.00 [bc]	105.00 ± 0.00 [bc]
T2W_L	0.89 ± 0.01 [ab]	54.00 ± 1.41 [b]	64.50 ± 0.72 [bc]	105.00 ± 0.00 [c]	105.00 ± 0.00 [c]	105.00 ± 0.00 [c]	105.00 ± 0.00 [c]	105.00 ± 0.00 [c]	105.00 ± 0.00 [bc]
T2W_M	0.62 ± 0.00 [m]	48.50 ± 0.72 [ef]	65.50 ± 0.72 [ab]	115.00 ± 0.00 [ab]	115.00 ± 0.00 [ab]	115.00 ± 0.00 [ab]	115.00 ± 0.00 [ab]	110.00 ± 0.00 [bc]	-
T2W_S	0.73 ± 0.00 [gh]	38.50 ± 0.72 [l]	56.50 ± 0.72 [hi]	120.00 ± 0.00 [a]	120.00 ± 0.00 [a]	120.00 ± 0.00 [a]	120.00 ± 0.00 [a]	120.00 ± 0.00 [a]	115.00 ± 0.00 [a]
CR_I	0.72 ± 0.01 [fg]	51.50 ± 0.72 [cd]	63.50 ± 0.72 [cd]	115.00 ± 0.00 [ab]	115.00 ± 0.00 [ab]	115.00 ± 0.00 [ab]	110.00 ± 0.00 [bc]	110.00 ± 0.00 [bc]	110.00 ± 0.00 [ab]
CR_L	0.90 ± 0.00 [a]	53.50 ± 0.72 [bc]	62.50 ± 0.72 [de]	110.00 ± 0.00 [bc]	110.00 ± 0.00 [bc]	110.00 ± 0.00 [bc]	110.00 ± 0.00 [bc]	105.00 ± 0.00 [c]	105.00 ± 0.00 [bc]
CR_M	0.69 ± 0.00 [k]	44.00 ± 1.41 [i]	55.50 ± 0.72 [ij]	115.00 ± 0.00 [ab]	115.00 ± 0.00 [ab]	115.00 ± 0.00 [ab]	115.00 ± 0.00 [ab]	115.00 ± 0.00 [ab]	105.00 ± 0.00 [bc]
CR_S	0.69 ± 0.01 [k]	43.50 ± 0.72 [ij]	62.50 ± 0.72 [de]	120.00 ± 0.00 [a]	120.00 ± 0.00 [a]	120.00 ± 0.00 [a]	120.00 ± 0.00 [a]	120.00 ± 0.00 [a]	110.00 ± 0.00 [ab]
T1R_I	0.72 ± 0.01 [hij]	46.50 ± 0.72 [fgh]	59.50 ± 0.72 [g]	115.00 ± 0.00 [ab]	115.00 ± 0.00 [ab]	115.00 ± 0.00 [ab]	115.00 ± 0.00 [ab]	115.00 ± 0.00 [ab]	110.00 ± 0.00 [ab]
T1R_L	0.86 ± 0.00 [c]	55.50 ± 0.72 [ab]	62.50 ± 0.72 [de]	-	-	-	-	-	-
T1R_M	0.65 ± 0.00 [l]	46.50 ± 0.72 [fgh]	61.50 ± 0.72 [efg]	120.00 ± 0.00 [a]	120.00 ± 0.00 [a]	120.00 ± 0.00 [a]	110.00 ± 0.00 [bc]	110.00 ± 0.00 [bc]	-
T1R_S	0.75 ± 0.00 [ef]	43.50 ± 0.72 [ij]	61.50 ± 0.72 [ef]	120.00 ± 0.00 [a]	120.00 ± 0.00 [a]	120.00 ± 0.00 [a]	115.00 ± 0.00 [ab]	115.00 ± 0.00 [ab]	115.00 ± 0.00 [a]
T2R_I	0.65 ± 0.00 [l]	50.00 ± 1.41 [de]	57.50 ± 0.72 [h]	115.00 ± 0.00 [ab]	115.00 ± 0.00 [ab]	115.00 ± 0.00 [ab]	115.00 ± 0.00 [ab]	115.00 ± 0.00 [ab]	100.00 ± 0.00 [c]
T2R_L	0.83 ± 0.00 [d]	55.00 ± 1.41 [ab]	64.00 ± 1.41 [bcd]	105.00 ± 0.00 [c]	105.00 ± 0.00 [c]	105.00 ± 0.00 [c]	105.00 ± 0.00 [c]	105.00 ± 0.00 [c]	105.00 ± 0.00 [bc]
T2R_M	0.75 ± 0.01 [ef]	44.50 ± 0.72 [hi]	60.50 ± 0.72 [fg]	115.00 ± 0.00 [ab]	115.00 ± 0.00 [ab]	115.00 ± 0.00 [ab]	115.00 ± 0.00 [ab]	-	-
T2R_S	0.76 ± 0.00 [e]	40.50 ± 0.72 [kl]	61.50 ± 0.72 [ef]	115.00 ± 0.00 [ab]	115.00 ± 0.00 [ab]	115.00 ± 0.00 [ab]	115.00 ± 0.00 [ab]	105.00 ± 0.00 [c]	105.00 ± 0.00 [bc]

CW—Control white sorghum, CR—control red sorghum, T1—dry heat treatment at 121 °C, T2—dry heat treatment at 140°C, L/M/S—particle sizes, I—integral. Each experiment was carried out in duplicate and data were reported as means ± standard deviation (SD). Means in the same column with different letters are significantly different ($p < 0.05$).

Flour foaming capacity was affected by particle size, dry heat treatment, and sorghum variety (Figure 1). Particle size reduction led to the increase of foaming capacity, the differences between varieties being not noticeable. On the other hand, dry heat treatment caused lower foaming capacity values compared to the controls and with temperature increase, excepting T2R_S.

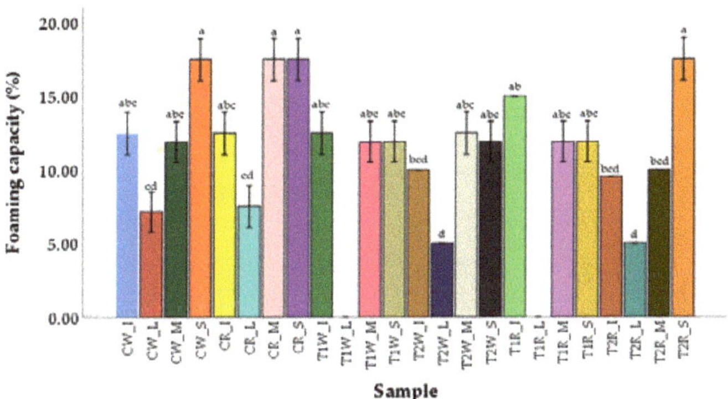

Figure 1. Effect of dry heat treatment on sorghum flour fractions foaming capacity: CW—Control white sorghum, CR—control red sorghum, T1—dry heat treatment at 121 °C, T2—dry heat treatment at 140 °C, L/M/S—particle sizes, I—integral. Means with distinct letters are significantly different ($p < 0.05$).

The variation of red and white sorghum flours foaming stability with dry heat treatment temperature increase and particle size decrease in time is presented in Table 3. Foaming stability registered an increase with particle size decrease, except for CW where an irregular trend was observed. Sorghum grains dry heat treatment caused an irregular variation of foaming stability, the lowest values being observed for T2W_L and T2R_L, the samples treated at 121 °C (T1W_L and T1R_L not presenting any foaming capacity). As expected, foaming stability decreased in time for all the studied samples.

3.3. Molecular Characteristics

FT-IR spectra of red and white untreated and treated sorghum flours with different particle sizes are shown in Figure 2. Many peaks can be observed on the FT-IR spectra that could be associated with the molecular bindings of sorghum chemical compounds such as starch, proteins, and polyphenols. Particle size reduction, sorghum variety, and dry heat treatment led to changes in the intensities of absorption band and the appearance of peaks in some regions. There can be observed two prominent peaks in the region 1700–1600 cm^{-1} (at 1649 cm^{-1}) and in the region 1060–960 cm^{-1} (997 cm^{-1}), the first one being ascribed to Amide I expression of proteins, while the second band to carbohydrates fingerprint [45]. The peaks found in 3800–3600 cm^{-1} may be attributed to the O-H groups of phenols intermolecular bonded [36], while the peak found at about 3296 cm^{-1} was given by the O-H stretching vibration [46].

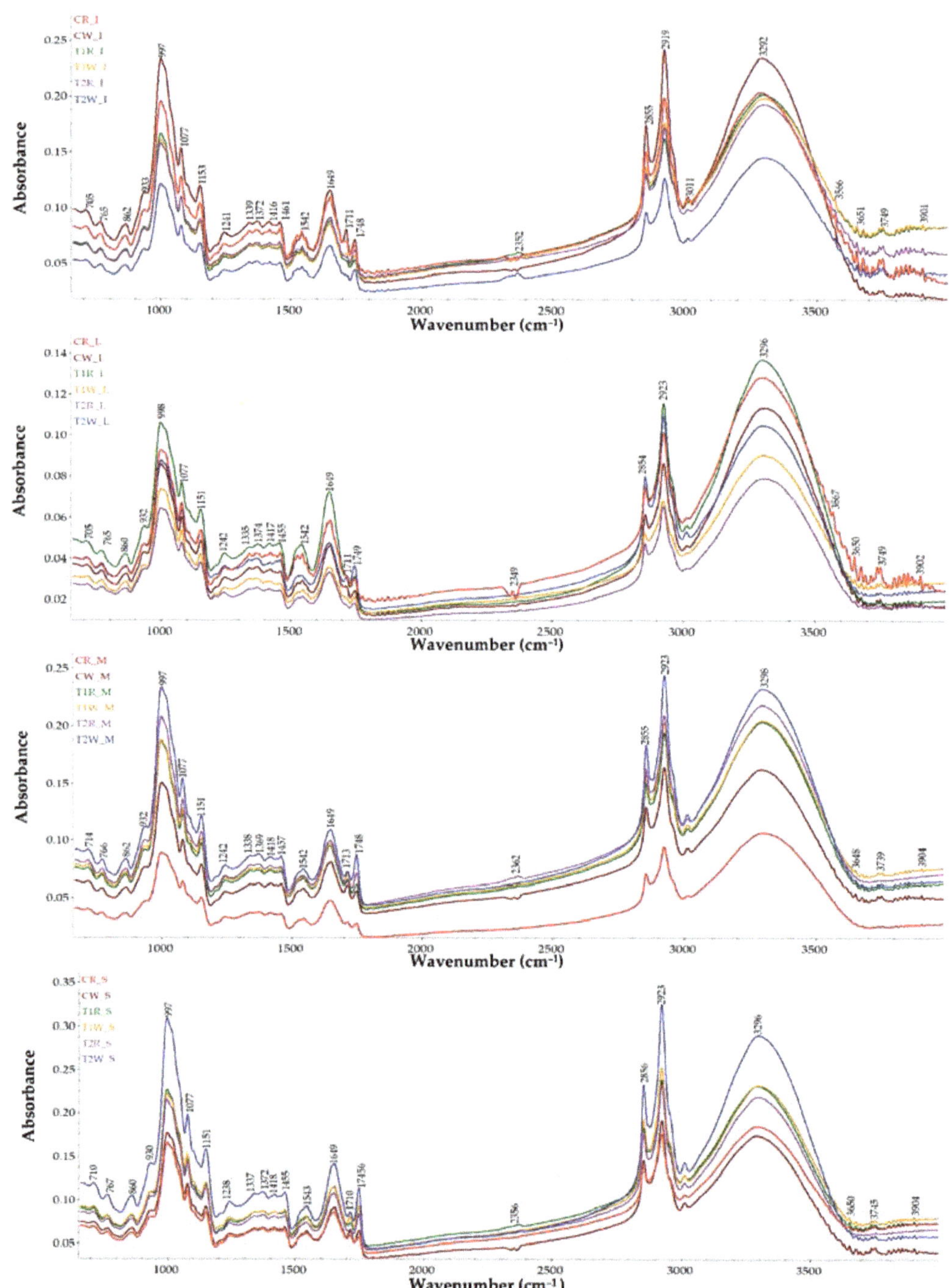

Figure 2. FT-IR spectra of untreated (C) and treated (T1, T2) red (R) and white (W) sorghum flours at different particle sizes: integral (I), large (L), medium (M), and small (S).

The absorption bands found at about 2923 and 2352 cm^{-1} could be due to the C–H stretching vibration and could suggest the presence of the alkane group and/or cis–olefinic group [2]. Interesting changes of FT-IR spectra were observed in the 2352 cm^{-1} region, in CR_L, CW_S, CW_M, T2R_M, and integral flours samples which showed higher peak absorbances compared to other samples (Figure 3). The absorption bands found at about 1151 and 1077 cm^{-1} could be attributed to the fiber fractions such as small hemicellulose and cellulose [45]. On the other hand, 1154 and 1416 could be due to the P–H, P–H banding, phosphine, and phosphoric acid which may be associated with the presence of phytic acid [43]. In the integral flours, the reduction of absorption intensity in 1154 and 1416 cm^{-1} was observed with the increase of sorghum grain dry heat treatment temperature. The peaks observed in 750–880 cm^{-1} region and 1150–1500 cm^{-1} zone could be associated with aromatic rings deformation, revealing the presence of phenols, with stretching given by them and C–C, C–H, O–H deformations suggesting the presence of phenolic acids and flavonoids [42]. The decrease of particle size led to higher absorbances in these regions, while white sorghum exhibited higher absorbances compared to the red one, except for L particle size. Dry heat treatment showed differential contribution in absorbances intensities, higher values compared to the controls being observed in the case of L, M, and S particle sizes. For the L particle size sample, the highest absorption intensities were observed for T1R_L, while for M and S particle sizes T2W samples exhibited the highest absorbances compared to the other studied samples.

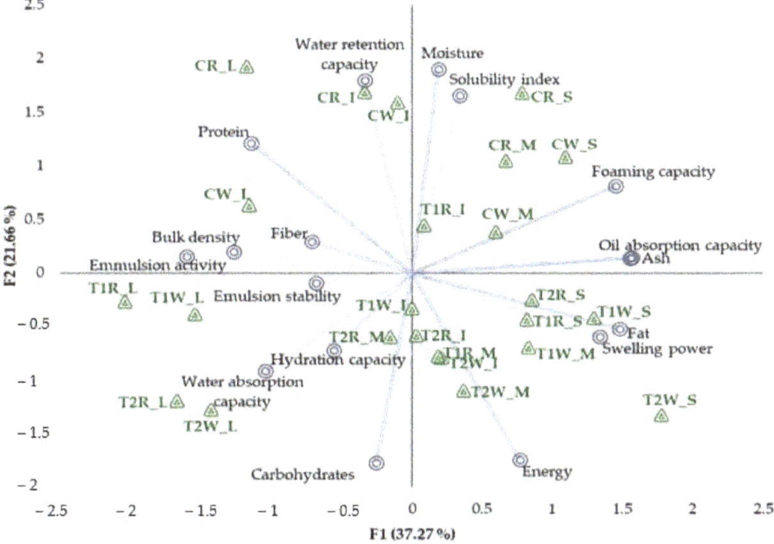

Figure 3. Principal Component Analysis (PCA) bi-plot: CW—Control white sorghum, CR—control red sorghum, T1—dry heat treatment at 121 °C, T2—dry heat treatment at 140 °C, L/M/S—particle sizes, I—integral.

3.4. Relations between Variables

The correlations between variables are presented in Table 4. Foaming capacity was positively correlated with the fat ($r = 0.52$, $p < 0.05$) and ash contents ($r = 0.78$, $p < 0.05$) and negatively with carbohydrates ($r = -0.37$, $p < 0.05$). Oil absorption capacity was correlated at $p < 0.05$ significance level with fat ($r = 0.67$), ash ($r = 0.84$), protein ($r = -0.50$) and foaming capacity ($r = 0.72$). Moderate correlations were observed between hydration capacity and ash ($r = -0.46$, $p < 0.05$), fat contents ($r = -0.48$, $p < 0.05$) and foaming capacity ($r = -0.50$, $p < 0.05$). Swelling power was positively correlated with foaming capacity ($r = 0.44$), fat

($r = 0.69$), ash ($r = 0.66$) and oil absorption capacity ($r = 0.60$) and negatively with protein content ($r = -0.51$) and fibers ($r = -0.48$), significant at $p < 0.05$ level. Moderate correlations were obtained between solubility index and flours moisture content ($r = 0.48$, $p < 0.05$), carbohydrates ($r = -0.36$, $p < 0.05$) and water retention capacity ($r = 0.45$, $p < 0.05$). Bulk density was positively correlated with the protein content ($r = 0.45$) and negatively with fat ($r = -0.50$), ash ($r = -0.59$) and oil absorption capacity ($r = -0.73$), significant at $p < 0.05$. Similar, but stronger correlations were obtained between emulsifying activity and protein content ($r = 0.64$), fat ($r = -0.77$), ash ($r = -0.80$), oil absorption capacity ($r = -0.84$) and hydration capacity ($r = 0.55$). Emulsion stability was moderately correlated with the protein content of flours ($r = 0.46$, $p < 0.05$). Significant negative ($p < 0.05$) correlations between carbohydrates and moisture ($r = -0.63$) and carbohydrates and fibers ($r = -0.40$) were obtained, while fat and ash contents were positively correlated ($r = 0.69$). The hydration capacity, swelling power and emulsifying activity were moderate correlated with fiber content ($p < 0.05$).

Table 4. Pearson's correlation coefficients.

Variables	FC	Protein	Fat	Ash	Moisture	Fiber	Carbohydrates	Energy	WAC	OAC	WRC	HC	SP	SI	BD	EA	ES
FC	1.00	−0.34	0.52	0.78	0.27	−0.16	−0.37	0.04	−0.26	0.72	0.11	−0.50	0.44	0.35	−0.63	−0.64	−0.20
Protein		1.00	−0.57	−0.57	0.33	0.19	−0.38	−0.53	0.09	−0.50	0.53	0.13	−0.51	0.18	0.45	0.64	0.46
Fat			1.00	0.69	−0.04	−0.40	−0.12	0.70	−0.23	0.67	−0.35	−0.48	0.69	−0.04	−0.50	−0.72	−0.17
Ash				1.00	0.12	−0.23	−0.26	0.23	−0.18	0.84	−0.14	−0.46	0.66	0.08	−0.59	−0.80	−0.17
Moisture					1.00	−0.23	−0.63	−0.62	−0.14	0.05	0.52	−0.33	0.00	0.48	0.05	−0.04	−0.08
Fiber						1.00	−0.40	−0.45	0.18	−0.07	−0.07	0.34	−0.48	0.00	0.22	0.31	0.08
Carbohydrates							1.00	0.54	0.08	−0.31	−0.32	0.27	0.06	−0.38	0.01	0.09	−0.11
Energy								1.00	−0.09	0.24	−0.52	−0.13	0.49	−0.36	−0.31	−0.38	−0.06
WAC									1.00	−0.15	−0.32	0.56	−0.09	−0.24	0.12	0.14	0.08
OAC										1.00	−0.20	−0.37	0.60	0.17	−0.73	−0.84	−0.36
WRC											1.00	−0.18	−0.37	0.45	0.18	0.19	0.01
HC												1.00	−0.29	−0.38	0.12	0.55	0.14
SP													1.00	−0.01	−0.55	−0.64	−0.16
SI														1.00	−0.15	−0.03	−0.33
BD															1.00	0.54	0.25
EA																1.00	0.46
ES																	1.00

Values in bold are significant at $p < 0.05$, FC—foaming capacity, WAC—water absorption capacity, OAC—oil absorption capacity, WRC—water retention capacity, HC—hydration capacity, SP—swelling power, SI—solubility index, BD—bulk density, EA—emulsifying activity, ES—emulsion stability.

Similarities and oppositions between variables were underlined by means of Principal Component Analysis (Figure 3), 58.93% of the total variance being explained. The first component (PC1) explained 37.27% of data variation, while the second one (PC2) explained 21.66% of the variance.

Emulsifying activity, bulk density, fiber content swelling power, fat content, oil absorption capacity, ash content, and foaming capacity were associated with PC1, while moisture content, solubility index, water retention capacity, carbohydrates content, and energetic value were associated with PC2. Emulsion stability, fiber content and hydration capacity position close to the origin underlined their smaller contribution to the data variation.

Control samples with different particle sizes were positioned in the upper part of the graphic and were associated with water retention capacity, moisture content, solubility index, and protein content, while treated sorghum flours with different particle sizes were placed on the lower side. Treated samples with L particle size were grouped and were associated with water absorption capacity, hydration capacity, and emulsifying activity. On the other hand, treated samples with M and S particle sizes were associated with energetic value, swelling power, fat and ash contents, foaming capacity, solubility index and oil absorption capacity.

4. Discussion

The nutritional and functional characteristics of food products are affected by structure-property relationships [24]. Modifications of chemical components proportions and functional properties determined by dry heat treatment and milling have been reported for cereals [24,25]. Protein content decrease with particle size decrease (Table 1) could be due to the localization of nutrients in the grain, being known that aleurone layer in the pericarp and the peripheral endosperm tissue are formed of cells with high amounts of proteins [47].

Fat and ash contents increases and carbohydrates decrease with particle size reduction agreed with the results reported by Alvarenga et al. [47]. In the case of a constant milling process, differences in dissociation between constituent parts of the sorghum grains are determined by the ability to remove the aleurone layer from the peripheral endosperm, the intercellular adherence at the interface of aleurone-endosperm layers being determined by the degree of bonding of arabinoxylan chains [48]. Another important aspect is led to the distribution of endosperm proteins such as kafirins and glutelins [48]. Our results showed that dry heat treatment induced the decrease of moisture and protein, and raised fat, carbohydrates and energetic value. Protein content decrease could be possibly attributed to the damage of amino acids as a result of heat [49]. He et al. [50] also reported increased metabolizable energy and net energy values induced by sorghum heat-related processing. The differences among red and white sorghum chemical composition were in line with those reported by Vargas-Solórzano et al. [48]. The significant ($p < 0.05$) correlations between, carbohydrates and ash, and between fat and ash contents were in agreement with those reported by Queiroz et al. [51] for sorghum chemical constituents. The results obtained for sorghum fractions fiber content were in line with those reported in the literature [52], the genotype and extraction rate playing a decisive role in fiber variation [26]. The dietary fiber of sorghum grains is mainly composed of cellulose and pentosan [53].

The water absorption capacity of flours decreased as the particle size was lower (Table 2) probably due to the loss of fiber which has great potential to hold water [54] and/or to the different chemical compounds of milling fractions and/or to the particle size distribution and its morphology [55]. For the acceptable food texture of baked goods, higher values for water absorption capacity are recommended [56]. Dry heat treatment caused the increase of water absorption capacity, similar to the trend reported by Adebowale et al. [57] for red sorghum starch treated by annealing and heat moisture treatment, probably due to the starch amylose and amylopectin chains reorganizations. Fat enhances the flavor retention and mouth feel of food products which means that oil retention capacity is an important quality characteristic to govern the ability of flour to physically entrap fat content [25]. Oil absorption capacity increase with particle size decrease could be possibly explained by the presence of lignin, its structure and surface characteristics, overall charge density, thickness, hydrophobic character, and particle size [58]. Grains milling may affect the absorption characteristics due to the increase in surface area. The differences between white and red sorghum could be due to the concentration of hydrophobic amino acids as lipid binding is determined by their concentration [54]. Water retention capacity showed the lowest values for M particle size samples (except for T2R). The hydrophilic compounds found in milling fractions such as polysaccharides present high water retention capacity, while the polar amino acid residues of proteins exert affinity for water molecules, enhancing thus the water binding capacity [55]. Dry heat treatment of sorghum grains caused the decrease of water retention capacity which may be explained by the changes induced by heat such as protein denaturation, starch gelatinization, and swelling of the crude fiber [59], a statement supported also by the correlations ($p < 0.05$) obtained between water retention capacity with protein and moisture content (Table 4). Flour swelling power increases with particle size reduction, probably as a result of starch damage during milling [60]. Swelling power is related to the intermolecular non-covalent linkages of starch, the degree of swelling being influenced by the molecular weight distribution, amylose-amylopectin ratio, and chain length [54]. Protein solubility decrease caused by dry heat treatment of sorghum grains could be responsible for solubility index decrease. Protein solubility is influenced by the intrinsic factors such as hydrophilic and hydrophobic properties of protein molecules, their dimension and charge, and the interaction with other grain components, and also by external factors such temperature [61]. Bulk density as an indicator for flour heaviness was influenced in an irregular way (Table 3) by particle size, similar to the results presented by Cairano et al. [60]. The presence of fat which could play a binder role in agglomeration of flour particles and the milling process conditions can affect flour granulometry and, thus, bulk density [60], a fact also supported by the significant correlation ($p < 0.05$) between bulk density and fat content (Table 4). In the case of sorghum flour samples with low lipid content (T1W_L, T2W_L, T1R_L, and T2R_L), the increase in bulk density was observed, and therefore using these flours may offer a significant advantage in terms of baked product volume [62]. The emulsion activity and stability values were in line with those reported by Siroha et al. [63] for millet flours and decreased with particle size reduction which may be related to the variations of the interfacial tension and surface hydrophobicity. Flour foaming capacity is determined by protein molecules structures and carbohydrates content [64], a fact confirmed also by the significant correlation ($p < 0.05$) obtained in this study. Particle size reduction led to the increase of foaming capacity and stability (Figure 1 and Table 3) probably due to the increase in proteins ability to form an elastic, flexible, and cohesive interfacial film that can catch and maintain air for enough time to slow down the coalescence rate [54]. Sorghum grains heat treatment induced the reduction of flour foaming capacity may be due to protein denaturation.

Regarding the molecular characteristics (Figure 3), the decrease of particle size determined proportional absorbances increases, dry heat treatment presenting differential contribution in absorbances intensities by inducing higher values compared to the controls in the case of L, M, and S particle sizes. The differences found at 3011 and 3292 cm^{-1} may be attributed to the changes in the activity of N-H of primary and secondary amines and/or O-H of either carboxylic acid, alcohols, or starch [36]. These modifications of the amine groups in 3292 cm^{-1} may also be assigned to their reactions with reducing sugars and α-dicarbonyls, resulting in pyrazinium radical cation and leading to the formation of brown pigments in the heat-treated sorghum flour [36]. The changes of FT-IR spectra intensities for CR_L, CW_S, CW_M, T2R_M samples at about 2352 cm^{-1} could be possibly attributed to the $-NH_3^+$ changes in amines or hydrohallides and $-PH$ in the phosphine functional groups [36]. Furthermore, Maillard reactions could determine the raise of unsaturated carbonyl compounds, the disintegration of amino acids to aldehydes, and their condensation with carbohydrates parts, furfurals, and other compounds to form chromophores and off-flavors and could be possibly led to the changes in 2300–2400 cm^{-1} region [36]. The changes in absorbances values at about 1151 and 1077 cm^{-1} could be related to the distribution of fiber fractions in the flours with different particle sizes [45]. The reduction of phytic acid content after sorghum grains dry heat treatment was observed by the decrease in absorbances at 1154 and 1416 [43], confirming thus the applicability of this kind of treatment for antinutrients reduction. The peaks found in the region 1200 to 1900 cm^{-1} are associated with different functional groups such as amides, amino acids, -C=O in aldehydes, C-O in esters, CO in anhydrides, =O in lactones, t-butyl groups, N-O pyridine groups, esters, lactones [36]. The production of chromophores and condensation reactions occurrence, such as reactions of furfurals or dehydro-reductones during Maillard reaction result in the formation of unsaturated brown nitrogenous polymers named melanoidins, depending on the heat treatment condition [36,65]. The changes observed in region 900–1100 cm^{-1} could be determined by the angular C-H linkage deformation in the sorghum flour, skeletal vibration of 1–4 glycosidic bonds (C–O–C), development of new groups, and stretching vibration of the C-O bond in the esters formed between the non-starch component such as –COOH in the protein and starch molecules [36].

5. Conclusions

Sorghum grain is an important alternative to conventional corps and its functionality and nutritional value can be modified by dry heat treatment, depending on the particle size. Dry heat treatment applied to white and red sorghum grains resulted in higher fat and fiber and lower protein, moisture, and ash contents. Particle size reduction caused protein content decrease and fat contents increase. The results of this study showed that red sorghum is richer in protein, ash, and fibers and less abundant in fat compared to the white variety. Dry heat treatment led to higher flour water absorption and lower water retention capacity, solubility index, and foaming capacity as the temperature increased. Particle size decrease induced the raise of flour oil absorption capacity, swelling power, and foaming capacity, while flour solubility index, emulsifying activity decreased. The molecular characteristics of sorghum flours showed significant differences among varieties, particle sizes, and dry heat treatment temperatures. Grains heat treatment led to lower intensities of peaks characteristics for phytic acid, suggesting the utility of this process in antinutrients diminishing. These results evidenced the opportunity to use dry heat treatment to enhance the nutritional and functional profiles of sorghum flours, at the same time underlying the importance of particle size and variety. This information could be helpful for processors to better decide the destination of sorghum flours to obtain further nutritious foods.

Author Contributions: All authors contributed equally to this research. All authors have read and agreed to the published version of the manuscript.

Funding: This research was funded by "Stefan cel Mare" University of Suceava.

Institutional Review Board Statement: Not applicable.

Informed Consent Statement: Not applicable.

Data Availability Statement: Not applicable.

Acknowledgments: This research was supported by "Stefan cel Mare" University of Suceava.

Conflicts of Interest: The authors declare no conflict of interest.

References

1. Ratnavathi, C.; Patil, L.V.; Chavan, U. *Sorghum Biochemistry: An Industrial Perspective*; Academic Press: Oxford, UK, 2016; ISBN 9780415475976.
2. Kamble, D.B.; Singh, R.; Rani, S.; Kaur, B.P.; Upadhyay, A.; Kumar, N. Optimization and characterization of antioxidant potential, in vitro protein digestion and structural attributes of microwave processed multigrain pasta. *J. Food Process. Preserv.* **2019**, *43*, e14125. [CrossRef]
3. Pérez, A.; Saucedo, O.; Iglesias, J.; Wencomo, H.B.; Reyes, F.; Oquendo, G.; Milián, I. Caracterización y potencialidades del grano de sorgo (*Sorghum bicolor* L. Moench). *Pastos Forrajes* **2010**, *33*, 1–25.
4. Beta, T.; Obilana, A.B.; Corke, H. Genetic diversity in properties of starch from Zimbabwean sorghum landraces. *Cereal Chem.* **2001**, *78*, 583–589. [CrossRef]
5. Tasie, M.M.; Gebreyes, B.G. Characterization of Nutritional, Antinutritional, and Mineral Contents of Thirty-Five Sorghum Varieties Grown in Ethiopia. *Int. J. Food Sci.* **2020**, *2020*, 8243617. [CrossRef] [PubMed]
6. Chávez, D.; Ascheri, J.; Martins, A.; Carvalho, C.; Bernardo, C.; Teles, A. Sorghum, an alternative cereal for gluten-free product. *Rev. Chil. Nutr.* **2018**, *45*, 169–177. [CrossRef]
7. Hager, A.S.; Wolter, A.; Jacob, F.; Zannini, E.; Arendt, E.K. Nutritional properties and ultra-structure of commercial gluten free flours from different botanical sources compared to wheat flours. *J. Cereal Sci.* **2012**, *56*, 239–247. [CrossRef]
8. Rashwan, A.K.; Yones, H.A.; Karim, N.; Taha, E.M.; Chen, W. Potential processing technologies for developing sorghum-based food products: An update and comprehensive review. *Trends Food Sci. Technol.* **2021**, *110*, 168–182. [CrossRef]
9. Weerasooriya, D.K.; Bean, S.R.; Nugusu, Y.; Ioerger, B.P.; Tesso, T.T. The effect of genotype and traditional food processing methods on in-vitro protein digestibility and micronutrient profile of sorghum cooked products. *PLoS ONE* **2018**, *13*, e0203005. [CrossRef] [PubMed]
10. Llopart, E.E.; Drago, S.R.; De Greef, D.M.; Torres, R.L.; González, R.J. Effects of extrusion conditions on physical and nutritional properties of extruded whole grain red sorghum (*sorghum* spp.). *Int. J. Food Sci. Nutr.* **2014**, *65*, 34–41. [CrossRef]
11. Shen, S.; Huang, R.; Li, C.; Wu, W.; Chen, H.; Shi, J.; Chen, S.; Ye, X. Phenolic Compositions and Antioxidant Activities Differ Significantly among Sorghum Grains with Different Applications. *Molecules* **2018**, *23*, 1203. [CrossRef]
12. Wolter, A. *Fundamental Studies of Sourdoughs Fermented with* Weissella cibaria *and* Lactobacillus plantarum: *Influence on Baking Characteristics, Sensory Profiles and In Vitro Starch Digestibility of Gluten-Free Breads*; National University of Ireland Coláiste: Cork, Ireland, 2013.
13. Schober, T.J.; Bean, S.R.; Boyle, D.L. Gluten-free sorghum bread improved by sourdough fermentation: Biochemical, rheological, and microstructural background. *J. Agric. Food Chem.* **2007**, *55*, 5137–5146. [CrossRef]
14. Winger, M.; Khouryieh, H.; Aramouni, F.; Herald, T. Sorghum Flour Characterization and Evaluation in Gluten-Free Flour Tortilla. *J. Food Qual.* **2014**, *37*, 95–106. [CrossRef]
15. Mary, C. *Fernholz Evaluation of Four Sorghum Hybrids through the Development of Sorghum Flour Tortillas*; Kansas State University: Manhattan, KS, USA, 2008.
16. Cisse, F.; Erickson, D.P.; Hayes, A.M.R.; Opekun, A.R.; Nichols, B.L.; Hamaker, B.R. Traditional malian solid foods made from sorghum and millet have markedly slower gastric emptying than rice, potato, or pasta. *Nutrients* **2018**, *10*, 124. [CrossRef]

17. Ferreira, S.M.R.; De Mello, A.P.; De Caldas Rosa Dos Anjos, M.; Krüger, C.C.H.; Azoubel, P.M.; De Oliveira Alves, M.A. Utilization of sorghum, rice, corn flours with potato starch for the preparation of gluten-free pasta. *Food Chem.* **2016**, *191*, 147–151. [CrossRef]

18. Suhendro, E.L.; Kunetz, C.F.; McDonough, C.M.; Rooney, L.W.; Waniska, R.D. Cooking characteristics and quality of noodles from food sorghum. *Cereal Chem.* **2000**, *77*, 96–100. [CrossRef]

19. Niba, L.L.; Hoffman, J. Resistant starch and β-glucan levels in grain sorghum (*Sorghum bicolor* M.) are influenced by soaking and autoclaving. *Food Chem.* **2003**, *81*, 113–118. [CrossRef]

20. Ofosu, F.K.; Elahi, F.; Daliri, E.B.M.; Yeon, S.J.; Ham, H.J.; Kim, J.H.; Han, S.I.; Oh, D.H. Flavonoids in Decorticated Sorghum Grains Exert Antioxidant, Antidiabetic and Antiobesity Activities. *Molecules* **2020**, *25*, 2854. [CrossRef]

21. Selma, M.V.; Espín, J.C.; Tomás-Barberán, F.A. Interaction between phenolics and gut microbiota: Role in human health. *J. Agric. Food Chem.* **2009**, *57*, 6485–6501. [CrossRef] [PubMed]

22. Raigar, R.K.; Prabhakar, P.K.; Srivastav, P.P. Effect of Different Thermal Treatments on Grinding Characteristics, Granular Morphology and Yield of Ready-to-Eat Wheat Grits. *J. Food Process Eng.* **2017**, *40*, e12363. [CrossRef]

23. Murthy, K.V.; Ravi, R.; Bhat, K.K.; Raghavarao, K.S.M.S. Studies on roasting of wheat using fluidized bed roaster. *J. Food Eng.* **2008**, *89*, 336–342. [CrossRef]

24. Schoeman, L.; du Plessis, A.; Verboven, P.; Nicolaï, B.M.; Cantre, D.; Manley, M. Effect of oven and forced convection continuous tumble (FCCT) roasting on the microstructure and dry milling properties of white maize. *Innov. Food Sci. Emerg. Technol.* **2017**, *44*, 54–66. [CrossRef]

25. Dayakar Rao, B.; Anis, M.; Kalpana, K.; Sunooj, K.V.; Patil, J.V.; Ganesh, T. Influence of milling methods and particle size on hydration properties of sorghum flour and quality of sorghum biscuits. *LWT Food Sci. Technol.* **2016**, *67*, 8–13. [CrossRef]

26. Trappey, E.F.; Khouryieh, H.; Aramouni, F.; Herald, T. Effect of sorghum flour composition and particle size on quality properties of gluten-free bread. *Food Sci. Technol. Int.* **2015**, *21*, 188–202. [CrossRef] [PubMed]

27. Rumler, R.; Bender, D.; Speranza, S.; Frauenlob, J.; Gamper, L.; Hoek, J.; Jäger, H.; Schönlechner, R. Chemical and physical characterization of sorghum milling fractions and sorghum whole meal flours obtained via stone or roller milling. *Foods* **2021**, *10*, 870. [CrossRef] [PubMed]

28. Zhao, Y. *Effect of Different Tempering Methods on Sorghum Milling*; Purdue University: West Lafayette, IN, USA, 2016.

29. Mahasukhonthachat, K.; Sopade, P.A.; Gidley, M.J. Kinetics of starch digestion in sorghum as affected by particle size. *J. Food Eng.* **2010**, *96*, 18–28. [CrossRef]

30. Palavecino, P.M.; Penci, M.C.; Calderón-Domínguez, G.; Ribotta, P.D. Chemical composition and physical properties of sorghum flour prepared from different sorghum hybrids grown in Argentina. *Starch Stärke* **2016**, *68*, 1055–1064. [CrossRef]

31. Zhu, F. Structure, physicochemical properties, modifications, and uses of sorghum starch. *Compr. Rev. Food Sci. Food Saf.* **2014**, *13*, 597–610. [CrossRef]

32. Wong, J.H.; Marx, D.B.; Wilson, J.D.; Buchanan, B.B.; Lemaux, P.G.; Pedersen, J.F. Principal component analysis and biochemical characterization of protein and starch reveal primary targets for improving sorghum grain. *Plant Sci.* **2010**, *179*, 598–611. [CrossRef]

33. Mohapatra, D.; Patel, A.S.; Kar, A.; Deshpande, S.S.; Tripathi, M.K. Effect of different processing conditions on proximate composition, anti-oxidants, anti-nutrients and amino acid profile of grain sorghum. *Food Chem.* **2019**, *271*, 129–135. [CrossRef]

34. Wu, G.; Ashton, J.; Simic, A.; Fang, Z.; Johnson, S.K. Mineral availability is modified by tannin and phytate content in sorghum flaked breakfast cereals. *Food Res. Int.* **2018**, *103*, 509–514. [CrossRef]

35. Ranganathan, V.; Nunjundiah, I.T.; Bhattacharya, S. Effect of roasting on rheological and functional properties of sorghum flour. *Food Sci. Technol. Int.* **2014**, *20*, 579–589. [CrossRef]

36. Sharanagat, V.S.; Suhag, R.; Anand, P.; Deswal, G.; Kumar, R.; Chaudhary, A.; Singh, L.; Singh Kushwah, O.; Mani, S.; Kumar, Y.; et al. Physico-functional, thermo-pasting and antioxidant properties of microwave roasted sorghum [*Sorghum bicolor* (L.) Moench]. *J. Cereal Sci.* **2019**, *85*, 111–119. [CrossRef]

37. Fao Chapter 4: Summary—Integration of Analytical Methods and Food Energy Conversion Factors. Available online: https://www.fao.org/3/y5022e/y5022e05.htm#bm5 (accessed on 9 December 2021).

38. Bordei, D.; Bahrim, G.; Pâslaru, V.; Gasparotti, C.; Elisei, A.; Banu, I.; Ionescu, L.; Codină, G. Quality Control in the Bakery Industry-Analysis Methods. *Galați Acad.* **2007**, *1*, 203–212.

39. Oladiran, D.A.; Emmambux, N.M. Nutritional and Functional Properties of Extruded Cassava-Soy Composite with Grape Pomace. *Starch* **2018**, *70*, 1700298. [CrossRef]

40. Elkhalifa, A.E.O.; Bernhardt, R. Combination Effect of Germination and Fermentation on Functional Properties of Sorghum Flour. *Curr. J. Appl. Sci. Technol.* **2018**, *30*, 1–12. [CrossRef]

41. Zhu, F.; Du, B.; Xu, B. Super fine grinding improves functional properties and antioxidant capacities of bran dietary fi bre from Qingke (hull-less barley) grown in Qinghai-Tibet Plateau, China. *J. Cereal Sci.* **2015**, *65*, 43–47. [CrossRef]

42. Sharma, R.; Sharma, S. Anti-nutrient & bioactive profile, in vitro nutrient digestibility, techno-functionality, molecular and structural interactions of foxtail millet (*Setaria italica* L.) as influenced by biological processing techniques. *Food Chem.* **2022**, *368*, 130815. [CrossRef]

43. Dave, G.; Modi, H. FT-IR method for estimation of phytic acid content during bread-making process. *J. Food Meas. Charact.* **2018**, *12*, 2202–2208. [CrossRef]

44. Yasar, S.; Tosun, R.; Sonmez, Z. Fungal fermentation inducing improved nutritional qualities associated with altered secondary protein structure of soybean meal determined by FTIR spectroscopy. *Meas. J. Int. Meas. Confed.* **2020**, *161*, 107895. [CrossRef]

45. González, M.; Vernon-Carter, E.J.; Alvarez-Ramirez, J.; Carrera-Tarela, Y. Effects of dry heat treatment temperature on the structure of wheat flour and starch in vitro digestibility of bread. *Int. J. Biol. Macromol.* **2021**, *166*, 1439–1447. [CrossRef]

46. Liu, F.Y.; Guo, X.N.; Xing, J.J.; Zhu, K.X. Effect of thermal treatments on: In vitro starch digestibility of sorghum dried noodles. *Food Funct.* **2020**, *11*, 3420–3431. [CrossRef] [PubMed]

47. Alvarenga, I.C.; Ou, Z.; Thiele, S.; Alavi, S.; Aldrich, C.G. Effects of milling sorghum into fractions on yield, nutrient composition, and their performance in extrusion of dog food. *J. Cereal Sci.* **2018**, *82*, 121–128. [CrossRef]

48. Vargas-Solórzano, J.W.; Carvalho, C.W.P.; Takeiti, C.Y.; Ascheri, J.L.R.; Queiroz, V.A.V. Physicochemical properties of expanded extrudates from colored sorghum genotypes. *Food Res. Int.* **2014**, *55*, 37–44. [CrossRef]

49. Desai, G.B.; Deshmukh, N.M.; Kshirsagar, R.B. Effect of Malting and Roasting of Grains on Nutritional and Sensory Quality of Sorghum Based Multigrain Roti. *Trends Biosci.* **2017**, *10*, 2235.

50. He, J.; Torres Lechuga, M.E.; Lei, Y.; Refat, B.; Prates, L.L.; Zhang, H.; Yu, P. Protein molecular structural, physicochemical and nutritional characteristics of warm-season adapted genotypes of sorghum grain: Impact of heat-related processing. *J. Cereal Sci.* **2019**, *85*, 182–191. [CrossRef]

51. Queiroz, V.A.V.; da Silva, C.S.; de Menezes, C.B.; Schaffert, R.E.; Guimarães, F.F.M.; Guimarães, L.J.M.; de Guimarães, P.E.; Tardin, F.D. Nutritional composition of sorghum [*Sorghum bicolor* (L.) Moench] genotypes cultivated without and with water stress. *J. Cereal Sci.* **2015**, *65*, 103–111. [CrossRef]

52. Xiong, Y.; Zhang, P.; Warner, R.D.; Fang, Z. Sorghum Grain: From Genotype, Nutrition, and Phenolic Profile to Its Health Benefits and Food Applications. *Compr. Rev. Food Sci. Food Saf.* **2019**, *18*, 2025–2046. [CrossRef]

53. Ragaee, S.; Abdel-Aal, E.S.M.; Noaman, M. Antioxidant activity and nutrient composition of selected cereals for food use. *Food Chem.* **2006**, *98*, 32–38. [CrossRef]

54. Sharma, R.; Sharma, S.; Dar, B.N.; Singh, B. Millets as potential nutri-cereals: A review of nutrient composition, phytochemical profile and techno-functionality. *Int. J. Food Sci. Technol.* **2021**, *56*, 3703–3718. [CrossRef]

55. Coțovanu, I.; Mironeasa, S. Buckwheat seeds: Impact of milling fractions and addition level on wheat bread dough rheology. *Appl. Sci.* **2021**, *11*, 1731. [CrossRef]

56. Gomez, M.; Martinez, M.M. Fruit and vegetable by-products as novel ingredients to improve the nutritional quality of baked goods. *Crit. Rev. Food Sci. Nutr.* **2018**, *58*, 2119–2135. [CrossRef] [PubMed]

57. Adebowale, K.O.; Olu-Owolabi, B.I.; Olayinka, O.O.; Lawal, O.S. Effect of heat moisture treatment and annealing on physico-chemical properties of red sorghum starch. *Afr. J. Biotechnol.* **2005**, *4*, 928–933. [CrossRef]

58. Zlatanović, S.; Kalušević, A.; Micić, D.; Laličić-Petronijević, J.; Tomić, N.; Ostojić, S.; Gorjanović, S. Functionality and storability of cookies fortified at the industrial scale with up to 75% of apple pomace flour produced by dehydration. *Foods* **2019**, *8*, 561. [CrossRef]

59. Caprita, R.; Caprita, A.; Cretescu, I. Effect of Heat Treatment and Digestive Enzymes on Cereal Water-Retention Capacity. *Sci. Pap. Anim. Sci. Biotechnol.* **2015**, *48*, 94–96.

60. Di Cairano, M.; Condelli, N.; Caruso, M.C.; Marti, A.; Cela, N.; Galgano, F. Functional properties and predicted glycemic index of gluten free cereal, pseudocereal and legume flours. *LWT* **2020**, *133*, 109860. [CrossRef]

61. Žilić, S.M.; Božović, I.N.; Savić, S.; Šobajić, S. Heat processing of soybean kernel and its effect on lysine availability and protein solubility. *Cent. Eur. J. Biol.* **2006**, *1*, 572–583. [CrossRef]

62. Joshi, A.U.; Liu, C.; Sathe, S.K. Functional properties of select seed flours. *Lwt* **2015**, *60*, 325–331. [CrossRef]

63. Siroha, A.K.; Sandhu, K.S.; Kaur, M. Physicochemical, functional and antioxidant properties of flour from pearl millet varieties grown in India. *J. Food Meas. Charact.* **2016**, *10*, 311–318. [CrossRef]

64. Culetu, A.; Susman, I.E.; Duta, D.E.; Belc, N. Nutritional and functional properties of gluten-free flours. *Appl. Sci.* **2021**, *11*, 6283. [CrossRef]

65. Bastos, D.M.; Monaro, É.; Siguemoto, É.; Séfora, M. *Maillard Reaction Products in Processed Food: Pros and Cons*; IntechOpen: London, UK, 2012; ISBN 9533079053. Available online: https://www.intechopen.com/chapters/29164 (accessed on 10 December 2021).

Article

Features of Bread Made from Different Amaranth Flour Fractions Partially Substituting Wheat Flour

Ionica Coțovanu * and Silvia Mironeasa *

Faculty of Food Engineering, Stefan cel Mare University of Suceava, 720229 Suceava, Romania
* Correspondence: ionica.cotovanu@usm.ro (I.C.); silviam@fia.usv.ro (S.M.);
  Tel.: +40-740-816-370 (I.C.); +40-741-985-648 (S.M.)

Featured Application: Particle size offers valuable information about the quality of amaranth flour, depending on milling fractions, and can be used to potentially predict the quality characteristics of new baked goods developed in the future.

Abstract: Amaranth flour (AF) is recognized as high-quality raw material regarding nutrients and bioactive compounds, essential in supplying human health benefits, compared with white flour (WF). In this study, the effects of factors, different particles sizes (large, medium, and small), and levels of AF (5, 10, 15, and 20%) substituting WF on the responses, empirical and dynamic dough rheological properties, and some quality parameters of bread were successfully modeled using predictive models. Finally, the optimization of a formulation to maximize the AF level whilst maintaining bread quality for each type of particle size (PS) was performed based on the response surface methodology models generated. The rheological properties of the composite flour formulated were evaluated using Mixolab, alveograph, rheofermentometer, and dynamic rheometer. In addition, bread quality parameters, loaf volume, instrumental texture features, and firmness were evaluated. The anticipation of the optimal value for each response in terms of dough rheological properties during mixing, protein weakening, starch gelatinization and retrogradation, biaxial extension, fermentation, viscoelastic moduli, and creep and recovery compliance depending on PS. The optimal addition level was determined by a multi-objective optimization approach. The optimal addition level was 9.41% for large, 9.39% for medium, and 7.89% for small PS. The results can help manufacturers to develop bread products with the desired particle size with optimal technological and physical features.

Keywords: amaranth flour; bread characteristics; dough rheology; particle size; wheat flour

Citation: Coțovanu, I.; Mironeasa, S. Features of Bread Made from Different Amaranth Flour Fractions Partially Substituting Wheat Flour. *Appl. Sci.* **2022**, *12*, 897. https://doi.org/10.3390/app12020897

Academic Editor: Anna Lante

Received: 21 December 2021
Accepted: 13 January 2022
Published: 17 January 2022

Publisher's Note: MDPI stays neutral with regard to jurisdictional claims in published maps and institutional affiliations.

1. Introduction

Bread is one of the most consumed foodstuffs that can satisfy daily requirements and its fortification would provide an opportunity to upgrade the nutritional level of the human diet [1]. The tendency to include pseudocereals to improve the nutritional value of bakery products has become more popular [2]. The addition of amaranth flour could improve the nutritional quality of bread and bakery products because amaranth is a pseudocereal very rich in protein and carbohydrates that fulfills the needs of people with celiac disease, individuals allergic to wheat, and vegetarians [3]. The protein content of amaranth seeds is characterized by a well-balanced amino acid composition. Furthermore, amaranth seeds are rich in lipids with high levels of unsaturated fatty acids, dietary fiber, minerals (zinc, copper, and manganese), vitamins (thiamine, niacin, riboflavin, and folate), and bioactive components (squalene, tocopherols, and phenolic compounds) [4–6] with important beneficial effects on human health [7]. Moreover, scientific investigations have demonstrated that amaranth has beneficial effects on health, such as decreasing hypocholesterolemic activity, improving the immune system, decreasing blood glucose levels, acting on liver functions, and decreasing hypertension, as well as having antitumor effects, anti-anemic effects, antioxidant activity, antiallergic actions, and beneficial effects on celiac disease [8]. The partial replacement of gluten with gluten-free flours represents a major technological challenge because gluten is the principal factor of structure-building protein, which is reflected in the quality attributes of many baked products. Furthermore, the gluten network is necessary for describing the rheological properties of dough, such as mixing ability, resistance to extension, elasticity, extensibility, and gas-holding capacity [9].

The rheological properties of dough can be evaluated by using empirical and fundamental tests. Fundamental dough rheological properties, such as oscillation, stress relaxation, and creep–recovery measurements have been previously studied to evaluate the mechanical behavior of wheat doughs [9–11], which influences the quality attributes of the end-product. In general, supplementation with non-wheat flour dilutes the gluten content in wheat flour which adversely affects the rheological properties of dough [12]. Some studies on white flour/pseudocereal composite flour highlighted that the bread properties (such as specific volume, crumb texture, and density) were positively related to dough rheological properties, and the bread's specific volume decreased as the doses of pseudocereal flour was increased [3,13,14]. Additionally, several studies demonstrated that incorporating pseudocereal flour significantly improved the nutritive values of wheat-based bakery products [15,16]. Compared with the continuous network and unique viscoelasticity of the wheat dough protein network, the protein matrix of pseudocereal flour was less desirable for bread making [17]. Including pulse flours into bread dough dilutes the gluten protein and affects both gluten development and starch–protein complexes, which are important to the dough rheology and the quality of the bread [18,19].

Mathematical modeling and statistical optimization in food processing manufacture represent a necessity in order to achieve a sustainable processing industry [20]. The response surface methodology (RSM) technique permits the study of the effects of the factors involved in processing and their interactions, using a reduced number of experimental runs, without being time-consuming.

This study aimed to optimize the formulation of amaranth flour addition levels for each particle size to enhance the dough's rheological and bread properties. To our knowledge, no other studies have examined amaranth flour's addition to wheat flour from a complex rheological properties point of view. The technological importance of each particle size could be highlighted by the chemical properties of each AF fraction, the rheological properties of the dough, and the technological properties of the bread obtained from the amaranth–wheat flour blend.

2. Materials and Methods

2.1. Materials

Wheat flour (flour yield 65%) from S.C. Mopan S.R.L., Suceava, România was used. Amaranth flours were obtained from amaranth seeds (acquired from S.C. Solaris Plant S.R.L., Ilfov, Romania). The flours were analyzed according to the international ICC standard methods [21]: moisture content was measured according to the gravimetric method (ICC 110/1); ash content was determined in a muffle oven by incineration at 900 °C (ICC 104/1); protein content was determined with a rapid Kjeldahl device, with digestion and steam distillation (VELP Scientifica, Usmate Velate (MB), Italy), and calculated with a general factor of 6.25 for wheat flour and 5.53 for amaranth flour fractions (ICC 105/2); fat content was determined with the Soxhlet method (VELP Scientifica, Usmate Velate (MB), Italy) (ICC 136); wet gluten content and the gluten deformation index were determined according to the SR 90:2007 method [22], as a percentage of the dried substances; and total carbohydrate content was calculated by difference [13]. Salt (S.C. Sanovita S.R.L., Vâlcea, Romania) and fresh *Saccharomyces cerevisiae* yeast (S.C. Rompak S.R.L., Pascani, România) acquired from the local market were used in the bread recipe.

2.2. Flours Formulation and Bread Manufacturing

Amaranth seeds were milled with a laboratory Grain Mill (KitchenAid, Whirlpool Corporation, Benton Harbor, MI, USA), then sieved with a Retsch Vibratory AS 200 basic (Haan, Germany) for 30 min at 70 Hz amplitude. Three different amaranth flour (AF) particle sizes were obtained: large (L > 300 µm), medium (M > 180 µm < 300 µm), and small fractions (S < 180 µm). Amaranth flours at each particle size were added at 0, 5, 10, 15, and 20% to refined wheat flour according to the experimental design (Table 1) and mixed for 30 min in a Yucebas Y21 mixer (Izmir, Turkey). For bread manufacturing, flour blend (0.3 kg), water (at water absorption capacities of the flours previously tested on the Mixolab device, Chopin, Tripette et Renaud, Paris, France), and yeast (1.8%) were used. The bread recipe followed a biphasic method, making leaven from the quantity of water and fresh yeast, and half of the quantity of WF–AF. This mixture was left to ferment until it doubled in volume, at 30 °C and relative humidity (85%), for approximately two hours in a fermenting chamber (PL2008, Piron, Cadoneghe, Padova, Italy). The leaven obtained, and the other part of WF–AF flours with salt, were kneaded in a laboratory mixer for another 10 min (Kitchen Aid, Whirlpool Corporation, Benton Harbor, MI, USA), and left for one hour for fermenting, for each dough formulation, in the same conditions [3]. At the end of the process, the dough was cut to 400 g/piece, molded, placed in aluminum trays for one hour to produce the final fermentation, and baked for 25 min, at 220 °C (oven Caboto PF8004D, Cadoneghe, Padova, Italy).

Table 1. Factors and their levels in the experimental design.

Run	A	Particle Size (µm)	B	Amaranth Flour (%)
	Coded Values	Real Values	Coded Values	Real Values
1	+1.00	380	0.00	10
2	−1.00	180	−0.50	5
3	0.00	280	−1.00	0
4	0.00	280	−0.50	5
5	−1.00	180	−1.00	0
6	+1.00	380	+1.00	20
7	−1.00	180	0.00	10
8	+1.00	380	−1.00	0
9	+1.00	380	−0.50	5
10	+1.00	380	+0.50	15
11	0.00	280	0.00	10
12	0.00	280	+1.00	20
13	−1.00	180	+1.00	20
14	0.00	280	+0.50	15
15	−1.00	180	+0.50	15

2.3. Empirical Dough Rheological Properties

The rheological behavior of the dough during the mixing and heating–cooling cycle was determined with the Mixolab equipment (Chopin, Tripette et Renaud, Paris, France) according to ICC standard method no. 173 (ICC, 2010) [21]. The mixing parameters from the registered Mixolab curves—water absorption (WA), dough development time (DT), dough stability (ST), torques related to protein weakening (C1-2), the starch gelatinization phase (C3-2), the stability of hot starch gel (C3-4), and the final starch paste viscosity after cooling at 50 °C (C5-4)—were reported.

The Falling Number index (FN) of the wheat flours and flour blends was determined by using a Falling Number device (FN 1305, Perten Instruments AB, Stockholm, Sweden) in order to determine the amylolytic activity. The dough rheological properties during extension were determined with an alveograph device (Chopin Technologies, Villeneuve-la-Garenne, France) according to ICC method no. 121 method at constant hydration to a 14% moisture basis. The determined parameters were: dough tenacity (P), dough extensibility (L), dough strength (W), and alveograph curve ratio (P/L).

The rheological properties of dough during fermentation were determined with the rheofermentometer device (Chopin Rheo, type F4, Villeneuve-La-Garenne, France) according to the AACC 89–01.01 method. The analyzed parameters were: maximum height of the gas release curve (H'm), the total volume of CO_2 production (VT), the volume of the gas retained in the dough at the end of the test (VR), and the retention coefficient (CR).

2.4. Fundamental Dough Rheological Properties

Dynamic oscillatory measurements as non-destructive tests were performed using HAAKE, MARS 40 (Thermo Scientific, Karlsruhe, Germany) with parallel plate–plates geometry. For this purpose, oscillatory frequency sweep, temperature sweep, and creep–recovery tests were performed, as reported by Iuga et al. [23]. Dough samples were preliminarily tested for the linear viscoelastic region (LVR) by applied strain sweep tests from 0.00 to 100 Pa at a constant oscillation frequency of 1 Hz. The dough, prepared at optimum water absorption capacity, but without the addition of the fresh yeast, was left to rest for 5 min before testing [23]. The excess dough was trimmed just before the measurement, and a layer of vaseline was applied to the exposed edge to avoid the evaporation of moisture during the resting period. The results, registered from a frequency sweep test applied in the LVR from 0.01 to 20 Hz, at 10 Pa stress and 20 °C, were evaluated by the elasticmodulus (G′) and the viscous modulus (G″) at 1 Hz, and the loss tangent (tan δ). A temperature sweep test was applied to determine the maximum gelatinization temperature (T_{max}), considered at the maximum G′ value by heating the dough from 20 to 100 °C at a rate of 4.0 ± 0.1 °C/min, a constant strain of 0.10%, and a frequency of 1 Hz.

A creep–recovery test at a constant shear stress of 25 Pa for a creep time of 60 s and a relaxation time of 180 s after removing the shear stress was applied to simulate different stresses during bread dough production [23]. Results were evaluated in terms of the maximum creep compliance (Jc_{max}), which corresponds to the maximum deformation at the end of the creep phase, and the maximum recovery compliance (Jr_{max}), associated with partial reformation after stress removal.

2.5. Bread Physical and Textural Parameters

After cooling, the bread samples obtained were analyzed for their bread volume (BV) according to the Romanian standard SR 90: 2007 by seed displacement method [22].

The textural parameter, bread firmness (BF), was evaluated based on the TPA mode by using a TVT-6700 texture analyzer (Perten Instruments, Hägersten, Sweden). The test speed of the probe was 1.0 mm/s. The compression strain was set at 20% while the auto-trigger force was 5.0 g, with an interval of 15 s between compressions. The firmness (BF) was recorded and processed by TexCalc 5 software (5.1.0.x. version, Perten Instruments, Hägersten, Sweden).

2.6. Factorial Design and Statistics

The study of PS and AF addition level effects on wheat dough rheological and bread characteristics and the optimization process were performed using Design-Expert software (Stat-Ease, Inc., Minneapolis, MN, USA). A full factorial design was used to study the main and interaction effects of two factors on 24 responses. The studied factors were three amaranth flour particle sizes (large, medium, and small) and five addition levels to wheat flour (0, 5, 10, 15, and 20%). The responses considered were the following: Falling Number (FN) index, the water absorption of the composite flour (WA), dough development time (DT), stability (DT), the consistency reached during the protein weakening stage (C1-2), the consistency reached during the starch gelatinization stage (C3-2), the consistency reached during the stability of hot starch gel (C3-4), the consistency during the starch retrogradation stage (C5-4), dough tenacity (P), dough extensibility (L), dough strength (W), alveograph configuration ratio (P/L), the maximum height of the gas release curve (H'm), the total volume of CO_2 production (VT), the volume of the gas retained in the dough at the end of the test (VR), the retention coefficient (CR), elastic modulus (G′), viscous modulus (G″), loss tangent (tan δ), maximum gelatinization temperature (T_{max}), maximum creep compliance (Jc_{max}), maximum recovery compliance (Jr_{max}), bread volume (BV), and bread firmness (BF).

An experimental design that consists of fifteen combinations was conducted and the coded versus the real values of the factors are presented in Table 1. The simultaneous effect of these two factors on the responses was investigated through response surface methodology (RSM). The effects of PS and AF addition levels on dough and bread properties were evaluated through mathematical modeling. The most suitable model for predicting data variation for each response was selected according to F-test results, the coefficient of determination (R^2), and adjusted coefficients of determination ($Adj.-R^2$). The effects of the factors and their interactions were underlined using Analysis of Variance (ANOVA), considering a significance level of 95%.

In order to establish the optimal value of the factors, amaranth flour particle size, and addition level, a multiple responses analysis was applied to the predictive models in conjunction with the desirability function approach. For the numerical optimization applied in this study, the desired goal established for each response included: addition level, ST, C3-4, H'm, VT, VR, CR, W, Jr_{max}, and BV at maximum value, while C1-2, C5-4, P/L were minimized, and the levels of all remaining responses which are considered in this study were kept within range.

3. Results

3.1. Flour Chemical Characteristics

The values of the chemical characteristics of the wheat flour and amaranth particle sizes, large (AL), medium (AM), and small (AS), are presented in Table 2. The wheat flour studied presented a low α amylase activity, 30.00% wet gluten content, and 6.00 mm gluten deformation index, characteristics that made it suitable for bread making according to Romanian standard SR 90:2007 [22].

Table 2. One-way analysis of variance (ANOVA) of the chemical composition of the wheat flour in comparison with amaranth flour fractions: large particle size (AL), medium particle size (AM) and, small particle size (AS).

		Amaranth Flour		
Parameters	**Wheat Flour**	**Particle Size**		
		AL	**AM**	**AS**
Moisture (%)	14.08 ± 0.08 [a]	10.61 ± 0.05 [b]	10.19 ± 0.06 [c]	9.35 ± 0.06 [d]
Ash (%)	0.69 ± 0.05 [e]	1.62 ± 0.02 [c]	3.54 ± 0.04 [b]	4.45 ± 0.04 [a]
Protein (%)	12.45 ± 0.15 [c]	10.18 ± 0.44 [d]	25.33 ± 0.25 [b]	29.36 ± 0.01 [a]
Fat (%)	1.41 ± 0.01 [d]	7.49 ± 0.02 [b]	8.09 ± 0.04 [a]	7.11 ± 0.01 [c]
Carbohydrates (%)	71.36 ± 0.02 [a]	71.16 ± 0.06 [a]	52.85 ± 0.53 [b]	49.74 ± 0.09 [c]
FN	312 ± 3.12	-	-	-
WGC	30.00 ± 0.15	-	-	-
GDI	6.00 ± 0.12	-	-	-

AL—amaranth large particle size; AM—medium particle size; AS—small particle size. FN—Falling Number index; WGC—wet gluten content; GDI—gluten deformation index. Lower-case letters ([a–e]) refer to the comparison of the same compound between the different particle size amaranth flour samples; results followed by the lowercase letter are significantly different according to Tukey's HSD post hoc test ($p < 0.05$).

A decrease in sample moisture can be observed as the particle size decreased, being significantly ($p < 0.05$) lower than wheat flour. The ash content varied between 1.62 and 4.45%, which was significantly higher than ash from wheat flour (0.69%), which increased when the particle size decreased, being significantly different between samples. The protein content from amaranth fractions ranged between 10.18 and 29.36% and followed the same upward trend, as did the ash content, and was significantly higher than wheat flour (12.45%). The fat content of amaranth particle size flours ranged between 7.11 and

8.09%, the highest values were observed in medium particle size (AM), with all samples showing significant ($p < 0.05$) differences between them, and all fractions being significantly higher than the fat content from wheat flour (1.41%). The total carbohydrate content for the amaranth particle fractions presented significant ($p < 0.05$) differences between all samples, the highest content being at large fractions, while the lowest carbohydrate content was present in small fractions (AS).

3.2. Influence of Particle Size and Addition Level of Amaranth Flour to Wheat Flour on FN Index, Dough Rheological Properties during Mixing and Heating-Cooling Cycle, and Dough Biaxial Extension

The accuracy test of the model (Table 3) showed that the quadratic and 2FI models properly predict the studied parameters as a function of the formulation factors. The data for the Falling Number index, Mixolab, and alveographic parameters were successfully fitted ($p < 0.05$) to the quadratic model, which explained proportions of 66–96% of the variation data, as the ANOVA results showed (Table 3). The 2FI mathematical model chosen for dough development time (DT) and the consistency reached during the starch gelatinization stage (C3-2) data fitting explained 68 and 95%, respectively, of the variation, and it was significant at $p < 0.05$ in both cases.

Table 3. The coefficients in the predictive models for the FN index, Mixolab, and alveograph parameters.

| | | | | | | Parameters | | | | | | | |
| Factors | Falling Number | | | | | Mixolab | | | | | Alveograph | | |
	FN (s)	WA (%)	DT (min)	ST (min)	C1-2 (Nm)	C3-2 (Nm)	C3-4 (Nm)	C5-4 (Nm)	P (mm)	L (mm)	W (10^{-4} J)	P/L
Constant	317.06	58.65	2.98	9.65	0.61	1.23	0.15	0.80	95.77	48.53	160.70	1.87
A	−1.96	−0.52 **	0.45	1.17 **	−0.05 **	0.05 **	0.08 **	−0.00	−6.38	1.88	6.08	−0.17
B	−2.98	0.39 **	1.53 **	−1.06 **	0.04 **	−0.15 ***	0.12 ***	−0.27 ***	9.62 **	−28.55 ***	−50.94 **	1.38 ***
A × B	−5.85 *	−0.77 **	0.47	0.77 *	−0.04 **	0.03 *	0.07 **	−0.03	−4.50	2.50	−5.30	−0.43
A^2	−2.40	−0.42 *		−0.55	0.00	-	−0.02	−0.03	2.40	−2.30	−5.30	0.08
B^2	−7.90 *	0.58 *		−0.52	0.03 *	-	0.03	0.07	1.33	13.43 **	33.90 *	0.29
R^2	0.71	0.93	0.68	0.83	0.93	0.95	0.96	0.89	0.69	0.96	0.85	0.86
Adj.-R^2	0.55	0.89	0.59	0.74	0.90	0.93	0.94	0.82	0.51	0.93	0.76	0.79
p-value	0.0265	<0.0001	0.0047	0.0028	<0.0001	<0.0001	<0.0001	0.0005	0.0352	<0.0001	0.0018	0.0011

*** $p < 0.001$; ** $p < 0.01$; * $p < 0.05$; A—particle size (μm); B—level of amaranth flour added to refined wheat flour (%); R^2, *Adj.-R^2*—measures of model fit; FN—Falling Number; WA—water absorption; DT—development time; ST—stability; C1-2—consistency reached during protein weakening stage; C3-2—consistency reached during starch gelatinization stage; C3-4—consistency reached during the stability of hot starch gel; C5-4—consistency during starch retrogradation stage; P—dough tenacity; L—dough extensibility; W—deformation energy; P/L—alveograph curve ratio.

The ANOVA for the quadratic model, as fitted to the experimental results, showed significance ($p < 0.05$). The FN index ranged from 289 to 321 s for the composite flours' formulation. The FN was significantly correlated ($p < 0.05$) with the interaction effect of the AF addition level to WF and particle size and, also, with the quadratic effect of the AF addition level, in a negative way. The effects of particle sizes and AF addition level is shown in Figure 1a, indicating a decrease in the FN index with an increase in the AF addition and PS.

During mixing, the dough was influenced by PS and AF addition to wheat flour. Water absorption registered a significant ($p < 0.05$) decrease (Figure 1b) when the PS increased and the AF addition level decreased, while the interaction between factors had a significant ($p < 0.05$) negative influence (Table 3). The dough development time (DT) showed a significant ($p < 0.05$) increase (Figure 1c) when the AF addition level was increased, while the PS had a non-significant effect ($p > 0.05$), which ranged between 1.33 and 5.75 min. The dough stability (ST) was significantly ($p < 0.05$) negatively affected by the AF addition level (Table 3), while the PS and the interaction between the factors had a positive significant influence on the ST, which ranged between 5.60 and 10.53 min. The rise in AF amounts led to proportionally lower dough stability (Figure 1d).

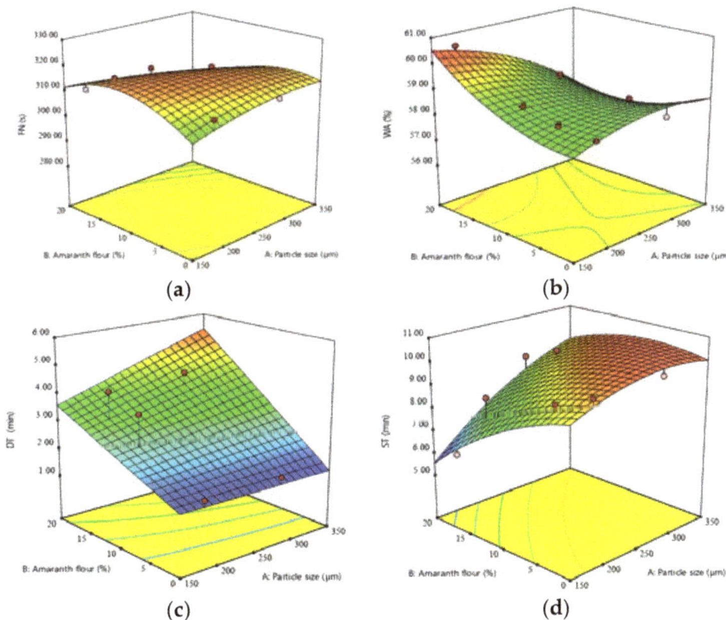

Figure 1. Three-dimensional response surface plots showing the interaction between amaranth flour particle size and addition level on (**a**) the Falling Number index (FN), (**b**) water absorption (WA), (**c**) development time (DT), and the (**d**) dough stability (ST) achieved during mixing.

Protein weakening (C1-2) values decreased significantly ($p < 0.05$) when the PS increased (Figure 2a), while this parameter was raised when the AF addition level was increased. The interaction between factors significantly affected protein weakening (C1-2 torque), indicating a decrease in protein weakening speed under the effect of temperature increase. The starch gelatinization stage showed a significant ($p < 0.05$) decrease as the addition level of AF was increased, while the PS and the interaction between factors significantly positively influenced C3-2. The effect of particle size and the AF addition level to wheat flour can be seen in Figure 2b, indicating an increase in C3-2 with particle size increase. The stability of hot starch gel (C3-4) was significantly positively affected by both factors and by the interaction between them. The effect of particle size and addition level on C3-4 is presented in Figure 2c, showing the capacity of the AF fractions, the addition level, and their interaction to increase the C3-4. The starch retrogradation during cooling was significantly ($p < 0.05$) decreased with AF addition (Figure 2d), while particle size and the interaction between factors showed a non-significant influence on C5-4 ($p > 0.05$).

The effects of factors on dough extension properties are presented in Figure 3. It can be seen from Table 3 that dough tenacity was significantly ($p < 0.05$) affected by the AF addition level, while particle size showed a non-significant effect ($p > 0.05$) on this parameter. It can be observed in Figure 3a that when the AF amount was increased, the dough tenacity was also increased. At the same time, when the dough tenacity increased with AF addition, the dough extensibility decreased significantly ($p < 0.05$) with an increase in this factor. The quadratic term of AF influences in a positive way the extensibility (Figure 3b). The quadratic regression model was fitted for the dough strength, which indicates that this alveograph parameter was significantly ($p < 0.05$) influenced by the linear and quadratic term of the AF addition level. It can be observed in Figure 3c that when the AF amount increased, the dough strength decreased. The alveograph configuration ratio presented a significant ($p < 0.05$) increase only with AF addition level increase (Figure 3d).

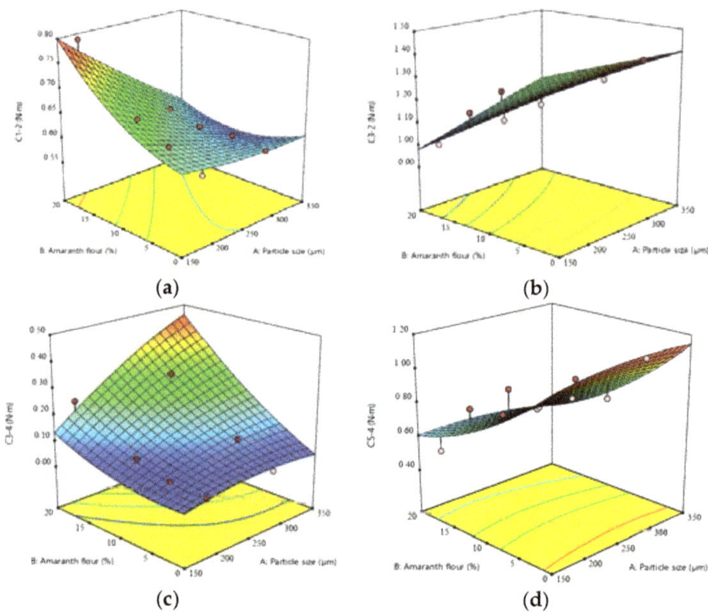

Figure 2. Three-dimensional response surface plots showing the interaction between amaranth flour particle size and addition level on: (**a**) protein weakening (C1-2; (**b**) starch gelatinization (C3-2); (**c**) the stability of hot starch gel (C3-4); and (**d**) starch retrogradation (C5-4).

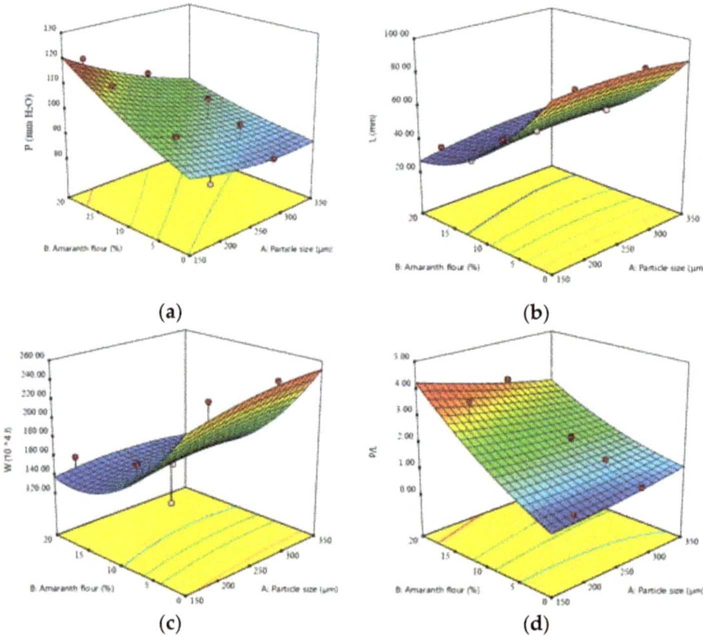

Figure 3. Three-dimensional response surface plots showing the interaction between amaranth flour particle size and addition level on: (**a**) dough tenacity (P); (**b**) dough extensibility (L); (**c**) dough strength (W); and (**d**) alveograph configuration ratio (P/L).

3.3. Influence of Particle Size and Addition Level of Amaranth Flour in Wheat Flour on Dough Fermentation, Dynamic Rheological Properties, and Bread Characteristics

The quadratic model successfully fitted ($p < 0.05$) the data for the maximum height of the gas release curve (H′m), the volume of gas retained in the dough at the end of the test (VR), the retention coefficient (CR), elastic modulus (G'), loss tangent (tan δ), maximum gelatinization temperature (T_{max}), maximum creep compliance (Jc_{max}), maximum recovery compliance (Jr_{max}), bread volume (BV), and bread firmness (BF). The variations were explained in proportions of 66 to 92% (Table 4). For total CO_2 volume production (VT) and viscous modulus (G'') data prediction, the 2FI model was found to be adequate ($p < 0.05$), with an explained variation of 70 and 71%, respectively (Table 4).

Table 4. The coefficients in the predictive models for dough during fermentation, dynamic rheological properties, and bread properties.

| | Parameters | | | | | | | | | | | |
| | Rheofermentometer | | | | Rheometer | | | | | | Bread Parameters | |
Factors	H′m (mL)	VT (mL)	VR (mL)	CR (%)	G' (Pa)	G'' (Pa)	tan δ (adim.)	T_{max} (°C)	Jc_{max} (Pa^{-1})	Jr_{max} (Pa^{-1})	BV (cm^3)	BF (g)
Constant	73.61	1268.45	1180.87	91.53	28,093.97	10,500.65	0.3476	79.32	20.27	13.40	360.83	951.01
A	0.25	35.40	16.96	−0.39	3939.00	1632.50 **	0.0144	−0.14	−7.52 **	−4.42 **	26.26 **	−365.40 *
B	6.30 **	84.58 **	89.36 **	1.20	8848.23 **	1463.54 *	−0.0375 **	−1.74 **	−1.46	−0.87	−40.73 ***	673.85 **
A × B	1.40	29.40	8.80	−1.15	2667.00	1837.65 *	0.0217	−0.57	−1.71	−1.59	7.70	202.60
A^2	0.44	-	−14.10	−0.99	−1172.50	-	−0.0205	0.52	3.68	1.77	−30.11 **	602.50 **
B^2	−4.53	-	−76.57 **	−4.36 **	9655.71 *	-	0.0275	1.67	2.14	1.70	−6.30	195.57
R^2	0.80	0.70	0.87	0.80	0.81	0.71	0.72	0.85	0.66	0.71	0.92	0.88
Adj.-R^2	0.68	0.62	0.80	0.69	0.71	0.64	0.57	0.77	0.47	0.55	0.87	0.81
p-value	0.0062	0.0032	0.0009	0.0054	0.0041	0.0025	0.0221	0.0015	0.0492	0.0268	0.0001	0.0007

*** $p < 0.001$; ** $p < 0.01$; * $p < 0.05$; A—particle size (μm); B—level of amaranth flour added to refined wheat flour (%); R^2, *Adj.-R^2*—measures of model fit; H′m—maximum height of the gas release curve; VT—total CO_2 volume production; VR—volume of the gas retained in the dough at the end of the assay; CR—retention coefficient; G'—elastic modulus; G''—viscous modulus; tan δ—loss tangent; T_{max}—maximum gelatinization temperature; Jc_{max}—maximum creep compliance; Jr_{max}—maximum recovery compliance; BV—bread volume; BF—bread firmness.

Table 4 shows the effects of PS and AF addition formulation factors on the maximum height of the gas release curve (H′m). The regression model indicates that the linear term of the addition level had a significant ($p < 0.05$) influence on the H′m parameter, while the PS had a non-significant effect ($p > 0.05$).

A response surface plot, showing the effect of AF level and PS on H′m, is represented in Figure 4a, and it can be seen that the H′m significantly increased as the AF level increased. The effect of the particle size and AF addition level on the total CO_2 volume production of amaranth–wheat composite flour dough as their corresponding regression coefficients in the 2FI model indicated is presented in Table 4. The VT varied from 1123 mL to 1421 mL, which was lower compared to the wheat control (1168 mL), but an increase in VT can be observed with the increase of AF addition, while the PS had no significant ($p > 0.05$) influence (Figure 4b). The volume of the gas retained in the dough at the end of the test (VR) was influenced significantly ($p < 0.05$) by the linear term of the AF addition level and the quadratic term of this factor (Table 4). The regression model for the VR (Table 4) showed a non-significant ($p > 0.05$) effect in linear terms of particle size and in terms of the interaction between particle size and AF addition level. The AF addition level positively affects VR, while the quadratic term of the AF addition level negatively influences it (Figure 4c). The effect of the PS and AF level on the CR, expressed as their corresponding regression coefficients in the quadratic regression model, is shown in Table 4. The CR was significantly influenced ($p < 0.05$) by the quadratic term of the AF level in a negative way. The response surface obtained for the CR (Figure 4d) showed that the increase in AF level decreased the CR parameter.

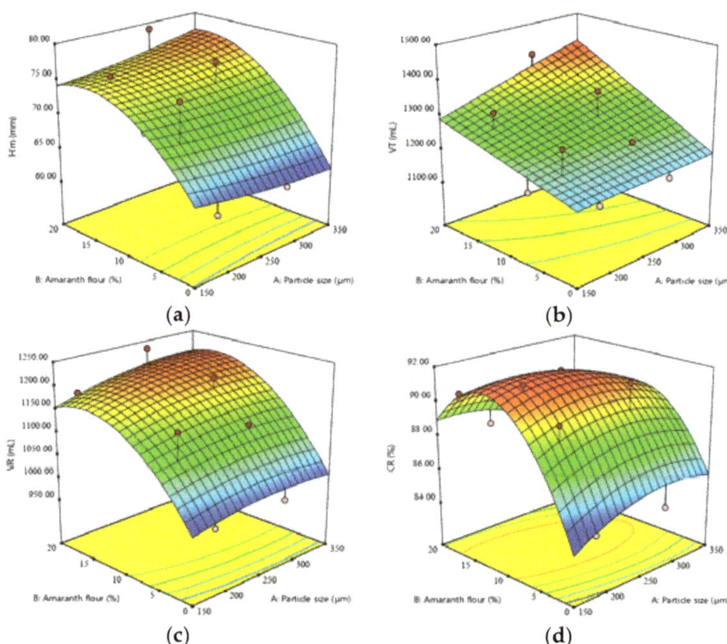

Figure 4. Three-dimensional response surface plots showing the interaction between amaranth flour particle size and addition level on: (**a**) maximum height of the gas release curve (H′m); (**b**) total CO_2 volume production (VT); (**c**) volume of the gas retained in the dough at the end of the test (VR); and the (**d**) retention coefficient (CR).

The dynamic dough rheological properties indicated different trends depending on particle sizes and AF addition levels. The elastic modulus (G′) was significantly influenced ($p < 0.05$) by the level of AF addition to WF, while the particle size had no significant influence ($p > 0.05$) (Table 4). G′ was influenced significantly ($p < 0.05$) by the linear term of the AF and the quadratic term of the AF addition level (Figure 5a), while the PS and the interactions between them presented a non-significant ($p > 0.05$) influence. The 2FI predictive model results from the regression analysis showed that both the PS and the addition level of AF and their interaction were significantly ($p < 0.05$) influenced and fitted well with the experimental data for G″ (Table 4). An increasing trend was observed with the increase in PS and AF addition levels (Figure 5b). The loss tangent (tan δ) decreased significantly ($p < 0.05$) as the amount of AF was increased (Figure 5c). Through ANOVA, the quadratic model was found to adequately represent the experimental data for the maximum gelatinization temperature (T_{max}), the R^2 value (0.85), confirming the adequacy of the model. The linear coefficient of the addition level indicated a significant ($p < 0.05$) negative influence on the T_{max}, while the linear term of the PS and the interaction between factors were not significant. The effect of the PS and AF addition level can be seen in Figure 5d, indicating a decrease in T_{max} with the increase in AF.

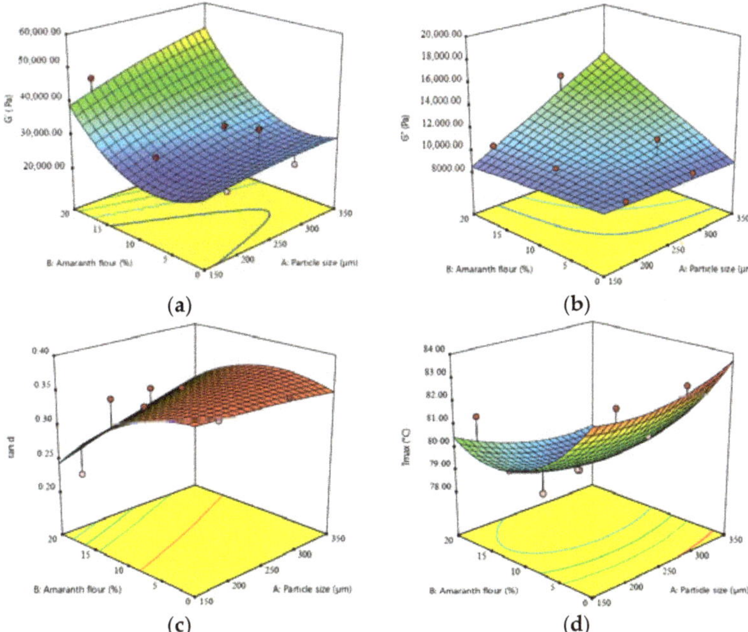

(a)   (b)

(c)   (d)

Figure 5. Three-dimensional response surface plots showing the interaction between amaranth flour particle size and addition level on: (**a**) elastic modulus (G′); (**b**) viscous modulus (G″); (**c**) loss tangent (tan δ); and (**d**) maximum gelatinization temperature (T_{max}).

The maximum creep compliance (Jc_{max}) was significantly influenced by the PS of amaranth flour, while the addition level did not influence this dynamic rheological parameter (Figure 6a). The recovery phase compliance was influenced ($p < 0.05$) by the particle size levels, while the AF added to wheat flour had a non-significant influence ($p > 0.05$). It was found that the PS had a highly significant negative effect ($p < 0.01$) on Jr_{max}, showing a decrease in Jr_{max} with the increase in PS (Figure 6b).

Bread firmness is an essential parameter in determining product quality, which determines its shelf-life. The crumb firmness was significantly influenced ($p < 0.05$) by the linear term of the PS, AF addition level, and the quadratic term of the PS. The response surface was generated (Figure 6c) to predict the bread firmness as a simultaneous function of the PS and AF added to wheat flour, and this physical parameter decreased significantly when the PS was increased. Otherwise, the bread firmness increased with greater AF amounts. The bread volume (BV) followed an inverse trend, being significantly positively affected by the PS, and negatively affected by the AF added to wheat flour. The results showed an increase in BV when the PS was higher, but it can be observed that this parameter decreased with the increase in the AF addition level (Figure 6d).

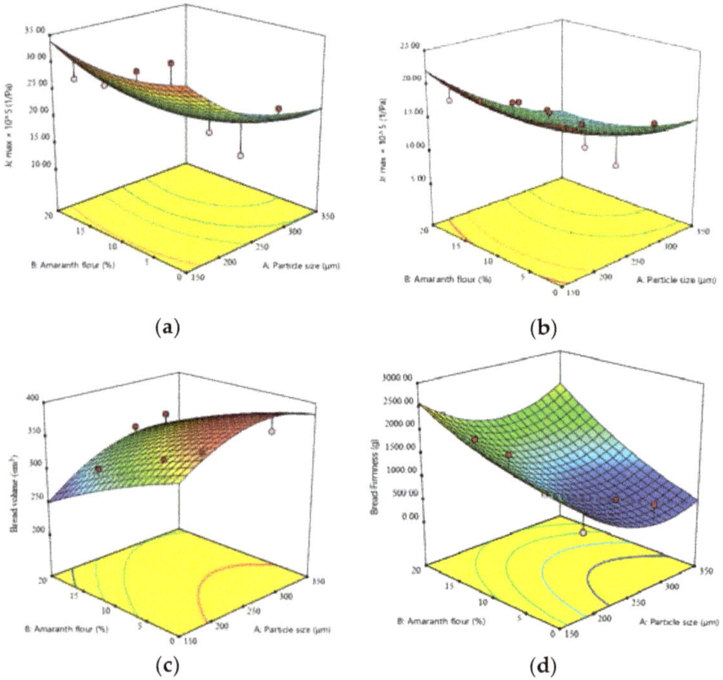

Figure 6. Three-dimensional response surface plots showing the interaction between amaranth flour particle size and addition level on: (**a**) maximum creep compliance (Jc_{max}); (**b**) maximum recovery compliance (Jr_{max}); (**c**) bread volume (BV); and (**d**) bread firmness (BF).

3.4. Optimal and Control Samples Properties

The optimal addition levels for each PS and the predicted values of the responses are presented in Table 5.

Table 5. Wheat flour dough and optimized composite flour for each amaranth flour particle size.

Parameters	Control Sample	O_AL	O_AM	O_AS
Addition Level	**100% WF**	**9.41%**	**9.39%**	**7.89%**
FN (s)	312.00	311.00	316.50	316.66
WA (%)	58.50	57.28	58.44	58.63
DT (min)	1.69	3.44	3.01	2.41
ST (min)	9.96	10.23	9.99	8.87
C1-2 (Nm)	0.61	0.56	0.60	0.64
C3-2 (Nm)	1.41	1.30	1.25	1.23
C3-4 (Nm)	0.05	0.22	0.17	0.07
C5-4 (Nm)	1.15	0.76	0.82	0.85
P (mm H_2O)	87.00	90.73	93.44	99.10
L (mm)	91.00	48.51	50.60	53.05
W $\times 10^{-4}$ (J)	253.00	162.92	165.33	165.28
P/L (adim.)	0.95	1.74	1.75	1.69
H'm (mm)	62.00	74.22	73.30	72.32
VT (mL)	1168.00	1307.73	1273.44	1230.20
VR (mL)	991.00	1173.33	1178.87	1141.19
CR (%)	84.80	89.34	91.25	90.70

Table 5. *Cont.*

Parameters	Control Sample	O_AL	O_AM	O_AS
Addition Level	**100% WF**	**9.41%**	**9.39%**	**7.89%**
G' (Pa)	26,370.00	30,587.03	28,624.29	23,720.87
G'' (Pa)	9488.00	12,411.70	10,869.10	9320.63
tan δ (adim.)	0.3600	0.3630	0.3550	0.3460
T_{max} (°C)	83.24	80.15	79.44	80.02
$Jc_{max} \times 10^{-5}$ (1/Pa)	24.50	16.91	18.47	27.48
$Jr_{max} \times 10^{-5}$ (1/Pa)	16.62	10.81	12.32	17.38
BV (cm^3)	372.20	345.73	368.28	337.11
BF (g)	786.00	1443.42	852.09	1398.75

O_AL—optimized samples with amaranth large particle size; O_AM—optimized samples with amaranth medium particle size; O_AS—optimized samples with amaranth small particle size; FN—Falling Number; WA—water absorption; DT—dough development time; ST—dough stability; C1-2—consistency reached during protein weakening stage; C3-2—consistency reached during starch gelatinization stage; C3-4—consistency reached during the stability of hot starch gel; C5-4—consistency during starch retrogradation stage; P—dough tenacity; L—dough extensibility; W—deformation energy; P/L—configuration ratio of the alveograph curve; H'm—maximum height of the gas release curve; VT—total CO_2 volume production, VR—volume of the gas retained in the dough at the end of the test; CR—retention coefficient; G'—elastic modulus; G''—viscous modulus; tan δ—loss tangent; T_{max}—maximum gelatinization temperature; Jc_{max}—maximum creep compliance; Jr_{max}—maximum recovery compliance; BV—bread volume; BF—bread firmness.

4. Discussion

4.1. Proximate Composition of Amaranth Particle Sizes

The higher shear force which was applied to the milling process can explain the lowest moisture in the finest amaranth particle flour. The biggest content of ash and protein content were in the small fractions, followed by medium PS, a fact that can be linked to the location of the protein in the germ (65%), and 35% in the endosperm of amaranth seed, compared to an average of 15 and 85%, respectively, in most cereals [24,25]. The protein content from the finest amaranth fractions is comparatively higher than the protein content from the same size of quinoa and buckwheat fractions [3]. Amaranth fractions contain higher lipids than most cereals [25].

4.2. Influence of PS and AF Addition Level on Falling Number, Mixolab, and Alveographic Parameters

The FN value is inversely correlated with α-amylase activity, so it may be concluded that with the increase in the AF addition, α-amylase activity increases. This increase can be correlated with the presence of calcium ions in amaranth [14,26,27]. High α-amylase activity is desired for improving the final product quality. The increase in WA with the increase in AF particle size and addition level may be explained by the gluten dilution, which requires less hydration, and therefore a lower amount of water is needed [28]. On the other hand, an increase in WA can be observed when the particle size decreases. This trend can be explained by the specific surface area of amaranth starch, which is larger than wheat starches, which can absorb more water compared to large and medium particle fractions. A similar result was reported by Iuga et al., Mironeasa et al., and Ahmed et al. [23,29,30], whereby small particle sizes of non-gluten flour increased the values of WA. Dough development time measures the dough strength and was significantly affected by the PS, demonstrating that large particles require a long time to reach the optimal elastic and viscosity characteristics. The addition of non-gluten flour affects gluten quality with an increased degree of softening, which is reflected in a higher DT and a lower ST. This behavior can be explained by the presence of larger amounts of crude fiber in the amaranth flour that tend to imbibe and retain water in the dough, affecting the mixing process parameters. Moreover, a longer development time could be linked with the difference in the rate of water absorption by wheat and amaranth flours, due to higher amounts of soluble proteins in amaranth flour and maybe also due to the water absorption characteristics of amaranth starch and non-starch polysaccharides. This may cause a delayed formation of the gluten network in the dough [31]. Otherwise, when the PS increases, the DT is improved, this fact being associated with amaranth albumins that interact with gluten proteins through disulfide bonds [31,32].

The protein weakening (C1-2) increased when the PS decreased, which can be explained by the changes in protein network structure, being more available for enzymatic attacking points and leading to a rise in the speed of protein weakening due to the heat [33]. Excessive mechanical impact weakens dough consistency, which could probably be explained by the high shear temperature from the milling process, which leads to water-binding by protein substances. Another factor might be the better retention of the hydration shell on the protein globules [34]. Starch gelatinization is a key factor in starch behavior, which occurs when the dough is heated at 60 °C. Lower values of C3-2 can be explained by the increase in the interactions between the low amount of amylose (1–5%) and higher long-chain amylopectin (20–25%) from amaranth starch, which generate a synergistic effect on the final viscosity and, thus, on starch retrogradation [35,36]. The combined effect of starch hydrolysis and a low FN index will lead to a decrease in viscosity [37], which will increase the water amount from the dough [38,39]. An increase in hot gel stability values may be related to the starch damage process. Our results indicate that amaranth flour addition can limit starch retrogradation, increasing bread freshness and shelf-life. The biaxial extension of the dough can be monitored with an alveograph analysis, which monitors vital parameters that can improve the fermentation process.

The dough tenacity (P), characterized by the force required for dough rupture, can predict the bread volume [40,41]. This alveographic parameter increased when the AF addition level increased, a fact that could be explained by the non-gluten flour incorporation, or could be a consequence of the higher level of protein and fiber from amaranth seeds [42,43]. The extensibility of dough (L) is the property of wheat flour dough to obtain the characteristic structure and volume of baked goods [44], can predict the handling properties of the dough [45], and was greatly reduced by the AF addition. This phenomenon can be explained by the small numbers of hydroxyl groups from fiber which has more water availability and the weakening of the gluten network [46]. Similar data were reported by Piga et al. [36] when amaranth flour was studied as a potential healthy ingredient for the development of an innovative gluten-free flatbread. Dough strength or deformation energy (W) decreased significantly when AF amounts were increased, while the PS did not influence this parameter, possibly due to the differences in protein content between wheat flour and amaranth flour [43]. The P/L ratio, which gives information about the elastic resistance and extensibility balance of dough, was augmented in doughs with amaranth flour.

4.3. Influence of PS and AF Addition Levels on Dough Fermentation, Dynamic Rheological Properties, and Bread Characteristics

The maximum height of the gas release curve (H'm) is a critical parameter in the fermentation process and is related to the maximum height of dough development and the height of dough development at the end of the test. The H'm parameter was closely related to the bread volume [47], and therefore it is a good attribute for predicting the final product. The dough, which contains non-diluting gluten, will be more elastic and will form bread with a continuous sponge structure after baking [48]. The gas retention was decreased, which may be due to dough permeability when the gluten network was weakened by the amylose and amylopectin hydrolysis on the presence of enzymes during the fermentation process. Martínez and Gómez (2017) [49] reported that the structure and morphology of the starch granules and flour particles were the major determinants of the dough changes produced during the fermentation and baking phases. Some authors reported that the greater milling damage to the starch level in superfine-ground flour gave rise to a decrease in total gas production during dough fermentation [50] due to the amylase inactivation, thus hindering the formation of fermentable sugars and reducing the capacity of yeasts to produce gas [51]. The compact particles of the flours produced dough with high consistency and bread with volume and textural properties lower than those obtained with large particles. These changes might be due to the competitiveness of water uptake between wheat gluten and AF components such as starch, protein, and fiber. The highest swelling capacity value was found for the small AF particle size due to an increase in damaged starch, which influences dough behavior [3].

Fundamental rheological tests can offer valuable information about the final product. An increase in elastic (G′) and viscous (G″) moduli values could be linked to the presence of binding agents from the composite dough and attractive forces between starch granules and amaranth fibers, this trend suggesting a solid, elastic-like behavior of the WF–AF composite dough. The higher values of G′ and G″ could probably also be a result of interactions between proteins and the formation of a protein–starch complex. The significant increase in G′ with the increased addition level of AF to wheat flour could be related to the increasing amounts of damaged starch in the composite flour. Hatcher et al. [52] found that the damaged starch in wheat flour greater affected the viscoelastic properties of noodles than the particle size, resulting in stiffer doughs, than did flour with low starch damage. Decreases in loss tangent (tan δ) values are typical for elastic and firm doughs. Similar results were found by Burešová et al. [53]. The tan δ values are influenced by the level of starch damage. The dough from flours with high starch damage presents significantly lower values than those with low or medium starch damage [52].

A decrease in the maximum gelatinization temperature with the increase in the AF addition could be due to the higher amount of water absorbed by the amaranth grain, and the greater swelling power and solubility of the amaranth starch granule compared with the wheat starch granule [54]. A higher amount of water absorbed could be related to the starch damage from the amaranth flour milling process.

The maximum creep compliance (Jc$_{max}$) and the maximum recovery compliance (Jr$_{max}$), measured at the end of the creep and recovery phases, respectively, represent the principal characteristics from the creep–recovery curves. The results of the creep–recovery measurements are significantly influenced by the protein or starch levels from amaranth fractions, while the AF addition level did not significantly affect these dynamic parameters. This strengthening phenomenon can be related to the hydroxyl groups from sugar compounds from amaranth seeds, which may directly interact with proteins, resulting in non-covalent or covalent bonding. Protein–polyphenol interactions modify proteins, influencing the quality and functional properties of a food [23,55]. The bread volume decreased as the AF addition level was increased, which could be explained by the poor baking quality of this flour, due to its lack of gluten proteins, which are present in wheat [25]. The proteins present in amaranth flour consist of three major fractions, albumins (51%), globulins (16%), and glutelins (24%), and a minor, alcohol-soluble fraction, prolamine, between 1.4 and 2.0% [56,57]. The albumin fraction is comparable with egg-white proteins and can be used as an egg substitute in different products [25]. The decrease in bread volume with the AF addition can be linked to the weakening of the gluten matrix and reduced gas retention of the dough (Table 4), which is predictable for the higher dough tenacity and lower extensibility (Table 3). Our results fall in line with those of Tömösközi et al. [31]. The high lipid content from small and medium amaranth fractions (Table 2) could have functionality as a gas stabilizing agent during breadmaking, which probably improves the bread's technological properties based on medium and small amaranth fractions (volume, elasticity) [58]. Some authors found a direct relation between dough elasticity/crumb chewiness and crumb firmness [59]. These results are in agreement with those reported by other authors [27,60–62].

The amaranth flour addition level negatively influenced the bread firmness; an increase in bread firmness was observed when the AF amount was increased. Regarding amaranth particle sizes, there was a decrease in bread firmness as the particle size was increased (especially for the 5–10% addition level), a phenomenon that can be explained by the albumin proteins from amaranth grain, which can act like gluten in the dough and have the capacity to interact with wheat glutenin protein through disulfide bonds, which does not weaken the gluten network overmuch. Similar results were obtained by Oszvald et al. and Miranda Ramos et al. [32,61].

4.4. Optimal PS and Addition Levels of AF

The numerical optimization procedure revealed that the most suitable composite flour, based on the large amaranth particle size (380 μm), would have 9.41% amaranth flour and 90.59% wheat flour; for the medium amaranth particle size (280 μm), a percent of 9.39% amaranth flour and 90.61% wheat flour would be most suitable; and for the small amaranth particle size (180 μm), the most suitable amaranth flour addition level is 7.89% to 92.11% wheat flour. According to the obtained results, both optimal and control flours present dough stability (ST) and protein weakening (C1-2) very close values. The development time (DT) was considerably higher than the control sample. In this case, the optimal dough formulations can retain gas easier than the wheat dough. The differences in the Falling Number, empirical, and dynamic rheological properties, and bread physical and textural properties between the optimal and control samples can be linked to the significance of proteins, lipids, carbohydrates, and minerals of the amaranth fractions and their interactions in the dough network.

5. Conclusions

This research studied the influence of amaranth flour particle size and addition level on the wheat dough rheological, technological, and textural bread properties, to optimize dough and bread quality in baking. Small fractions, followed by medium ones, presented the highest content of protein and ash, while the large particle sizes have a high content of carbohydrates.

The results obtained revealed that particle size and AF addition level modified dough behavior during mixing, extension, and fermentation, as well as the bread's physical and textural parameters. Dough rheology and bread parameters were significantly influenced by particle size and the addition level of amaranth flour, all the regression models obtained for responses being significant ($p < 0.05$) and with high coefficients of determination ($R^2 = 0.66$–0.96).

The combined effect of AF particle size and addition level on wheat flour led to a decrease in the Falling Number, starch gelatinization, the final starch paste viscosity after cooling, dough extensibility, baking strength, maximum gelatinization temperature, maximum creep and recovery compliance, and bread volume, while dough development and stability, protein weakening, the stability of hot starch gel, dough tenacity, the alveograph curve ratio, all rheofermentometer parameters (maximum height of the gas release curve, the total volume of CO_2 production, the volume of the gas retained in the dough at the end of the test, the retention coefficient), visco-elastic moduli, loss tangent, and bread firmness increased proportionally with the amount of AF used.

The optimization of twenty-four responses allowed us to obtain, for each particle size, the optimal amaranth flour level which can be substituted into wheat flour to obtain dough and bread with the best rheological and technological properties. It can be concluded that particle size is very important for obtaining the desired rheological properties together with the appropriate bread volume and firmness. These optimal composite flours do not alter the dough network but improve the technological parameters desired to enrich bread and other bakery products due to particle size and approximate composition.

Author Contributions: Conceptualization, I.C. and S.M.; methodology, I.C.; software, I.C.; validation, I.C. and S.M.; formal analysis, I.C. and S.M.; investigation, I.C.; resources, I.C. and S.M.; data curation, I.C. and S.M.; writing—original draft preparation, I.C.; writing—review and editing, I.C. and S.M.; visualization, I.C. and S.M.; supervision, S.M.; funding acquisition, S.M. The authors contributed equally to this research. All authors have read and agreed to the published version of the manuscript.

Funding: This research was funded by "Stefan cel Mare" University of Suceava.

Institutional Review Board Statement: Not applicable.

Informed Consent Statement: Not applicable.

Data Availability Statement: Not applicable.

Acknowledgments: This work was funded by the Ministry of Research, Innovation and Digitalization within Program 1—Development of national research and development system, Subprogram 1.2—Institutional Performance—RDI excellence funding projects, under contract no. 10PFE/2021.

Conflicts of Interest: The authors declare no conflict of interest.

References

1. Rosell, C.M.; Garzon, R. *Chemical Composition of Bakery Products, Handbook of Food Chemistry*; Springer: Berlin/Heidelberg, Germany, 2015; pp. 191–224.
2. Alvarez-Jubete, L.; Auty, M.; Arendt, E.K.; Gallagher, E. Baking properties and microstructure of pseudocereal flours in gluten-free bread formulations. *Eur. Food Res. Technol.* **2010**, *230*, 437–445. [CrossRef]
3. Coțovanu, I.; Mironeasa, S. Impact of different amaranth particle sizes addition level on wheat flour dough rheology and bread features. *Foods* **2021**, *10*, 1539. [CrossRef]
4. Chauhan, A.; SaxenA, D.C.; Singh, S. Total dietary fibre and antioxidant activity of gluten free cookies made from raw and germinated amaranth (*Amaranthus* spp.) flour. *LWT-Food Sci. Technol.* **2015**, *63*, 939–945. [CrossRef]
5. Inglett, G.E.; Chen, D.; Liu, S.X. Physical properties of gluten-free sugar cookies made from amaranth–oat composites. *LWT-Food Sci. Technol.* **2015**, *63*, 214–220. [CrossRef]
6. Venskutonis, P.R.; Kraujalis, P. Nutritional components of amaranth seeds and vegetables: A review on composition, properties, and uses. *Compr. Rev. Food Sci. Food Saf.* **2013**, *12*, 381–412. [CrossRef] [PubMed]
7. Kalinova, J.; Dadakova, E. Rutin and total quercetin content in amaranth (*Amaranthus* spp.). *Plant Foods Hum. Nutr.* **2009**, *64*, 68–74. [CrossRef] [PubMed]
8. Caselato-Sousa, V.M.; Amaya-Farfán, J. State of knowledge on amaranth grain: A comprehensive review. *J. Food Sci.* **2012**, *77*, 93–104. [CrossRef]
9. Lazaridou, A.; Biliaderis, C.G. Gluten-free doughs: Rheological properties, testing procedures–methods and potential problems. In *Gluten-Free Food Science and Technology*; Gallagher, E., Ed.; Wiley-Blackwell: Chichester, UK, 2009; pp. 52–82.
10. Skendi, A.; Papageorgiou, M.; Biliaderis, C.G. Effect of barley β-glucan molecular size and level on wheat dough rheological properties. *J. Food Eng.* **2009**, *91*, 594–601. [CrossRef]
11. Edwards, N.M.; Mulvaney, S.J.; Scanlon, M.G.; Dexter, J.E. Role of gluten and its components in determining durum semolina dough viscoelastic properties. *Cereal Chem.* **2003**, *80*, 755–763. [CrossRef]
12. Bloksma, A.H.; Bushuk, W.; Pomeranz, Y. Wheat chemistry and technology. *Rheol. Chem. Dough* **1988**, *2*, 131–200.
13. Coțovanu, I.; Mironeasa, S. Buckwheat Seeds: Impact of Milling Fractions and Addition Level on Wheat Bread Dough Rheology. *Appl. Sci.* **2021**, *11*, 1731. [CrossRef]
14. Coțovanu, I.; Ungureanu-Iuga, M.; Mironeasa, S. Investigation of Quinoa Seeds Fractions and Their Application in Wheat Bread Production. *Plants* **2021**, *10*, 2150. [CrossRef]
15. Alvarez-Jubete, L.; Arendt, E.K.; Gallagher, E. Nutritive value and chemical composition of pseudocereals as gluten-free ingredients. *Int. J. Food Sci. Nutr.* **2009**, *60*, 240–257. [CrossRef]
16. Martínez-Villaluenga, C.; Peñas, E.; Hernández-Ledesma, B. Pseudocereal grains: Nutritional value, health benefits and current applications for the development of gluten-free foods. *Food Chem. Toxicol.* **2020**, *137*, 111178. [CrossRef] [PubMed]
17. Sluková, M.; Levková, J.; Michalcova, A.; Skřivan, P.; Skřivan, P. Effect of the dough mixing process on the quality of wheat and buckwheat proteins. *Czech J. Food Sci.* **2017**, *35*, 522–531. [CrossRef]
18. Zhang, Y.; Hu, R.; Tilley, M.; Siliveru, K.; Li, Y. Effect of Pulse Type and Substitution Level on Dough Rheology and Bread Quality of Whole Wheat-Based Composite Flours. *Processes* **2021**, *9*, 1687. [CrossRef]
19. Millar, K.A.; Barry-Ryan, C.; Burke, R.; McCarthy, S.; Gallagher, E. Dough properties and baking characteristics of white bread, as affected by addition of raw, germinated and toasted pea flour. *Innov. Food Sci. Emerg. Technol.* **2019**, *56*, 102189. [CrossRef]
20. Mironeasa, S.; Mironeasa, C. Dough bread from refined wheat flour partially replaced by grape peels: Optimizing the rheological properties. *J. Food Process Eng.* **2019**, *42*, e13207. [CrossRef]
21. ICC. *Standard Methods of the International Association for Cereal Chemistry. Methods 104/1, 110/1, 136, 105/2, 171, 121, 107/1*; International Association for Cereal Chemistry: Vienna, Austria, 2010.
22. *SR 90:2007*; Wheat Flour. Analysis Method. Romanian Standards Association: Bucharest, Romania, 2007.
23. Iuga, M.; Mironeasa, C.; Mironeasa, S. Oscillatory rheology and creep-recovery behaviour of grape seed-wheat flour dough: Effect of grape seed particle size, variety and addition level. *Bull. Univ. Agric. Sci. Vet. Med. Cluj-Napoca Food Sci. Technol.* **2019**, *76*, 40–51. [CrossRef]
24. Stallknecht, G.F.; Schulz-Schaeffer, J.R. *Amaranth Rediscovered. New Crops*; Wiley: New York, NY, USA, 1993; pp. 211–218.
25. Singh, N.; Singh, P.; Shevkani, K.; Virdi, A.S. Amaranth: Potential source for flour enrichment. In *Flour and Breads and Their Fortification in Health and Disease Prevention*; Academic Press: Cambridge, MA, USA, 2019; pp. 123–135. [CrossRef]
26. Sundarram, A.; Murthy, T.P.K. α-amylase production and applications: A review. *J. Appl. Environ. Microbiol.* **2014**, *2*, 166–175. [CrossRef]
27. Sanz-Penella, J.M.; Wronkowska, M.; Soral-Smietana, M.; Haros, M. Effect of whole amaranth flour on bread properties and nutritive value. *LWT-Food Sci. Technol.* **2013**, *50*, 679–685. [CrossRef]

28. Mohammed, I.; Ahmed, A.R.; Senge, B. Effects of chickpea flour on wheat pasting properties and bread making quality. *J. Food Sci. Technol.* **2014**, *51*, 1902–1910. [CrossRef] [PubMed]

29. Mironeasa, S.; Iuga, M.; Zaharia, D.; Mironeasa, C. Optimization of grape peels particle size and flour substitution in white wheat flour dough. *Sci. Stud. Res. Chem. Chem. Eng. Biotechnol. Food Ind.* **2019**, *20*, 29–42.

30. Ahmed, J.; Al-Attar, H.; Arfat, Y.A. Effect of particle size on compositional, functional, pasting and rheological properties of commercial water chestnut flour. *Food Hydrocoll.* **2016**, *52*, 888–895. [CrossRef]

31. Tömösközi, S.; Gyenge, L.; Pelcéder, A.; Abonyi, T.; Lásztity, R. The effects of flour and protein preparations from amaranth and quinoa seeds on the rheological properties of wheat-flour dough and bread crumb. *Czech J. Food Sci.* **2011**, *29*, 109–116. [CrossRef]

32. Oszvald, M.; Tamás, C.; Rakszegi, M.; Tömösközi, S.; Békés, F.; Tamás, L. Effects of incorporated amaranth albumins on the functional properties of wheat dough. *J. Sci. Food Agric.* **2009**, *89*, 882–889. [CrossRef]

33. Tester, R.F.; Karkalas, J.; Qi, X. Starch structure and digestibility enzyme-substrate relationship. *Poult. Sci. J.* **2004**, *60*, 186–195. [CrossRef]

34. Derkanosova, N.; Stakhurlova, A.; Pshenichnaya, I.; Ponomareva, I.; Peregonchaya, O.; Sokolova, S. Amaranth as a bread enriching ingredient. *Foods Raw Mater.* **2020**, *8*, 223–231. [CrossRef]

35. Corke, H.; Cai, Y.Z.; Wu, H.X. *Amaranth: Overview In Encyclopedia of Food Grains*; Wrigley, C., Corke, H., Seetharaman, K., Faubion, J., Eds.; Academic Press: Oxford, UK, 2016; pp. 287–296.

36. Piga, A.; Conte, P.; Fois, S.; Catzeddu, P.; Del Caro, A.; Sanguinetti, A.M.; Fadda, C. Technological, Nutritional and sensory Properties of an Innovative Gluten-Free Double-Layered Flat Bread Enriched with Amaranth Flour. *Foods* **2021**, *10*, 920. [CrossRef] [PubMed]

37. De Souza, P.M.; Magalhães, P.D.O. Application of microbial α-amylase in industry—A review. *Braz. J. Microbiol.* **2010**, *41*, 850–861. [CrossRef]

38. Mironeasa, S.; Codină, G.G. The mixolab rheological properties and dough microstructure of defatted mustard seed–wheat composite flours. *J. Food Process. Preserv.* **2017**, *41*, e13130. [CrossRef]

39. Pahwa, A.; Kaur, A.; Puri, R. Influence of hydrocolloids on the quality of major flat breads: A review. *J. Food Process.* **2016**, *2016*. [CrossRef]

40. Dobraszczyk, B.J.; Salmanowicz, B.P. Comparison of predictions of baking volume using large deformation rheological properties. *J. Cereal Sci.* **2008**, *47*, 292–301. [CrossRef]

41. Kenny, S.; Wehrle, K.; Dennehy, T.; Arendt, E.K. Correlations between empirical and fundamental rheology measurements and baking performance of frozen bread dough. *Cereal Chem.* **1999**, *76*, 421–425. [CrossRef]

42. Kurek, M.A.; Karp, S.; Wyrwisz, J.; Niu, Y. Physicochemical properties of dietary fibers extracted from gluten-free sources: Quinoa (*Chenopodium quinoa*), amaranth (*Amaranthus caudatus*) and millet (*Panicum miliaceum*). *Food Hydrocoll.* **2018**, *85*, 321–330. [CrossRef]

43. Sánchez-Marrowín, A.; Domingo, M.V.; Maya, S.; Saldana, C. Amaranth flour blends and fractions for baking applications. *J. Food Sci.* **1985**, *50*, 789–794. [CrossRef]

44. Kieffer, R. The role of gluten elasticity in the baking quality of wheat. In *Future of Flour—A Compendium of Flour Improvement*; Popper, L., Schäfer, W., Freund, W., Eds.; Verlag Agrimedia: Clenze, Germany, 2006; pp. 169–178.

45. Rosell, C.M.; Rojas, J.A.; De Barber, C.B. Influence of hydrocolloids on dough rheology and bread quality. *Food Hydrocoll.* **2001**, *15*, 75–81. [CrossRef]

46. Ungureanu-Iuga, M.; Atudorei, D.; Codină, G.G.; Mironeasa, S. Rheological Approaches of Wheat Flour Dough Enriched with Germinated Soybean and Lentil. *Appl. Sci.* **2021**, *11*, 11706. [CrossRef]

47. Gandikota, S.; MacRitchie, F. Expansion capacity of doughs: Methodology and applications. *J. Cereal Sci.* **2005**, *42*, 157–163. [CrossRef]

48. Hosney, R.C. *Principles of Cereal Science and Technology*; American Association of Cereal Chemists: Saint Paul, MN, USA, 1994.

49. Martínez, M.M.; Gómez, M. Rheological and microstructural evolution of the most common gluten-free flours and starches during bread fermentation and baking. *J. Food Eng.* **2017**, *197*, 78–86. [CrossRef]

50. Xu, X.; Xu, Y.; Wang, N.; Zhou, Y. Effects of superfine grinding of bran on the properties of dough and qualities of steamed bread. *J. Cereal Sci.* **2018**, *81*, 76–82. [CrossRef]

51. Wang, Q.; Li, L.; Zheng, X. A review of milling damaged starch: Generation, measurement, functionality and its effect on starch-based food systems. *Food Chem.* **2020**, *315*, 126267. [CrossRef]

52. Hatcher, D.W.; Anderson, M.J.; Desjardins, R.G.; Edwards, N.M.; Dexter, J.E. Effects of flour particle size and starch damage on processing and quality of white salted noodles. *Cereal Chem.* **2002**, *79*, 64–71. [CrossRef]

53. Burešová, I.; Buňka, F.; Kráčmar, S. Rheological Characteristics of Gluten-Free Dough. *J. Microbiol. Biotechnol. Food Sci.* **2014**, *2021*, 195–198.

54. Perez-Rea, D.; Antezana-Gomez, R. The functionality of pseudocereal starches. In *Starch in Food*; Woodhead Publishing: Sawston, UK, 2018; pp. 509–542. [CrossRef]

55. Baxter, N.J.; Lilley, T.H.; Haslam, E.; Williamson, M.P. Multiple interactions between polyphenols and a salivary proline-rich protein repeat result in complexation and precipitation. *Biochemistry* **1997**, *36*, 5566–5577. [CrossRef]

56. Gorinstein, S.; Denue, I.A.; Arruda, P. Alcohol soluble and total proteins from amaranth seeds and their comparison with other cereals. *J. Agric. Food Chem.* **1991**, *39*, 848–850. [CrossRef]

57. Martinez, E.N.; Castellani, O.F.; Anon, M.C. Common molecular features among amaranth storage proteins. *J. Agric. Food Chem.* **1997**, *45*, 3832–3839. [CrossRef]
58. D'Amico, S.; Schoenlechner, R.; Tömösközi, S.; Langó, B. Proteins and amino acids of kernels: Chemistry and technology. In *Pseudocereals: Chemistry and Technology*, 1st ed.; Haros, M., Schoenlechner, R., Eds.; John Wiley & Sons Ltd.: Oxford, UK, 2017; pp. 94–99.
59. Meullenet, J.-F.; Lyon, B.G.; Carpenter, J.A.; Lyon, C.E. Relationship between sensory and instrumental texture profile Attributes. *J. Sens. Stud.* **1998**, *13*, 77–93. [CrossRef]
60. Amare, E.; Grigoletto, L.; Corich, V.; Giacomini, A.; Lante, A. Fatty Acid Profile, Lipid Quality and Squalene Content of Teff (*Eragrostis teff* (Zucc.) Trotter) and Amaranth (*Amaranthus caudatus* L.) Varieties from Ethiopia. *Appl. Sci.* **2021**, *11*, 3590. [CrossRef]
61. Miranda-Ramos, K.C.; Sanz-Ponce, N.; Haros, C.M. Evaluation of technological and nutritional quality of bread enriched with amaranth flour. *LWT* **2019**, *114*, 108418. [CrossRef]
62. Almeida, E.; Chang, Y.; Steel, C. Dietary fibre sources in bread: Influence on technological quality. *LWT-Food Sci. Technol.* **2013**, *50*, 545–553. [CrossRef]

applied
sciences

Article

Amaranth Seed Polyphenol, Fatty Acid and Amino Acid Profile

Ovidiu Procopet and Mircea Oroian *

Faculty of Food Engineering, Stefan cel Mare University of Suceava, 720229 Suceava, Romania;
ovidiu.procopet@fia.usv.ro
* Correspondence: m.oroian@fia.usv.ro

Abstract: In this paper, the extraction of polyphenols from amaranth seed using a Box–Benhken design using four factors—ultra-turrax speed, solid-to-liquid ratio (RSL), methanol concentration and extraction time—were studied. There were two responses studied for the model: total phenolic content (TPC) and total flavonoid content (TFC). The factors which influenced the most the extraction of the TPC and TFC were the RSL, methanol concentration and ultra-turrax speed. Twelve phenolic acids (rosmarinic acid, *p*-coumaric acid, chlorogenic acid, vanillic acid, caffeic acid, *p*-hydroxybenzoic acid, protocatechuic acid and gallic acid) and flavonoids (kaempferol, quercetin, luteolin and myricetin) were studied, and the most abundant one was kaempferol followed by myricetin. The amaranth seed is a valuable source of fatty acids, and 16.54% of the total fatty acids determined were saturated fatty acids, while 83.45% of the fatty acids were unsaturated ones. Amaranth seed is a valuable source of amino acids, with 9 essential amino acids being reported: histidine, isoleucine, leucine, lysine, methionine, phenylalanine, threonine, tryptophan and valine.

Keywords: amaranth seed; extraction; polyphenols; fatty acids; amino acids

1. Introduction

Amaranth is a pseudo-cereal which is known to have a high nutritional value thanks to its amino acids, fiber, trace elements, vitamins, flavonoids, phenolic compounds and polyunsaturated fatty acids [1,2]. Amaranth seed is rich in different bioactive and nutritional compounds such as tocopherols, sterols, squalene, fatty acids, phenolic acids, flavonoids (as free molecules of conjugated molecules with sugars or linked to proteins in a complex matrix through covalent or non-covalent links), dietary fibers and peptides [3–6].

The use of amaranth seeds has increased over the last 10 years both in the ordinary diet as well as in the diets of people suffering from certain diseases and in the diets of the people allergic to several types of cereals [7]. Thus, amaranth seeds, considered to be pseudo-cereals, are characterized by high nutritional and functional values, associated with the antioxidant quality and quantity of proteins and lipids [2,8,9]. The amaranth seed flour is a well-known source of polyphenols, and it is recommended by doctors to be used in balanced diets [8]. The polyphenols are easily degraded in the intestine of humans and animals because of the abundance of beta-glucosidase enzyme, which liberates the aglyconic moiety of the molecules [10]. Amaranth seed production has become more and more popular among the people interested in their own health and especially among high performance athletes as a consequence of a change in their eating habits [2].

Antioxidants have triggered considerable interest in food technology research. Their availability in various diets and their high potential role in the fight against different diseases such as cancer and neurodegenerative or cardiovascular diseases has been outlined in a variety of studies [11,12].

Some research has shown that amaranth seeds and leaves have antioxidant capabilities when tested in vitro [13]. Thus, the presence of polyphenols in the amaranth seeds has been reported [14], and so has the presence of flavonoids and betalaines [15]. In other research, *Amaranthus hypochondriacus* seeds have been analyzed, and phenolic compounds

Citation: Procopet, O.; Oroian, M. Amaranth Seed Polyphenol, Fatty Acid and Amino Acid Profile. *Appl. Sci.* **2022**, *12*, 2181. https://doi.org/10.3390/app12042181

Academic Editor: Massimo Lucarini

Received: 24 January 2022
Accepted: 17 February 2022
Published: 19 February 2022

Publisher's Note: MDPI stays neutral with regard to jurisdictional claims in published maps and institutional affiliations.

with antioxidant properties were discovered [10]. The amaranth can be considered a source for the development of new edible products, as it has a high nutritional value, and there is a trend in recent years regarding vegan diets. It is well known that vegetable proteins present functional properties with high potential for human health [16,17].

Extraction of compounds from the matrices is an important process for achieving the structural composition or a fundamental step for obtaining natural medicines. This process is influenced by many factors, such as the particle size of the matrices, solvents, time, technique and temperature [18]. There are various compound extraction methods for vegetable products. One extraction method uses solvents like ethanol, methanol, ethyl acetate or hexane among other solvents. Conventional solvents like methanol and hexane are recognized for their high extraction yield, but for environmental safety, ethanol is a safer alternative. Various extraction techniques can be used, such as mechanical agitation, ultrasounds, homogenization, Soxhlet-type extraction, maceration and supercritical fluid extraction [8,10,13,15]. The particle size of the matrix is important because the smaller the particle is, the better the extraction process is. The ultra-turrax was used for its high versatility in the homogenization process in the case of two immiscible systems (solvent and plant material) [19]. Extraction assisted by an ultra-turrax can be used for bioactive compound extraction from plant matrices.

The extraction of compounds from plant matrices using an ultra-turrax has many advantages, such as narrow and uniform particle size distribution, being a fast and low-cost extraction method and being able to generate high extraction rates at a normal temperature through crushing the matrices into smaller particles [18,20]. Moreover, under partial negative pressure permeation, high solubility of the compounds from the matrix into the solvent can be obtained. The extraction rate is increased because of the particle reduction, and drastic stirring adds dynamic molecular permeation [20]. The ultra-turrax was used for the extraction of triterpenoid saponins from *Pulsatilla* herbs [20], organic acids from honeysuckle [18], lignans from *Schisandra chinensis* fruit [21] and flavonoids from the seeds of *Oroxylum indicum* [22].

The aim of this study was to evaluate the influence of the ultra-turrax speed, solid-to-liquid ratio, methanol concentration and extraction time on the total phenolic content and total flavonoid content from amaranth seed. In addition, the most concentrated solution was used for the identification and quantification of 12 phenolics. The amino acids and fatty acids from the amaranth seed were determined using a GC-MS method.

2. Materials and Methods

2.1. Materials

Ecological amaranth seed (*Amaranthus cruentus*) was procured from Driedfruits supplies (Timisoara, Romania). The seeds were authenticated based on their botanical features, and the seeds were kept in paper bags until extraction.

Methanol, Folin Ciocalteu reagent, sodium carbonate, gallic acid, rosmarinic acid, *p*-coumaric acid, chlorogenic acid, vanillic acid, caffeic acid, *p*-hydroxybenzoic acid, protocatechiuc acid, kaempferol, quercetin, luteolin, myricetin, n-hexane, BF_3 and NaCl were purchased from Sigma Aldrich (Steinheim am Albuch, Germany). Fatty acid methyl esters (FAMEs) were purchased from Restek (Bellefonte, PA, USA, 35077).

2.2. Methods

2.2.1. Extraction of Bioactive Compounds

The extraction of bioactive compounds was carried out according to the Box–Behnken design presented in Table 1. The amount mentioned for each experiment was transferred into a flask, filled to 30 mL with the corresponding solvent and submitted to ultraturrax. After extraction, the solution was passed to a centrifuge tube and centrifugated for 10 min at 5000 rpm. The supernatant was collected in a 50-mL flask, filled up with the solvent and kept at $-20\ °C$ prior analysis.

Table 1. Box–Behnken design of the TPC and TFC from amaranth seed.

No.	Ultra-Turrax Speed (rpm)	RSL	Methanol (%)	Time (s)
1	10,000	0.02	40	60
2	10,000	0.02	60	30
3	10,000	0.01	60	60
4	10,000	0.03	60	60
5	10,000	0.02	60	90
6	10,000	0.02	80	60
7	20,000	0.02	40	30
8	20,000	0.01	40	60
9	20,000	0.03	40	60
10	20,000	0.02	40	90
11	20,000	0.01	60	30
12	20,000	0.03	60	30
13	20,000	0.02	60	60
14	20,000	0.01	60	90
15	20,000	0.03	60	90
16	20,000	0.02	80	30
17	20,000	0.01	80	60
18	20,000	0.03	80	60
19	20,000	0.02	80	90
20	30,000	0.02	40	60
21	30,000	0.02	60	30
22	30,000	0.01	60	60
23	30,000	0.03	60	60
24	30,000	0.02	60	90
25	30,000	0.02	80	60

2.2.2. Total Phenolic Content Determination

A total of 0.1 mL of extract, according to the design presented in Table 1, was mixed in 1.9 mL water and 0.1 mL of Folin Ciocalteu reagent. The mixture was then homogenized for 2 min, and after that, 0.8 mL of 5% sodium carbonate was added. The mixture was kept at 40 °C for 20 min and cooled down in an ice bath to stop the reaction. The total phenolic content was determined at 750 nm, based on a gallic acid calibration curve expressing the results as a gallic acid equivalent (mg GAE/g) [23].

2.2.3. Total Flavonoid Content Determination

A total of 0.2 mL of extract, according to the design presented in Table 1, was mixed with 2 mL of methanol and 0.1 mL of 5% AlCl3 (prepared in methanol). The solution was left for 30 min at room temperature, and its absorbance was measured at 425 nm. The concentration was determined based on a quercetin calibration curve expressing the results as mg quercetin equivalent/L (mg QE/L) [23].

2.2.4. Individual Phenolic Determination

The phenolic acids (rosmarinic acid, *p*-coumaric acid, chlorogenic acid, vanillic acid, caffeic acid, *p*-hydroxybenzoic acid, protocatechuic acid and gallic acid) and flavonoids (kaempferol, quercetin, luteolin and myricetin) were determined from the methanolic extract using a Shimadzu high-performance liquid chromatograph (Kyoto, Japan) with a diode array detector. The separation was carried out on a Zorbax SP-C18 column 150 mm in length, and a 4.6-mm i.d. 5 μm-diameter particle was used for the separation. Elution was carried out with a solvent system consisting of 0.1% acetic acid in water (solvent A) and acetonitrile (solvent B) at a flow rate of 1 mL·min^{-1}. The determined phenolic compounds of gallic acid, vanillic acid, protocatechuic acid and *p*-hydroxibenzoic acid were determined at 280 nm, and chlorogenic acid, *p*-coumaric acid, caffeic acid, rosmarinic acid, myricetin, quercetin, luteolin and kaempherol were determined at 320 nm [24]. Figure 1 presents an HPLC-DAD chromatogram at 280 nm and 320 nm for the standards (100 mg/L).

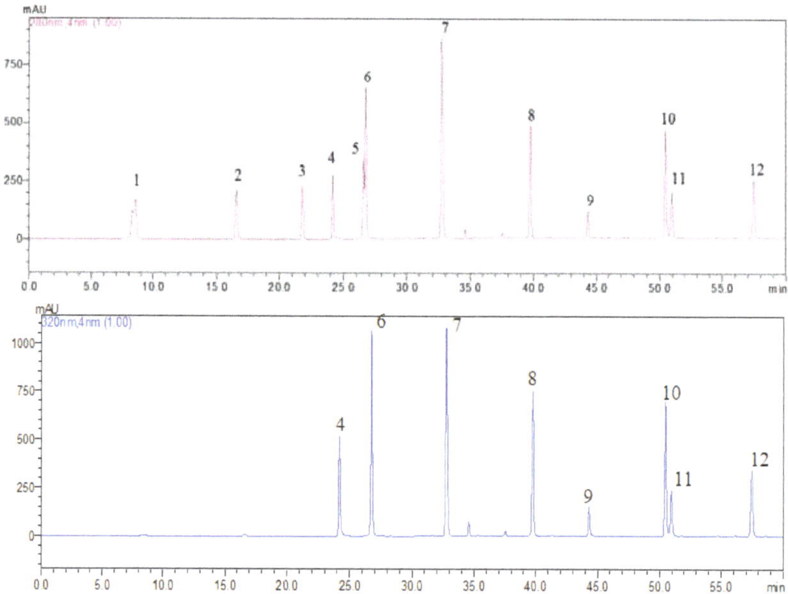

Figure 1. HPLC-DAD chromatogram at 280 nm and 320 nm for the standard (100 mg/L) for gallic acid (peak 1 (8.578 min)), protocatechuic acid (peak 2 (16.6 min)), *p*-hydroxybenzoic acid (peak 3 (21.8 min)), caffeic acid (peak 4 (24.2 min)), vanillic acid (peak 5 (26.6 min)), chlorogenic acid (peak 6 (26.7 min)), *p*-coumaric acid (peak 7 (32.7 min)), rosmarinic acid (peak 8 (39.8 min)), myricetin (peak 9 (44.3 min)), luteolin (peak 10 (50.5 min)), quercetin (peak 11 (50.9 min)) and kaempferol (peak 12 (57.4 min)).

2.2.5. Fatty Acid Profile

Prior to analysis, the amaranth seed flour was extracted for 48 h with n-hexane at room temperature. The oil (0.1 g) was mixed with 1 mL of n-hexane and 1 mL of 15% BF_3 in methanol. The mixture was kept for 15 min at 60 °C in a water bath. The mixture was cooled to 20 °C and mixed with 5 mL of an NaCl-saturated solution, and the solution was centrifuged for 5 min at 3000 rpm afterward. The supernatant was filtered with a 0.45-μm nylon filter and kept at −20 °C prior to analysis. The separation of the fatty acid methyl esters (FAMEs) was carried out on a SUPELCOWAX 10 column (60 m × 0.25 mm i.d., 0.25-μm film thickness; Supelco Inc., Bellefonte, PA, USA) using a Shimadzu GC-MS instrument (GC MS-QP 2010 Plus, Shimadzu, Kyoto, Japan) equipped with an AOC-01 auto-injector that was used to perform the gas chromatographic–mass spectrometric analyses. Identification of FAMEs was performed by comparing their retention times to those of the known standards (37 component FAME Mix, Restek, Bellefonte, PA, USA, 35077) and the resulting mass spectra to the ones from our database (NIST MS Search 2.0) [24]. The FAMEs were quantified by comparison with the standard curves of the FAME 37 mixture, which was made with different concentrations from 0.1 to 1 mg·mL^{-1}. The results were expressed as mg of fatty acid per 100 g of amaranth seed.

2.2.6. Determination of Free Amino Acids

For the extraction and identification of free amino acids, 1.75 ± 0.1 g of amaranth flour was mixed with 15 mL of 15% trichloroacetic acid (TCA). The pH of the mixture was adjusted to 2.2 with 15% trichloroacetic acid (TCA) (isoelectric precipitation point of the proteins), and the extract was further diluted to 25 mL with 15% TCA [25]. Then, the supernatant was collected and filtered using 0.45-μm microfilters, and 100 μL of the filtered supernatant was subjected to the determination of organic components using the EZfaast GC-MS kit, following the protocol given by the manufacturer.

2.2.7. Statistical Analysis

Experimental Design and Statistical Analysis

The three-level Box–Behnken design was adapted in this study to investigate and optimize the effect of the independent variables of the ultra-turrax speed (X_1), solid/liquid ratio (X_2), methanol concentration (X_3) and extraction time (X_4) on the TPC and TFC. The coded levels of the variables are shown in Table 1. All calculations and graphics were carried out using the statistical software Design Expert 11 (trial version, Minneapolis, MN, USA). Triplicate experiments were conducted in order to validate the optimal extraction conditions.

The Box–Behnken design was based on a second-order (quadratic) polynomial response surface model using the following equation:

$$y = b_0 + \sum_{i=1}^{n}(b_i x_i) + \sum_{i=1}^{n}\left(b_{ii} x_{ii}^2\right) + \sum_{ij=1}^{n}(b_{ij} x_i x_j) \tag{1}$$

where y is the predicted response (TPC or TFC), x_i stands for the coded levels of the design variable (solvent concentration, time, temperature and ultrasonic frequency (Table 1)), b_0 is a constant, b_i is the linear effects, b_{ii} is the quadratic effects and b_{ij} is the interaction effects.

3. Results and Discussion

3.1. Box–Behnken Design and Model Fitting

The Box–Behnken design was applied to study the combined effects of the four parameters (ultra-turrax speed, solid liquid ratio, methanol concentration and time) on the total phenolic content (TPC) and total flavonoid content (TFC) from amaranth seed flour. The regression coefficients of the two proposed models are presented in Table 2. The two models were significant ones ($p < 0.05$), and the F-value was 10.31 for the TPC and 10.29 for the TFC. As can be seen, the two models were significant in terms of variance analysis, and the regression coefficients were higher than 0.90. For the two parameters studied (TPC and TFC), three parameters (ultra-turrax speed, solid-to-liquid ratio (RSL) and methanol concentration) presented significant linear effects ($p < 0.05$), and the linear effects were higher in the case of the TPC than the TFC. The interaction effects between the parameters studied were not significant ones, while one quadratic effect (methanol concentration) was a significant one.

Table 2. Analysis of variance of the total phenolic content (TPC) and total flavonoid content (TFC) modeling using the Box–Benhken design.

Source	TPC					TFC				
	Sum of Squares	Df	Mean Square	F-Value	p-Value	Sum of Squares	Df	Mean Square	F-Value	p-Value
Model	8.012×10^6	14	5.723×10^5	10.31	<0.0001	6.570×10^5	14	46,931.81	10.29	<0.0001
A-Ultraturax Speed	4.979×10^5	1	4.979×10^5	8.97	0.0096	40,973.81	1	40,973.81	8.98	0.0096
B-RSL	1.193×10^6	1	1.193×10^6	21.49	0.0004	97,906.84	1	97,906.84	21.46	0.0004
C-Methanol	1.042×10^6	1	1.042×10^6	18.78	0.0007	85,711.02	1	85,711.02	18.79	0.0007
D-Time	1.569×10^5	1	1.569×10^5	2.83	0.1149	12,878.07	1	12,878.07	2.82	0.1151
AB	6387.21	1	6387.21	0.1151	0.7394	524.40	1	524.40	0.1149	0.7396
AC	91,827.18	1	91,827.18	1.65	0.2192	7465.19	1	7465.19	1.64	0.2217
AD	2838.76	1	2838.76	0.0512	0.8243	233.07	1	233.07	0.0511	0.8245
BC	2.364×10^5	1	2.364×10^5	4.26	0.0581	19,406.33	1	19,406.33	4.25	0.0582
BD	12,818.77	1	12,818.77	0.2310	0.6382	1052.44	1	1052.44	0.2307	0.6385
CD	9980.01	1	9980.01	0.1799	0.6779	819.37	1	819.37	0.1796	0.6782
A^2	511.49	1	511.49	0.0092	0.9249	29.23	1	29.23	0.0064	0.9373
B^2	312.19	1	312.19	0.0056	0.9413	34.40	1	34.40	0.0075	0.9320
C^2	4.430×10^6	1	4.430×10^6	79.83	<0.0001	3.624×10^5	1	3.624×10^5	79.43	<0.0001
D^2	5095.40	1	5095.40	0.0918	0.7663	386.17	1	386.17	0.0846	0.7754
Residual	7.769×10^5	14	55,490.70			63,877.95	14	4562.71		
Lack of Fit	7.769×10^5	10	77,686.99			63,874.49	10	6387.45	7387.71	<0.0001
Pure Error	0.0000	4	0.0000			3.46	4	0.8646		
Cor Total	8.789×10^6	28				7.209×10^5	28			
Model Stdev			235.57					67.55		
Model R^2			0.9116					0.9014		
Model Adjusted R^2			0.8232					0.8128		

3.2. Fitting of Second-Order Polynomial Equations

The relationship between the extraction parameters and experimental data (TPC and TFC) was described using the Box–Behnken design using second-order polynomial equations. The effects which were not significant were eliminated from the equations, and the final equations are presented below:

$$\mathbf{TPC} = 506.2 + 203.7 \cdot \mathbf{X}_1 + 315.2 \cdot \mathbf{X}_2 + 294.7 \cdot \mathbf{X}_3 + 826.4 \cdot \mathbf{X}_3^2 \tag{2}$$

$$\mathbf{TFC} = 145.7 + 58.4 \cdot \mathbf{X}_1 + 90.3 \cdot \mathbf{X}_2 + 84.5 \cdot \mathbf{X}_3 + 236.4 \cdot \mathbf{X}_3^2 \tag{3}$$

3.3. Surface Response 3D Plots

The surface response methodology is a statistical method used to depict the relationship between two parameters and the experimental data [26]. In our study, the Box–Benhken design was used for the optimization of the TPC and TFC extraction from amaranth seed flour using four parameters (ultra-turrax speed, solid-to-liquid ratio, methanol concentration and time). In Figures 2 and 3, the evolution of the TPC and TFC as a function of the parameters studied are presented.

In the case of the TPC, a positive combined effect of a high RSL and high ultra-turrax speed was observed, and a high speed corelated with a high amount of matrices influenced the extraction of phenolic compounds positively. Regarding the combined effect of the methanol concentration and ultra-turrax speed, it can be observed that a high ultra-turrax speed and a high methanol concentration generated a high TPC concentration. Furthermore, at a low methanol concentration, there was a maximum TPC, and one possible explanation for this may be the complex chemical composition of amaranth that includes other polyphenols more soluble in high water solvent concentrations. The combined effect of methanol and RSP showed a maximum point at a high methanol concentration and high RSL, respectively. In the case of time combined with the ultra-turrax speed and RSL, a direct correlation between the TPC concentration and time/RSL can be observed. The combination of the methanol concentration and time shows that the time did not influence the evolution significantly. In the case of the TFC, a positive combined effect of a high RSL and high ultra-turrax speed was observed, and a high speed corelated with high amount of matrices influenced the extraction of phenolic compounds positively. Regarding the combined effect of the methanol concentration and ultra-turrax speed, it can be observed that a high TFC concentration was achieved at the maximum speed and concentration, and at a low methanol concentration, there was a maximum. One possible explanation may be that because of the complex chemical composition of the amaranth, there can be other flavones more soluble in high water solvent concentrations. The combination of the methanol concentration and time shows that the time did not influence the evolution significantly. The combined effect of methanol and RSP showed a maximum point at a high methanol concentration and high RSL, respectively. In the case of time, it can be observed that its combination with the ultra-turrax speed and RSL had a direct correlation between the TPC concentration and time/RSL.

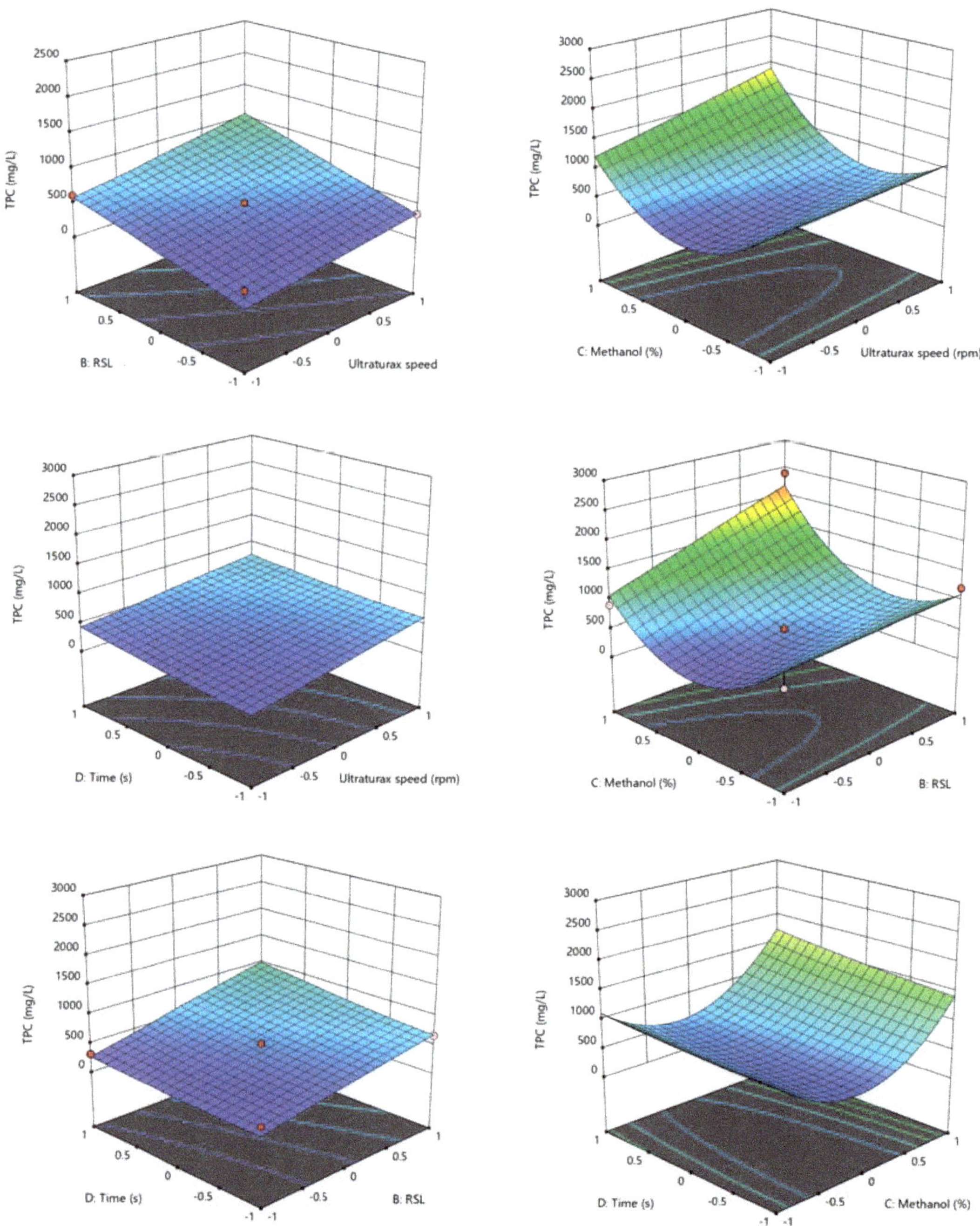

Figure 2. 3D evolution of TPC according to the Box–Benhken design. −1 the low value of the parameter, +1 the highest value of the parameter according to data from Table 1.

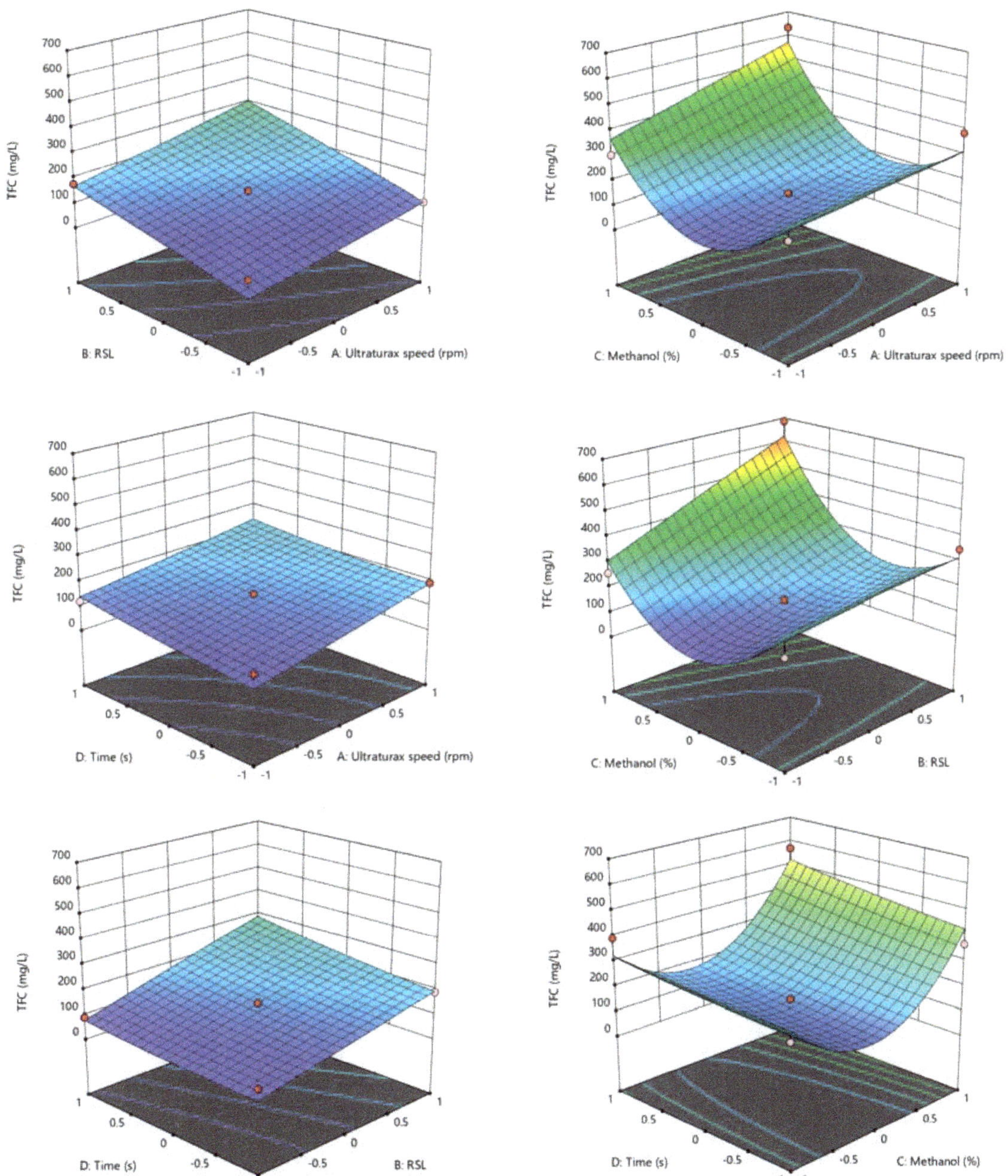

Figure 3. 3D evolution of TFC according to the Box–Benhken design. −1 the low value of the parameter, +1 the highest value of the parameter according to data from Table 1.

3.4. Individual Phenolic Profile

Table 3 presents the concentrations of individual phenolics (expressed as mg/kg amaranth seed) determined using the chromatographic method. From the 12 compounds considered, 8 were determined (vanillic acids, caffeic acid, chlorogenic acid, *p*-coumaric acid, rosmarinic acid, myricetin, luteolin and kaempferol), while 4 were under the limit of detection (gallic acid, protocatechuic acid, 4-hydroxybenzoic acid and quercetin). From the phenolic acids determined, vanillic acid was the most abundant, followed by rosmarinic acid, chlorogenic acid, *p*-coumaric acid and caffeic acid. The most abundant flavonoid was kaempferol, followed by myricetin and luteolin. Barba de la Rosa et al. determined three polyphenols in amaranth seed—rutin (10.1 mg/g), nicotiflorin (7.2 mg/g) and iso-quercetin (0.5 mg/g)—and regarding phenolic acids, they quantified 4-hydroxybenzoic acid (2.2 mg/g), syringic acid (0.8 mg/g) and vanillic acid (1.8 mg/g) [10]. The polyphenols in amaranth seed present high antioxidative activity and anti-inflammatory capacity as well as miscellaneous effects like anti-fibrotic, anti-bacterial, anti-hypertensive, anti-atherogenic and anti-proliferative effects [27]. Kaempferol, the major phenolic observed in our study, was reported to have a nephroprotective effect, preventing DOX-induced cardiac damage, hippocampal damage and neural and lung ischemia injury as well as an antioxidative effect [28,29]. Myricetin, the second-largest phenolic observed, possesses antimicrobial activity and has a positive effect on microbiota [30]. The myricetin supplementation decreases hepatic lipid synthesis and inflammation by modulating the gut microbiota [31].

Table 3. Phenolic profile of amaranth seed.

Phenolic	mg/kg Amaranth Seed
Gallic acid	ND
Protocatechuic acid	ND
4-hydroxybenzoic acid	ND
Vanillic acid	10.33
Caffeic acid	3.37
Chlorogenic acid	4.00
p-coumaric acid	3.63
Rosmarinic acid	7.68
Myricetin	105.05
Luteolin	39.15
Quercitin	ND
Kaempferol	1041.50
Total	1214.71

ND = not detected.

3.5. Fatty Acid Composition

The fatty acid composition of the amaranth seed was found based on the oil extracted from the seeds (7.01% oil content in amaranth seed). In Table 4, the fatty acid composition of the amaranth seed is presented. As can be observed, the fatty acid in the highest amount was reported to be oleic acid (23.78 mg/100 g amaranth), representing 42.68% of the total fatty acids, followed by *cis,cis*-9,12-linoleic acid (C18:2), palmitic acid C16:0 and *trans,trans*-9,12-linoleic acid (C18:2). The saturated fatty acids represented 16.54%, while unsaturated fatty acids represented 83.45% (poly-unsaturated fatty acids represented 37.56%). In a previous study, a higher concertation of saturated fatty acids (26.72–27.28%) and lower concentration of unsaturated fatty acids (72.72–73.28%) in amaranth seed were reported [32]. Bozorov et al. reported that the predominant saturated fatty acid in amaranth seed was palmitic acid, while the predominant unsaturated fatty acids were linoleic acids and oleic acids [32]. According to He et al., the major fatty acids in the oil from 11 genotypes of 4 grain amaranth species were palmitic (19.1–23.4%), oleic (18.7–38.9%) and linoleic (36.7–55.9%) acids [33]. The linoleic acids represented in amaranth oil are essential fatty acids which cannot be synthetized by the human body and play an important role due to their antiadipogenic, antidiabetogenic, anti-carcinogenic and anti-therosclerotic activities [34]. Based on the fatty acid profile of the amaranth oil, amaranth can be considered a promising food and an important raw material for pharma.

Table 4. Fatty acid composition of amaranth seed.

Fatty Acid	mg/100 g Amaranth
Dodecanoic acid C12:0	0.30
Myristoleic acid C14:0	0.52
cis-9-Octadecenoic acid C18:1	0.74
Pentadecanoic acid C15:0	0.08
Palmitic acid C16:0	7.38
cis-9-Palmitoleic acid C16:1	0.63
Heptadecanoic Acid C17:0	0.16
cis-10-Heptadecenoic acid C17:1	0.06
Oleic acid C18:1	23.78
trans,trans-9,12-linoleic acid (C18:2)	3.18
cis,cis-9,12-linoleic acid (C18:2)	13.53
Arachidic acid (C20:0)	0.30
Gamma linolenic acid (C18:3)	3.05
cis-11-Eicosenoic acid (C20:1)	0.36
Linolenic acid (C18:3)	0.71
Heneicosanoic acid (C21:0)	0.08
Eicosadienoic acid (C20:2, *cis*-11,14)	0.16
Behenic acid (C22:0)	0.34
Eicosatetranoic acid (20:3)	0.05
cis-5,8,11,14-arachidonic acid (C20:4)	0.19
Tricosanoic acid (C23:0)	0.06
Nervonic acid (C24:1)	0.08
TOTAL	55.71
Saturated fatty acids (%)	16.54
Unsaturated fatty acids (%)	83.45
Poly-unsaturated fatty acids	37.56

3.6. Amino Acid Profile

Amaranth is a protein-rich pseudo-cereal (~14.45%), and the sum of the essential amino acids in the seeds is approximately equal to the recommended FAO and WHO standard [35]. Table 5 presents the composition of amino acids in amaranth seed. There are nine essential amino acids, namely leucine, isoleucine, valine, phenylalanine, threonine, tryptophan, methionine, histidine and lysine [36]. Figure 4 presents a typical chromatogram for the amino acid profile of amaranth seed. As can be observed, all 9 essential amino acids known were reported: histidine, isoleucine, leucine, lysine, methionine, phenylalanine, threonine, tryptophan and valine in a total concentration of 29.67 g/100 g protein, with the major essential amino acid being lysine, followed by leucine and histidine. According to the FAO and WHO standard, the tryptophan concentration is 1.0 g/100 g protein, methionine's is 3.5 g/100 g protein, threonine's is 4.0 g/100 g protein, isoleucine's is 4.0 g/100 g, valine's is 5.0/100 g protein, lysine's is 5.5 g/100 g protein, leucine's is 7.0 g/100 g protein and cysteine's is 3.5 g/100 g protein. In our study, there were some negative differences with the FAO and WHO standard in terms of threonine, isoleucine and valine and positive differences in terms of lysine, leucine, cysteine, methionine and tryptophan. Our results are in agreement with those reported in the literature [37,38]. The essential amino acids play an important role in the human body, promoting protein synthesis, influencing the human metabolism, regulating multiple biological processes and influencing body weight and energy balance [36,39,40].

Table 5. Amino acids present in amaranth seed.

Amino Acid	g/100 g Protein
Alanine	1.79
Asparagine	2.87
Aspartic acid	3.72
Cystathionine	5.09
Cystine	6.06
Glutamic acid	4.95
Glutamine	0.97
Glycine	1.03
Histidine *	6.83
3-hydroxyproline	3.34
Isoleucine *	2.87
Leucine *	7.85
Lysine *	10.10
Methionine *	4.69
Phenylalanine *	3.80
Proline	3.83
Hydroxyproline dipeptide	2.58
Serine	3.40
Thioproline	2.40
Threonine *	3.05
Tryptophan *	5.51
Tyrosine	5.30
Valine *	3.11
α-aminopimelic acid	4.87

* Essential amino acids.

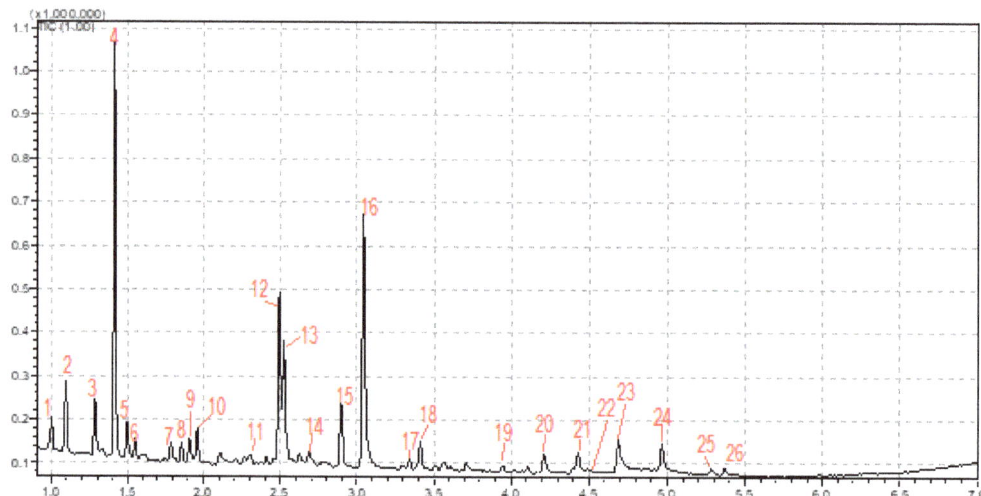

Figure 4. Amaranth seed amino-acid profile: 1 = alanine, 2 = glycine, 3 = valine, 4 = internal standard, 5 = leucine, 6 = isoleucine, 7 = threonine, 8 = serine, 9 = proline, 10 = asparagine, 11 = thioproline, 12 = aspartic acid, 13 = methionine, 14 = 3-hydroxyproline, 15 = phenylalanine, 16 = glutamic acid, 17 = α-aminopimelic acid, 18 = glutamine, 19 = 4-hydroxylserine, 20 = hydroxyproline dipeptide, 21 = histidine, 22 = lysine, 23 = tyrosine, 24 = tryptophan, 25 = cystationine, and 26 = cystine.

4. Limitations and Future Perspectives

Our study's limitations are related to the time of extraction. An increase in the extraction time could lead to a higher yield for the TPC and TFC, but the high shear rates of the ultra-turrax could cause a decrease in the TPC and TFC. Future studies will cover the hydrothermal impact (e.g., sous-vide treatment) on the amaranth properties (e.g., mass transfer, TPC, TFC and amino acid composition) and the extraction of bioactive compounds from amaranth seed using other techniques (e.g., ultrasonication extraction and microwave extraction).

5. Conclusions

Amaranth seed can be considered a valuable source of polyphenols (phenolic acids and flavonoids) and fatty acids. The parameters which influenced the extraction of polyphenols from amaranth seed the most were the solid-to-liquid ratio, methanol concentration and ultra-turrax speed. The Box–Benhken design was suitable for predicting the TPC and TFC using the response surface method. The two models were significant ones ($p < 0.05$), and the F-value was 10.31 for the TPC and 10.29 for the TFC. From the phenolic acids determined, vanillic acid was the most abundant, followed by rosmarinic acid, chlorogenic acid, *p*-coumaric acid and caffeic acid. The most abundant flavonoid was kaempferol, followed by myricetin and luteolin. The fatty acid reported to be of the highest amount was oleic acid (23.78 mg/100 g amaranth), representing 42.68% of the total fatty acids, followed by *cis,cis*-9,12-linoleic acid (C18:2), palmitic acid C16:0 and *trans,trans*-9,12-linoleic acid (C18:2). The saturated fatty acids represented 16.54% of the total, and unsaturated fatty acids represented 83.45% (poly-unsaturated fatty acids represented 37.56%). Nine essential amino acids were reported in amaranth seed—histidine, isoleucine, leucine, lysine, methionine, phenylalanine, threonine, tryptophan and valine—at a total concentration of 29.67 g/100 g protein, with the major essential amino acid being lysine followed by histidine and methionine. The limits of our experimental work were related to the particle size of the materials, as it is known that the extraction efficiency is influenced by the particle size. In the future, we intend to study the influence of thermal treatments on the phenolic compounds from amaranth seed.

Author Contributions: Conceptualization, O.P. and M.O.; methodology, O.P.; software, M.O.; validation, O.P. and M.O.; formal analysis, O.P. and M.O.; investigation, O.P.; resources, O.P. and M.O.; data curation, O.P. and M.O.; writing—original draft preparation, O.P.; writing—review and editing, O.P. and M.O.; visualization, O.P. and M.O.; supervision, M.O. The authors contributed equally to this research. All authors have read and agreed to the published version of the manuscript.

Funding: This paper was supported by "DECIDE—Development through entrepreneurial education and innovative doctoral and post doctoral research, project code POCU/380/6/13/125031, project co-financed from the European Social Fund through the 2014–2020 Operational Program Human Capital".

Institutional Review Board Statement: Not applicable.

Informed Consent Statement: Not applicable.

Data Availability Statement: Not applicable.

Conflicts of Interest: The authors declare no conflict of interest.

References

1. Lamothe, L.M.; Srichuwong, S.; Reuhs, B.L.; Hamaker, B.R. Quinoa (*Chenopodium quinoa* W.) and amaranth (*Amaranthus caudatus* L.) provide dietary fibres high in pectic substances and xyloglucans. *Food Chem.* **2015**, *167*, 490–496. [CrossRef] [PubMed]
2. Paśko, P.; Sajewicz, M.; Gorinstein, S.; Zachwieja, Z. Analysis of selected phenolic acids and flavonoids in Amaranthus cruentus and *Chenopodium quinoa* seeds and sprouts by HPLC. *Acta Chromatogr.* **2008**, *20*, 661–672. [CrossRef]
3. Rodríguez, M.; Tironi, V.A. Polyphenols in amaranth (*A. manteggazianus*) flour and protein isolate: Interaction with other components and effect of the gastrointestinal digestion. *Food Res. Int.* **2020**, *137*, 109524. [CrossRef]
4. Venskutonis, P.R.; Kraujalis, P. Nutritional Components of Amaranth Seeds and Vegetables: A Review on Composition, Properties, and Uses. *Compr. Rev. Food Sci. Food Saf.* **2013**, *12*, 381–412. [CrossRef]
5. Sandoval-Sicairos, E.S.; Milán-Noris, A.K.; Luna-Vital, D.A.; Milán-Carrillo, J.; Montoya-Rodríguez, A. Anti-inflammatory and antioxidant effects of peptides released from germinated amaranth during in vitro simulated gastrointestinal digestion. *Food Chem.* **2021**, *343*, 128394. [CrossRef]
6. Vacek, J.; Ulrichová, J.; Klejdus, B.; Imánek, V. Analytical methods and strategies in the study of plant polyphenolics in clinical samples. *Anal. Methods* **2010**, *2*, 604–613. [CrossRef]

7. Berti, C.; Riso, P.; Brusamolino, A.; Porrini, M. Effect on appetite control of minor cereal and pseudocereal products. *Br. J. Nutr.* **2005**, *94*, 850–858. [CrossRef]

8. Gorinstein, S.; Vargas, O.J.M.; Jaramillo, N.O.; Salas, I.A.; Ayala, A.L.M.; Arancibia-Avila, P.; Toledo, F.; Katrich, E.; Trakhtenberg, S. The total polyphenols and the antioxidant potentials of some selected cereals and pseudocereals. *Eur. Food Res. Technol.* **2007**, *225*, 321–328. [CrossRef]

9. Paśko, P.; Bartoń, H.; Fołta, M.; Gwizdz, J. Evaluation of antioxidant activity of amaranth (*Amaranthus cruentus*) grain and by-products (flour, popping, cereal). *Rocz. Państw. Zakł. Hig.* **2007**, *58*, 35–40.

10. Barba de la Rosa, A.P.; Fomsgaard, I.S.; Laursen, B.; Mortensen, A.G.; Olvera-Martínez, L.; Silva-Sánchez, C.; Mendoza-Herrera, A.; González-Castañeda, J.; De León-Rodríguez, A. Amaranth (*Amaranthus hypochondriacus*) as an alternative crop for sustainable food production: Phenolic acids and flavonoids with potential impact on its nutraceutical quality. *J. Cereal Sci.* **2009**, *49*, 117–121. [CrossRef]

11. Manach, C.; Scalbert, A.; Morand, C.; Rémésy, C.; Jiménez, L. Polyphenols: Food sources and bioavailability. *Am. J. Clin. Nutr.* **2004**, *79*, 727–747. [CrossRef] [PubMed]

12. Scalbert, A.; Johnson, I.T.; Saltmarsh, M. Polyphenols: Antioxidants and beyond. *Am. J. Clin. Nutr.* **2005**, *81*, 215S–217S. [CrossRef] [PubMed]

13. Ozsoy, N.; Yilmaz, T.; Kurt, O.; Can, A.; Yanardag, R. In vitro antioxidant activity of *Amaranthus lividus* L. *Food Chem.* **2009**, *116*, 867–872. [CrossRef]

14. Alvarez-Jubete, L.; Wijngaard, H.; Arendt, E.K.; Gallagher, E. Polyphenol composition and in vitro antioxidant activity of amaranth, quinoa buckwheat and wheat as affected by sprouting and baking. *Food Chem.* **2010**, *119*, 770–778. [CrossRef]

15. Repo-Carrasco-Valencia, R.; Hellström, J.K.; Pihlava, J.M.; Mattila, P.H. Flavonoids and other phenolic compounds in Andean indigenous grains: Quinoa (*Chenopodium quinoa*), kañiwa (Chenopodium pallidicaule) and kiwicha (Amaranthus caudatus). *Food Chem.* **2010**, *120*, 128–133. [CrossRef]

16. Barbin, D.F.; Natsch, A.; Müller, K. Improvement of functional properties of rapeseed protein concentrates produced via alcoholic processes by thermal and mechanical treatments. *J. Food Process. Preserv.* **2011**, *35*, 369–375. [CrossRef]

17. Adiamo, O.Q.; Gbadamosi, O.S.; Abiose, S.H. Functional Properties and Protein Digestibility of Protein Concentrates and Isolates Produced from Kariya (Hildergadia bateri) Seed. *J. Food Process. Preserv.* **2016**, *40*. [CrossRef]

18. Xu, W.J.; Zhai, J.W.; Cui, Q.; Liu, J.Z.; Luo, M.; Fu, Y.J.; Zu, Y.G. Ultra-turrax based ultrasound-assisted extraction of five organic acids from honeysuckle (*Lonicera japonica* Thunb.) and optimization of extraction process. *Sep. Purif. Technol.* **2016**, *166*, 73–82. [CrossRef]

19. Scholz, P.; Keck, C.M. Nanoemulsions produced by rotor-stator high speed stirring. *Int. J. Pharm.* **2015**, *482*, 110–117. [CrossRef]

20. Jin, M.M.; Zhang, W.D.; Song, G.S.; Xu, Y.M.; Du, Y.F.; Guo, W.; Cao, L.; Xu, H.J. Discrimination and Chemical Phylogenetic Study of Four Pulsatilla Herbs Using UPLC-ESI-MS/MS Combined with Hierarchical Cluster Analysis. *J. Chromatogr. Sci.* **2018**, *56*, 216–224. [CrossRef]

21. Cheng, Z.; Song, H.; Yang, Y.; Zhou, H.; Liu, Y.; Liu, Z. Smashing Tissue Extraction of Five Lignans from the Fruit of Schisandra chinensis. *J. Chromatogr. Sci.* **2016**, *54*, 246–256. [CrossRef]

22. Yin, X.S.; Zhong, Z.F.; Bian, G.L.; Cheng, X.J.; Li, D.Q. Ultra-rapid, enhanced and eco-friendly extraction of four main flavonoids from the seeds of Oroxylum indicum by deep eutectic solvents combined with tissue-smashing extraction. *Food Chem.* **2020**, *319*, 126555. [CrossRef] [PubMed]

23. Pauliuc, D.; Dranca, F.; Oroian, M. Antioxidant activity, total phenolic content, individual phenolics and physicochemical parameters suitability for Romanian honey authentication. *Foods* **2020**, *9*, 306. [CrossRef] [PubMed]

24. Oroian, M.; Ursachi, F.; Dranca, F. Ultrasound-assisted extraction of polyphenols from crude pollen. *Antioxidants* **2020**, *9*, 322. [CrossRef]

25. Urcan, A.C.; Criste, A.D.; Dezmirean, D.S.; Bobiș, O.; Bonta, V.; Dulf, F.V.; Mărgăoan, R.; Cornea-Cipcigan, M.; Campos, M.G. Botanical origin approach for a better understanding of chemical and nutritional composition of beebread as an important value-added food supplement. *LWT* **2021**, *142*, 111068. [CrossRef]

26. Tabaraki, R.; Heidarizadi, E.; Benvidi, A. Optimization of ultrasonic-assisted extraction of pomegranate (*Punica granatum* L.) peel antioxidants by response surface methodology. *Sep. Purif. Technol.* **2012**, *98*, 16–23. [CrossRef]

27. Antoniewska, A.; Rutkowska, J.; Pineda, M.M.; Adamska, A. Antioxidative, nutritional and sensory properties of muffins with buckwheat flakes and amaranth flour blend partially substituting for wheat flour. *LWT -Food Sci. Technol.* **2018**, *89*, 217–223. [CrossRef]

28. El-kott, A.F.; Abd-Lateif, A.E.K.M.; Khalifa, H.S.; Morsy, K.; Ibrahim, E.H.; Bin-Jumah, M.; Abdel-Daim, M.M.; Aleya, L. Kaempferol protects against cadmium chloride-induced hippocampal damage and memory deficits by activation of silent information regulator 1 and inhibition of poly (ADP-Ribose) polymerase-1. *Sci. Total Environ.* **2020**, *728*, 138832. [CrossRef]

29. Alagal, R.I.; AlFaris, N.A.; Alshammari, G.M.; ALTamimi, J.Z.; AlMousa, L.A.; Yahya, M.A. Kaempferol attenuates doxorubicin-mediated nephropathy in rats by activating SIRT1 signaling. *J. Funct. Foods* **2022**, *89*, 104918. [CrossRef]

30. Fan, L.; Zhao, X.; Tong, Q.; Zhou, X.; Chen, J.; Xiong, W.; Fang, J.; Wang, W.; Shi, C. Interactions of Dihydromyricetin, a Flavonoid from Vine Tea (*Ampelopsis grossedentata*) with Gut Microbiota. *J. Food Sci.* **2018**, *83*, 1444–1453. [CrossRef]

31. Sun, W.L.; Li, X.Y.; Dou, H.Y.; Wang, X.D.; Li, J.D.; Shen, L.; Ji, H.F. Myricetin supplementation decreases hepatic lipid synthesis and inflammation by modulating gut microbiota. *Cell Rep.* **2021**, *36*, 109641. [CrossRef] [PubMed]

32. Bozorov, S.S.; Berdiev, N.S.; Ishimov, U.J.; Olimjonov, S.S.; Ziyavitdinov, J.F.; Asrorov, A.M.; Salikhov, S.I. Chemical composition and biological activity of seed oil of amaranth varieties. *Nova Biotechnol. Chim.* **2018**, *17*, 66–73. [CrossRef]

33. He, H.P.; Cai, Y.; Sun, M.; Corke, H. Extraction and purification of squalene from *Amaranthus grain. J. Agric. Food Chem.* **2002**, *50*, 368–372. [CrossRef] [PubMed]

34. Zhang, Z.S.; Kang, Y.J.; Che, L. Composition and thermal characteristics of seed oil obtained from Chinese amaranth. *LWT* **2019**, *111*, 39–45. [CrossRef]

35. Thoufeek Ahamed, N.; Singhai, R.S.; Kulkarni, P.R.; Pal, M. A lesser-known grain, *Chenopodium quinoa*: Review of the chemical composition of its edible parts. *Food Nutr. Bull.* **1998**, *19*. [CrossRef]

36. Xiao, F.; Guo, F. Impacts of essential amino acids on energy balance. *Mol. Metab.* **2021**, *57*, 101393. [CrossRef] [PubMed]

37. López, D.N.; Galante, M.; Robson, M.; Boeris, V.; Spelzini, D. Amaranth, quinoa and chia protein isolates: Physicochemical and structural properties. *Int. J. Biol. Macromol.* **2018**, *109*, 152–159. [CrossRef]

38. Thanapornpoonpong, S.N.; Vearasilp, S.; Pawelzik, E.; Gorinstein, S. Influence of various nitrogen applications on protein and amino acid profiles of amaranth and quinoa. *J. Agric. Food Chem.* **2008**, *56*, 11464–11470. [CrossRef]

39. Tashiro, Y.; Han, Q.; Tan, Y.; Sugisawa, N.; Yamamoto, J.; Nishino, H.; Inubushi, S.; Higuchi, T.; Aoki, T.; Murakami, M.; et al. Oral recombinant methioninase prevents obesity in mice on a high-fat diet. *In Vivo (Brooklyn)* **2020**, *34*, 489–494. [CrossRef]

40. Drummen, M.; Tischmann, L.; Gatta-Cherifi, B.; Adam, T.; Westerterp-Plantenga, M. Dietary protein and energy balance in relation to obesity and co morbidities. *Front. Endocrinol. (Lausanne)* **2018**, *9*, 443. [CrossRef]

applied sciences

Article

Effect of Hemp Seed Oil Addition on the Rheological Properties of Dough and Bread

Sorina Ropciuc *, Laura Carmen Apostol, Cristina Damian and Ancuța Elena Prisacaru *

Faculty of Food Engineering, Stefan cel Mare University, Universitatii Street 13, 720229 Suceava, Romania; laura.apostol@fia.usv.ro (L.C.A.); cristinadamian@fia.usv.ro (C.D.)
* Correspondence: sorina.ropciuc@fia.usv.ro (S.R.); ancuta.prisacaru@fia.usv.ro (A.E.P.)

Abstract: The aim of the study was to determine the influence of hemp seed oil on the rheological properties of the dough and the properties of white and black wheat flour bread. In this work, the dough was obtained from wheat flour types 550, 650, and 1350 to which hemp seed oil was added in percentages ranging from 0–15% (flour-based). The empirical and fundamental rheological properties of the dough were characterization using an alveograph test and a rotational rheometer. The rheological properties determined by the alveograph test indicated a correlation between the dough extensibility index and the bread volume. The obtained results led to the conclusion that the addition of oil in the dough, at a maximum percentage of 15%, modifies rheological properties by decreasing the modulus of viscosity and elasticity. Textural properties were used to determine hardness, springiness, cohesiveness, and adhesiveness. The hardness and softness of the bread decreased as the percentage of added oil increased and the elasticity of the bread samples increased with the addition of more oil. Supplementing bread with oil has led to improved textural features and sensory scores. Generally, the best quality was obtained at the optimum usage level of 5–10% hemp oil.

Keywords: hempseed oil; bread; dough; rheological properties; textural properties

Citation: Ropciuc, S.; Apostol, L.C.; Damian, C.; Prisacaru, A.E. Effect of Hemp Seed Oil Addition on the Rheological Properties of Dough and Bread. *Appl. Sci.* **2022**, *12*, 2764. https://doi.org/10.3390/app12062764

Academic Editor: Suyong Lee

Received: 20 January 2022
Accepted: 5 March 2022
Published: 8 March 2022

Publisher's Note: MDPI stays neutral with regard to jurisdictional claims in published maps and institutional affiliations.

1. Introduction

Bread is considered a staple food all over the world because of its nutritional value, its ability to be eaten with other foods, and its low price [1].

In the composition of bakery products, oil is an auxiliary component, but with an important functional and rheological role in the dough [2]. Adding a small amount of fat to yeast dough helps the gluten to spread, resulting in bread with a higher volume. Fats are used in bread making to improve gas retention in the dough and thereby increase its volume and softness, to lubricate, aerate, and help heat transference in the dough, imparting desirable texture [3]. The flow and deformation behavior of the dough is recognized as essential to the successful production of baked goods [4]. Since the doughs are viscoelastic, oscillating dynamic testing has been widely used in recent years to characterize the rheological properties of dough [5,6]. The amount of oil added is determined by the type of flour and the capacity of the proteins in the flour to bind these lipids during kneading [7]. The percentage of added oil should not exceed 5–10%.

The rheological properties of wheat flour dough, supplemented with different amounts of hemp seed oil, have been studied previously and it has been shown that water absorption decreased as the amount of oil added increased. Other authors who have studied the effect of adding flaxseed oil to wheat flour dough have reported increased water absorption and dough stability [8,9]. They reported that the use of ground flaxseed at a level of 10% significantly increased the volume of the bread, the specific volume of the bread, and delayed the aging of the bread.

The sensory qualities of baked bread, such as flavor, softness, volume, palatability, and appearance are improved, while oxidation reactions, staling, and moisture migration are reduced by the addition of oil to bread recipes [10].

The current trend in the industry is to introduce healthier fats into food products. As a result, a decrease in monounsaturated fatty acids (MUFA) are expected alongside an increase in polyunsaturated fatty acids (PUFAs) in consumer foods [11–13].

Hemp seed oil contains high concentrations of polyunsaturated fatty acids (PUFA) (70–90%) and is a good dietary source of essential fatty acids and linoleic acids [14,15]. Linoleic acid (50–70%) is the major fatty acid in seed oil. Hemp seed oil is considered healthy because of its desirable omega-6: omega-3 fatty acid ratio, 3:1. This value is considered optimal for human health, as it has a positive influence on health conditions through reducing the risk of cardiovascular diseases [16,17].

The main purpose of this research was to investigate the effects hemp seed oil had upon the rheological properties of dough obtained from wheat flour, with specific attention paid to three types of flour (550, 650, and 1350), as well as its effects upon the quality of the bread.

2. Materials and Methods

2.1. Materials

Wheat flour was bought from the local producer, Mopan S.A, and have with document's verifying its quality. The following three types of flour were used: white flour (type 550), white flour (type 650), and black wheat flour (type 1350). Wheat flour was characterized according to ICC methods in terms of its composition as follows: Type 550, 14.0 g/100 g moisture; 0.55 g/100 g ash; 13.2 g/100 g protein; 31 g/100 g wet gluten. Wheat flour type 650, 14.5 g/100 g humidity; 0.65 g/100 g ash; 10.5 g/100 g protein, 27 g/100 g wet gluten. Wheat flour type 1350, 14.5 g/100 g humidity; 1.35 g/100 g ash; 10.5 g/100 g protein; 26 g/100 g wet gluten.

The results obtained were duplicated and expressed as average values. The hemp seed oil was purchased from a producer in Romania (Pronat SRL, Romania). The acidity of the oil at its time of use in the formulation of dough was 1.16 and the iodine number of hemp oil was 152 g/100 g. The percentage of hemp seed oil added was as follows: S0 control test–0%; sample 1–5%; sample 2–10%; sample 3–15%. A total of twelve bread samples were obtained for analysis.

2.2. Methods

2.2.1. Dough and Bread Preparation Method

The bread recipe was as follows: 250 g of wheat flour of each type: 550, 650, and 1350 in g of yeast was added, 3.75 g salt and hemp seed oil in percentages ranging from 0% to 15%. Water was added depending on the hydration capacity of the dough. The dough was kneaded using the Kitchen Aid Heavy Duty mixer by mixing until full dough development. After kneading, the dough was split, re-mixed, and fermented for 30 min at a temperature of 35 °C and 80% relative humidity using a fermentation cabinet Piron. When the fermentation was complete, the bread was baked in a Mistral TTR electric convection oven at 180–210 °C for 30 min.

2.2.2. Rheological Analysis

The rheological analysis of the dough was performed on a dynamic rheometer (HAAKE RheoWin Mars 40, Karlsruhe, Germany). All measurements were performed at a temperature of 20 °C, using rotor/plate geometry with a rotor diameter of 40 mm and a gap of 2 mm. A frequency range of 1–20 Hz and a temperature of 20 °C were used for the elastic modulus (G') and the viscous modulus (G''). Frequency recovery tests were performed in the linear viscoelasticity range at a constant stress of 50 Pa. The whistle time was 60 s and the recovery stage was 180 s. The viscoelastic properties of the dough samples also appreciated under the influence of temperature. G' modulus, "storage" or "elastic", and

G" module, "loss" or "plastic", were analyzed at temperatures between 20 °C–100 °C. The determination was made using plate-plate geometry, with a rotor diameter of 40 mm and a distance of 2 mm [18,19].

The empirical dough's rheological properties were documented using an alveograph device (Chopin Technologies, type F4, Villeneuve-La-GarenneCedex, France). The following characteristics were defined: dough extensibility (mm) was indicated by the length of the alveogram curve (L), the maximum pressure (P) represented the peak height (mm), baking strength (W), index of swelling (G), and the configuration ratio of the alveograph curve (P/L). The alveograph test was done according to ICC 121, AACC 54–30A, and ISO 5530/4 approval with 14.0–14.5 g/100 g of moisture [20–22].

2.2.3. Evaluation of Bread Quality

Texture Analysis

The texture profile evaluation test aims to mimic the mastication process, which is run in the form of double compression cycles. Through this test, a number of properties relating to the bread's texture can be evaluated, such as the following: hardness, springiness, cohesiveness, adhesiveness. The textural characteristics of the bread samples were determined using a texturometer device (TVT-6700, Perten Instruments) using a 45 mm cylindrical probe, which operates at a speed of 5.0 mm/s and a trigger force of 5 g. To highlight the textural parameters of the bread samples, they were cut into 50 mm high slices. Compression was performed up to a displacement equal to 75% of the sample height, with a pressing speed of 4 mm/s. The test parameters were as follows: 10.0 mm/s pre-test speed, 5.0 mm/s test speed, 5.0 mm/s post-test speed, and 40% strain. Thus, the device software has been programmed to record forces and displacements, as well as to calculate texture parameters. These analyses were conducted 2 h after baking [22–26].

Sensory Analysis

The 12 bread samples were proposed for sensory analysis to a group of 11 specially trained panelists; a selection of students from the Faculty of Food Engineering. This group evaluated the bread samples, giving grades from 1 (lowest intensity) to 5 (highest intensity), for the following sensory attributes: crust color, crumb pore uniformity (size of pores in section (small/big), crumb softness, specific aroma, and after-taste. The coded samples were offered to the tasters and the sensory evaluation was performed using the preference test. Additionally, overall acceptability was determined using a 9-point hedonic scale (from 9 = like extremely to 1 = dislike extremely). Differences were considered to be significantly different at a value of $p < 0.05$ [25,26].

Statistical Analysis

The obtained results were statistically analyzed using XLSTAT 2021, software (Addinsoft, NY, USA) for the one-way analysis of variance, followed by the Tukey test to determine the significant differences between means at $p < 0.05$.

PCA analysis enabled the multivariate distribution of the data sets obtained in determining the alveograph parameters (dough extensibility (L), the maximum pressure (P), baking strength (W), and swelling (G). The data sets were separated into quadrants indicating the strength of the links that are established between the analyzed characteristics. The greater the strength of the connections, the shorter the distance from the point of origin. This method was used to distribute rheological characteristics and to determine the correlations between rheological, textural, and sensory characteristics.

Rheological variables (dough extensibility, maximum pressure, baking strength, swelling, elastic modulus, and viscous modulus) and textural variables (hardness, springiness, cohesiveness, and adhesiveness) were correlated with sensory characteristics (crust color, crumb pore, crumb softness, specific aroma, and after-taste) and then ana-lyzed using the Matlab 2021b program (MathWorks, Natick, MA, USA).

3. Results and Discussion

3.1. Results in Obtaining Bread Samples

A total of 12 bread samples (Figure 1a) were obtained using three types of flour (type 550, type 650, and type 1350) to which hemp seed oil was incorporated in kneading in percentages of 0%—Control, 5% (P1), 10% (P2), and 15% (P3). The samples in the section are presented in Figure 1b.

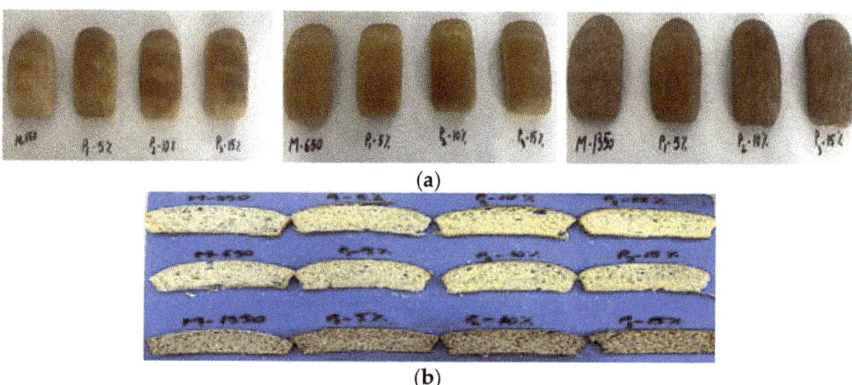

(a)

(b)

Figure 1. (**a**) Bread samples obtained from the three types of flour (external appearance); (**b**) Section appearance of bread samples.

3.2. Rheological Analysis Results

Figure 2a shows the values of elastic and viscous modulus, depending on the frequency, for the dough obtained from white flour type 550 with different amounts of hemp seed oil. A visual inspection of the data suggests that all samples are dependent on the amount of oil added. The rheological properties of the 550-flour dough demonstrate the same linear decreasing tendency of both the elastic and viscous modes. Figure 2b describes the viscoelastic properties of the flour dough 650. A significant decrease in the elastic and viscous mode is observed with the addition of 15% hemp oil. This behavior can be attributed to the high protein content of white flour, but also to the size of the starch granules and the bonds that form between the starch and the incorporated oil. In general, the incorporation of oil into white flour dough decreases viscoelastic modules, G′ and G″, leading to lower values than the control dough [27,28].

Figure 2c describes the frequency and viscous modulus, depending on the frequency, for the dough obtained from black wheat flour (type 1350) with the addition of hemp seed oil. The same tendency to decrease the elasticity and viscosity of the dough is also noticeable in this dough sample. The decrease is more visible as the percentage of added oil increases. The loss modulus (G″) presents lower values than the storage modulus (G′) in all formulated dough samples, suggesting a more elastic than viscous characteristic in dough samples. This result influences the development of the dough during fermentation, increasing its capacity to retain fermentation gases and thus, increasing the elasticity of the bread crumb and volume [29,30].

Figure 3a–c describe the behavior of dough obtained from wheat flour with the addition of hemp seed oil in proportions of 0–15% under the influence of temperature variations. There has a tendency to increase the viscosity of the dough as the temperature increases. The flour dough sample 550 has the effect of increasing the viscous and elastic modulus by 5% and, by increasing the amount of oil, there is a considerable decrease in elastic and viscous behavior [31,32]. The dough obtained from 650 flour with the addition of hemp seed oil behaves differently due to the manner with which it reactions to temperature. The elastic modulus decreases significantly with a 5% addition of oil, especially when compared to the sample control, as well as with 10% and 15% oil additions.

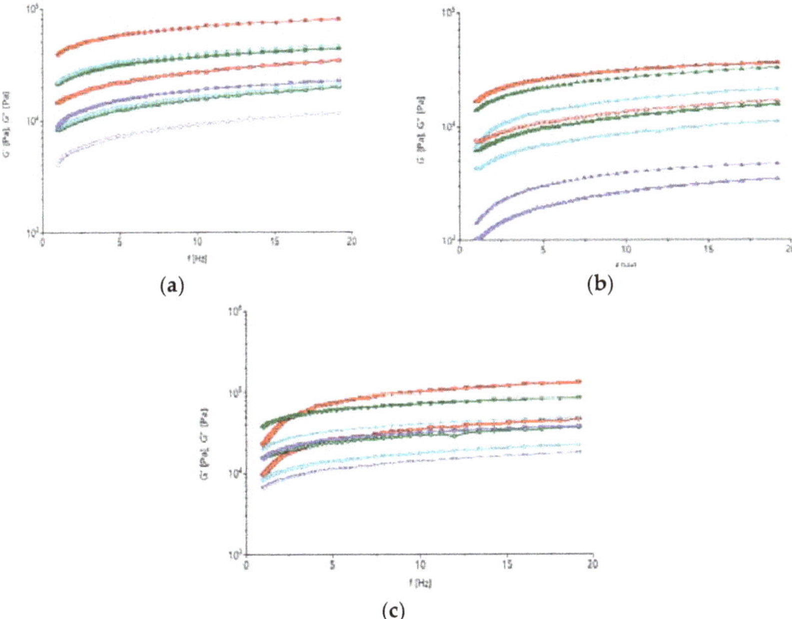

Figure 2. (**a**) G′ G″ dough modulus (points) and power curves (lines) depending on frequency at 550 flour dough samples for four different added oil contents: 0% (red point), 5% (green point), 10% (blue point), and 15% (purple point); (**b**) G′ G″ dough modulus (triangle with tip up) and power curves (lines) depending on frequency at 650 flour dough samples for four different added oil contents: 0% (red triangle), 5% (green triangle), 10% (blue triangle), and 15% (purple triangle); (**c**) G′ G″ dough modulus (triangle with tip down) and power curves (lines) depending on frequency at 1350 flour dough samples for four different added oil contents: 0% (red triangle), 5% (green triangle), 10% (blue triangle), and 15% (purple triangle).

The temperature ranges for starch gelatinization are lower for dough samples with higher oil additions. Figure 3c shows the behavior of the dough obtained from black flour (1350) under the influence of temperature. The G′ G″ modules show a linear trend for the control sample, compared to the dough sample which had an added percentage of 15% oil. The difference is significant between the two samples, this aspect will be observed in the results obtained when assessing the quality of the bread [33,34].

The alveograph data are shown in Table 1.

Analyzing the alveographic parameters highlights the bread's extensibility and the effects of increasing the elasticity of the dough as the percentage of added oil increases. The characteristics were determined: the extensibility of the dough (mm) indicated by the length (L) of the alveogram curve, the maximum pressure (P) representing the peak height (mm), the baking strength (W), the swelling index (G), and the configuration ratio of alveograph curve (P/L). The swelling index (G) increases in all dough samples in proportion to the amount of oil added and the baking strength decreases inversely in proportion to the addition of oil [35,36]. The incorporation of oil decreases water absorption, the mixing tolerance, and the tenacity of the doughs, but increases extensibility. The incorporation of oil tends to reduce the extensibility of the doughs and the energy of deformation, measured with the alveograph which can cause problems in the use of the doughs or in the retention of gas during fermentation.

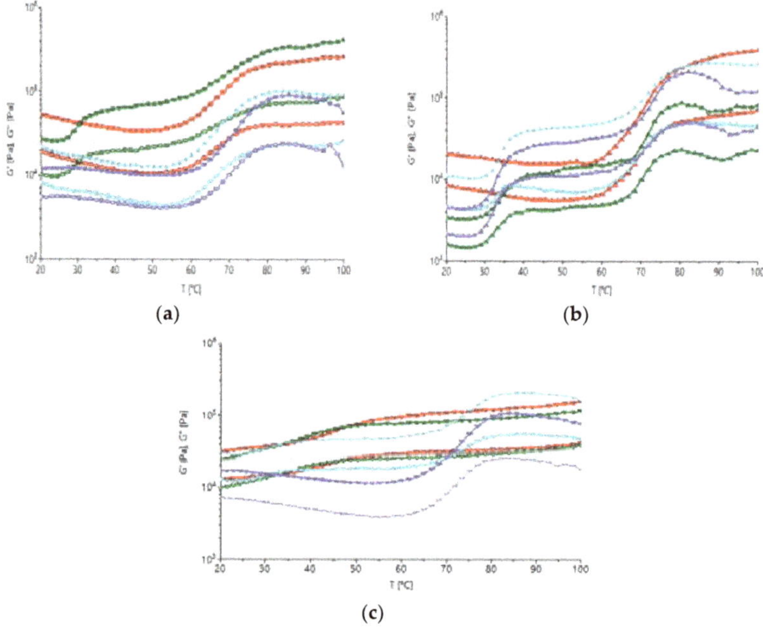

Figure 3. (a) G′ G″ dough modulus (points) and power curves (lines) depending on temperature at 550 flour dough samples for four different amounts of added oil: 0% (red point), 5% (green point), 10% (blue point), and 15% (purple point); (b) G′ G″ dough modulus (triangle with tip up) and power curves (lines) depending on temperature at 650 flour dough samples for four different amounts of added oil: 0% (red triangle), 5% (green triangle), 10% (blue triangle), and 15% (purple triangle); (c) G′ G″ dough modulus (triangle with tip up) and power curves (lines) depending on temperature at 1350 flour dough samples for four different amounts of added oil: 0% (red triangle), 5% (green triangle), 10% (blue triangle), and 15% (purple triangle).

Table 1. Results of alveographer characteristics for wheat flour with the addition of hemp seed oil.

Sample	P, mm	L, mm	G, mm	W, J	P/L
M_550	99 ± 0.01 [a]	79 ± 0.02 [a]	19.8 ± 0.02 [a]	283 ± 0.02 [c]	1.47 ± 0.01 [a]
P1_550	89 ± 0.01 [a]	110 ± 0.01 [c]	23.3 ± 0.02 [c]	262 ± 0.02 [b]	1.81 ± 0.02 [c]
P2_550	106 ± 0.02 [b]	99 ± 0.01 [ab]	22.1 ± 0.03 [c]	179 ± 0.02 [a]	1.70 ± 0.02 [b]
P3_550	122 ± 0.02 [c]	96 ± 0.01 [b]	21.8 ± 0.01 [b]	179 ± 0.01 [a]	2.56 ± 0.02 [d]
M_650	105 ± 0.01 [b]	97 ± 0.02 [c]	21.9 ± 0.01	298 ± 0.01 [c]	1.08 ± 0.02 [c]
P1_650	95 ± 0.01 [a]	86 ± 0.01 [a]	20.6 ± 0.01	221 ± 0.01 [b]	0.94 ± 0.01 [b]
P2_650	112 ± 0.01 [c]	96 ± 0.01 [b]	21.8 ± 0.02	184 ± 0.01 [a]	0.68 ± 0.03 [a]
P3_650	120 ± 0.01 [d]	79 ± 0.02 [c]	19.8 ± 0.03	122 ± 0.01 [ab]	0.63 ± 0.02 [a]
M_1350	102 ± 0.01 [a]	27 ± 0.01 [a]	11.6 ± 0.02	271 ± 0.01 [c]	8.0 ± 0.03 [d]
P1_1350	108 ± 0.02 [a]	24 ± 0.01 [a]	10.9 ± 0.01	183 ± 0.02 [b]	7.13 ± 0.03 [c]
P2_1350	117 ± 0.03 [b]	22 ± 0.02 [a]	10.4 ± 0.01	135 ± 0.02 [b]	6.27 ± 0.03 [b]
P3_1350	128 ± 0.02 [c]	21 ± 0.01 [a]	10.2 ± 0.01	116 ± 0.02 [a]	6.05 ± 0.03 [a]

[a–d] Mean values in the same column (corresponding to the same parameters) indicated statistically significant differences ($p < 0.05$).

Figure 4 expresses the following through a three-dimensional graph: the distribution of swelling index, the extensibility of the dough, and the maximum pressure which describes the empirically determined rheological properties. A homogeneous distribution of the results obtained for the following characteristics was observed: the extensibility of the dough (L), the maximum pressure (P), and the swelling index (G). The values are evenly distributed because the 550 and 650 flour assortments have similar rheological behavior, with the exception of the 1350 flour, which has a different behavior both in the description of the dynamic rheology and in empirical determinations.

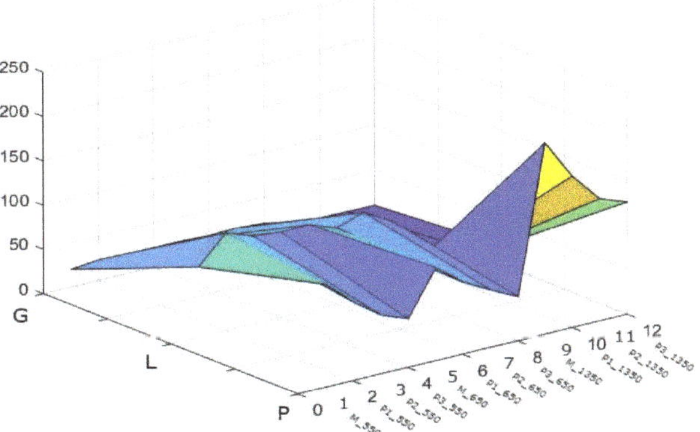

Figure 4. Graphic representation of the main alveograph characteristics.

Principal Component Analysis (PCA) was applied to highlight the effects of adding oil to white and black flour with the alveograph determined rheological characteristics of the dough. The loads of the variables studied with regards to the first two main components, PC1 (74.96%) and PC2 (24.30%) (Figure 4) described 99.27% of the total variant. Figure 5 represents the distribution of the quality attributes determined by the alveograph determinations and the dough samples, with which they are in a significantly positive correlation. PC1 was associated with dough samples with 0–5% added oil with rheological characteristics (extensibility, baking strength, maximum pressure, and the configuration ratio of the alveograph curve). PC2 was associated with dough samples with additions of 10–15% of hemp seed oil. There is a good relationship between the extensibility and the swelling index in the control sample and the dough with an addition of 5% hemp seed oil. The samples with 10–15% were not associated with the analyzed rheological characteristics [37,38].

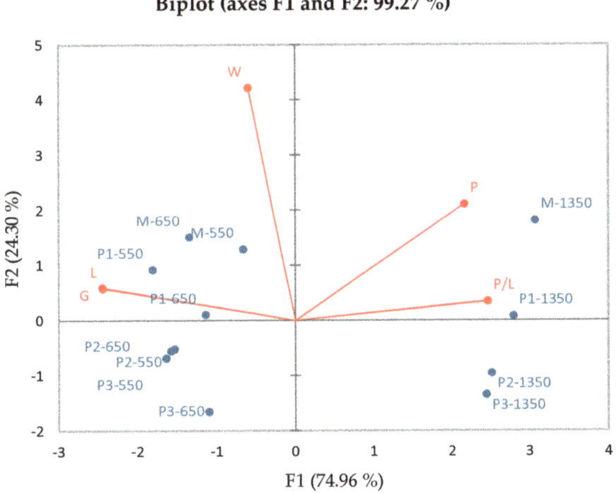

Figure 5. Principal component analysis (PCA) graph showing the distribution of rheological variables for the dough samples analyzed.

It can be seen that dough's of 650 and 550 flour with a maximum of 5% added hemp oil had the best characteristics, having qualities such as high extensibility, swelling index and high resistance to baking. The 1350 flour doughs (control and P1) correlate with the maximum pressure.

3.3. Evaluation of Bread Quality

The results regarding the analysis of the textural characteristics are presented in Table 2.

Table 2. Texture parameters of the bread samples with different percentages of hemp seed oil added.

Sample	Hardness (N)	Springiness (Adimensional)	Cohesiveness (Adimensional)	Gumminess (N)	Resilience (Adimensional)
M_550	6.95 ± 0.02 [b]	3.86 ± 0.01 [a]	0.9 ± 0.01 [b]	6.28 ± 0.00 [d]	2.3 ± 0.12 [d]
P1_550	6.37 ± 0.03 [b]	5.58 ± 0.02 [b]	0.8 ± 0.02 [a]	5.19 ± 0.00 [b]	1.8 ± 0.11 [a]
P2_550	5.36 ± 0.01 [a]	8.42 ± 0.03 [d]	0.8 ± 0.03 [a]	4.14 ± 0.01 [a]	2.3 ± 0.14 [d]
P3_550	5.60 ± 0.02 [c]	6.20 ± 0.02 [c]	0.8 ± 0.03 [a]	5.39 ± 0.01 [c]	2.5 ± 0.02 [d]
M_650	6.45 ± 0.02 [b]	3.68 ± 0.03 [a]	0.9 ± 0.03 [c]	5.54 ± 0.00 [c]	2.5 ± 0.01 [b]
P1_650	6.31 ± 0.00 [b]	4.83 ± 0.00 [c]	0.9 ± 0.01 [c]	5.76 ± 0.01 [c]	3.0 ± 0.11 [c]
P2_650	5.53 ± 0.02 [a]	4.12 ± 0.01 [b]	0.7 ± 0.01 [a]	4.09 ± 0.01 [a]	1.7 ± 0.10 [a]
P3_650	6.21 ± 0.03 [c]	3.67 ± 0.01 [a]	0.8 ± 0.03 [b]	5.09 ± 0.00 [b]	2.7 ± 0.04 [b]
M_1350	9.35 ± 0.01 [c]	2.95 ± 0.11 [c]	0.8 ± 0.03 [b]	7.52 ± 0.10 [c]	1.9 ± 0.04 [a b]
P1_1350	9.01 ± 0.01 [c]	2.74 ± 0.15 [c]	0.8 ± 0.02 [b]	6.85 ± 0.01 [b]	1.7 ± 0.03 [a]
P2_1350	8.71 ± 0.02 [b]	1.94 ± 0.21 [b]	0.8 ± 0.03 [b]	6.82 ± 0.01 [b]	2.2 ± 0.03 [b]
P3_1350	8.99 ± 0.03 [a]	1.86 ± 0.20 [a]	0.7 ± 0.02 [a]	6.41 ± 0.01 [a]	1.6 ± 0.04 [a]

[a–d] Mean values in the same column (corresponding to the same parameters) indicate statistically differ significantly ($p < 0.05$).

Determining the texture of white bread when supplemented with oil demonstrated that a decrease in hardness occurred when the amount of oil added amounted to 10%. When an oil percentage of 15% was added to the dough, it increased the hardness value for all resulting bread; both white and black flour samples [39,40]. This increase in hardness comes about during the process of changing the structure of the dough. The absorption of water in the starch granule is reduced, the proteins in the flour are covered with oil, and the gluten network is inhibited by the 15% addition of oil. The proteins' glutenin and gliadin require water to combine and form gluten strands, which gives the dough its strength, stretchability, and ability to trap fermentation gasses, preventing them from rising. This increase in hardness when adding 15% oil also has a negative effect on the elasticity of the crumb. This value decreases proportionally with the increase in the percentage of incorporated oil. The fermentation of the dough is slower with the addition of high percentages of oil, which leads to an increase in the fermentation time, a decrease in the retention capacity of the fermentation gases, and finally, a reduction in the volume of the bread [41,42]. The gumminess of the crumb is correlated with the addition of oil to the bread samples. This characteristic is associated with chewability. Chewing properties (chewing time or number of chews) and the sensory perception of fat are attributes that the consumer does not like [43]. Comparable findings have been described by Puerta et al. (2020), and Jourdren, et al. (2016), correlating the perceived sensations to the appreciation of the textural characteristics of bread by panelists [44,45]. The recommendation for consumers is that the bread samples with high gumminess can be toasted before consumption.

The characteristics of the structure and texture of the bread were analyzed (Figure 6) to demonstrate the overall effect of the distribution of quality attributes. Two components accounted for 81% of the total variation in data; 29% and 52% of the variations found in main components 1 and 2, respectively. It is evident that the 650 and 550 flour bread samples correlate with the cohesiveness and extensibility properties of the dough. In the right quadrant are grouped bread samples obtained from 1350 flour, the control sample, and the sample with 15% oil that demonstrated increased hardness.

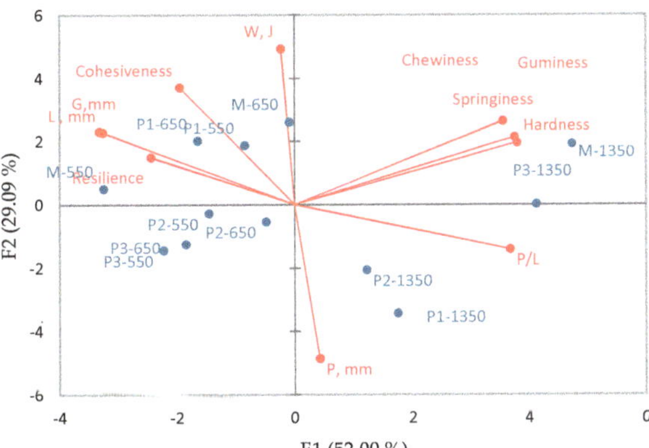

Figure 6. Analysis of the main components for the correlations between the bread samples and the rheological and textural characteristics.

The graph represented in Figure 7 expresses the distribution of bread samples in relation to all quality attributes that described the fundamental rheological characteristics, empirical and textural characteristics. On axis 1, at the distance of origin, are the bread samples obtained from black flour (1350) with the addition of 10% and 15% hemp seed oil. On axis 2 are the white flour bread samples, type 650 and 550 (P3_550 and P3_550), which have a strong positive ratio. Observing distribution on the graph reveals significant correlations between samples P2_550 and P2_650. In conclusion, the addition of oil in the dough influences the rheological and textural properties, depending on the percentage added and the type of flour.

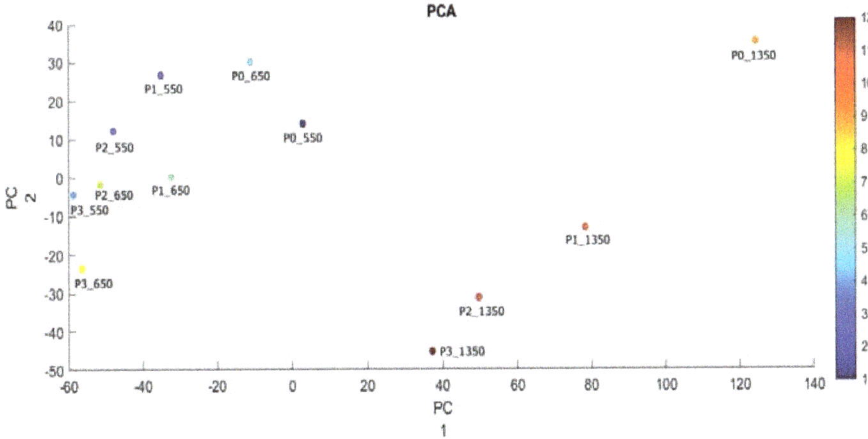

Figure 7. Graph of the distribution of bread samples in relation to the quality characteristics and the proportion of added hemp seed oil.

3.4. Evaluation of Sensory Quality of Bread

Sensory evaluation (Figure 8) of bread samples was performed 24 h after baking. The bread samples received up to 9 points regarding general acceptability with the following average scores: the bread samples with the addition of 5% hemp oil scored 7.66; samples with the addition of 10% hemp oil scored 8.20; samples with the addition of 15% scored

7.12. On average, the evaluators said, with regards to the addition of 5–15% hemp oil, "I like it very much" and "I like it moderately". The evaluators appreciated the samples of black bread with the addition of 5% and 10% for its crumb pore and specific aroma. Although using hemp seed oil benefits is beneficial to bread quality, it cannot be used in excess of 10%, as, according to the evaluators' ratings, this causes the taste to worsen. Related to this, is that other several studies have also evaluated consumer satisfaction with oils, such as sesame, chia, pumpkin, and flax, finding that 5–10% additions did not significantly alter sensory properties.

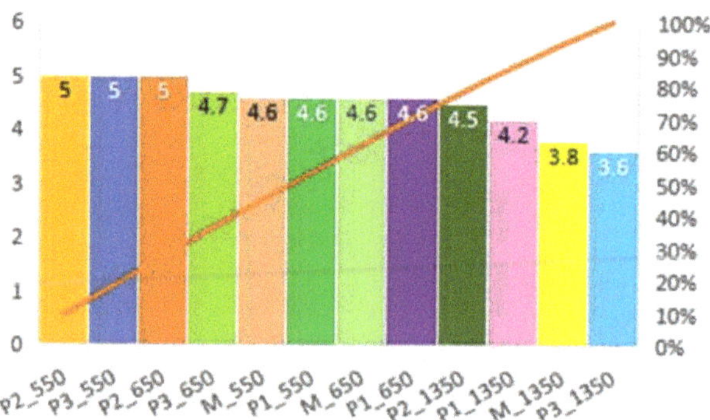

Figure 8. Sensory properties of bread samples.

4. Conclusions

This study demonstrates the possibility of using hemp seed oil in the formulation of bread dough for both white and black bread.

Alveograph parameters were used to analyze and highlight the extensibility properties and the effect of increasing the elasticity of the dough as the percentage of added oil increases. The results indicate that the presence of hemp seed oil changes the rheological properties of the dough, increasing storage and loss values.

The loss modulus (G″) presents lower values than the storage modulus (G′) in all formulated dough samples, suggesting a more elastic than viscous characteristic in said dough samples.

The rheological properties of the dough have been improved, as have the textures in the samples obtained from black flour with the specification that the maximum oil addition is 10%. An increase in the hardness of the bread samples was noticed in bread with the addition of 15% hemp seed oil. The effect of incorporating oils in bread depends on the quantity added. In small amounts, it can be beneficial, while in larger quantities, they weaken the dough and have negative effects on the springiness. Therefore, depending on the flour type and the amount of oil added, the effects on the quality of bread may vary.

Author Contributions: Conceptualization, S.R., L.C.A. and C.D.; methodology, S.R. and A.E.P.; software, S.R.; validation, A.E.P., C.D. and L.C.A.; formal analysis, S.R. and A.E.P.; investigation, S.R. and L.C.A.; resources, A.E.P. and S.R.; data curation, S.R. and C.D.; writing—original draft preparation, A.E.P. and S.R.; writing—review and editing, S.R. and C.D.; visualization S.R. and L.C.A.; supervision, S.R.; funding acquisition, A.E.P. All authors have read and agreed to the published version of the manuscript.

Funding: This work was funded by Ministry of Research, Innovation and Digitalization within Program 1—Development of national research and development system, Subprogram 1.2—Institutional Performance—RDI excellence funding projects, under contract no. 10PFE/2021.

Institutional Review Board Statement: Not applicable.

Informed Consent Statement: Not applicable.

Data Availability Statement: Not applicable.

Conflicts of Interest: The authors declare no conflict of interest.

References

1. Pareyt, B.; Finnie, S.M.; Putseys, J.A.; Delcour, J. Lipids in bread making: Sources, interactions, and impact on bread quality. *J. Cereal Sci.* **2011**, *54*, 266–279. [CrossRef]
2. Manzocco, L.; Calligaris, S.; Da Pieve, S.; Marzona, S.; Nicoli, M.C. Effect of monoglyceride-oil–water gels on white bread properties. *Food Res. Int.* **2012**, *49*, 778–782. [CrossRef]
3. Tarancón, P.; Salvador, A.; Sanz, T. Sunflower oil-watercellulose ether emulsions as trans-fatty acid-free fat replacers in biscuits: Texture and acceptability study. *Food Bioprocess Technol.* **2013**, *6*, 2389–2398. [CrossRef]
4. Chin, N.L.; Rahman, R.A.; Hashim, D.M.; Kowng, S.Y. Palm Oil Shortening Effects on Baking Performance of White Bread. *J. Food Process Eng.* **2009**, *33*, 413–433. [CrossRef]
5. Skrbic, B.; Filipčev, B. Nutritional and sensory evaluation of wheat breads supplemented with oleic-rich sunflower seed. *Food Chem.* **2008**, *108*, 119–129. [CrossRef]
6. Rios, R.V.; Pessanha, M.D.F.; de Almeida, P.F.; Viana, C.L.; da Silva Lannes, S.C. Application of fats in some food products. *Food Sci. Technol.* **2014**, *34*, 3–15. [CrossRef]
7. Osuna, M.B.; Romero, C.A.; Romero, A.M.; Judis, M.A.; Bertola, N.C. Proximal composition, sensorial properties and effect of ascorbic acid and α-tocopherol on oxidative stability of bread made with whole flours and vegetable oils. *LWT* **2018**, *98*, 54–61. [CrossRef]
8. Leonard, W.; Zhang, P.; Ying, D.; Fang, Z. Hempseed in food industry: Nutritional value, health benefits, and industrial applications. *Compr. Rev. Food Sci. Food Saf.* **2019**, *19*, 282–308. [CrossRef]
9. El-Adawy, T.A. Effect of sesame seed proteins supplementation on the nutritional, physical, chemical and sensory properties of wheat flour bread. *Mater. Veg.* **1995**, *48*, 311–326. [CrossRef]
10. Austria, J.A.; Aliani, M.; Malcolmson, L.J.; Dibrov, E.; Blackwood, D.P.; Maddaford, T.G.; Guzman, R.; Pierce, G.N. Daily choices of functional foods supplemented with milled flaxseed by a patient population over one year. *J. Funct. Foods* **2016**, *26*, 772–780. [CrossRef]
11. Montesano, D.; Blasi, F.; Simonetti, M.S.; Santini, A.; Cossignani, L. Chemical and Nutritional Characterization of Seed Oil from *Cucurbita maxima* L. (var. Berrettina) Pumpkin. *Foods* **2018**, *7*, 30. [CrossRef] [PubMed]
12. Hyvärinen, H.K.; Pihlava, J.-M.; Hiidenhovi, J.A.; Hietaniemi, V.; Korhonen, A.H.J.T.; Ryhänen, E.-L. Effect of Processing and Storage on the Stability of Flaxseed Lignan Added to Bakery Products. *J. Agric. Food Chem.* **2005**, *54*, 48–53. [CrossRef] [PubMed]
13. Gomez, M.; Ronda, F.; Blanco, C.; Caballero, P.A.; Apesteguía, A. Effect of dietary fibre on dough rheology and bread quality. *Eur. Food Res. Technol.* **2003**, *216*, 51–56. [CrossRef]
14. Moreira, R.; Chenlo, F.; Torres, M. Effect of chia (*Sativa hispanica* L.) and hydrocolloids on the rheology of gluten-free doughs based on chestnut flour. *LWT* **2013**, *50*, 160–166. [CrossRef]
15. Mansour, E.H.; Dworschák, E.; Pollhamer, Z.; Gergely, Á.; Hóvári, J. Pumplin and canola seed proteins and bread quality. *Acta Aliment.* **1999**, *28*, 59–70.
16. Jourdren, S.; Panouillé, M.; Saint-Eve, A.; Déléris, I.; Forest, D.; Lejeune, P.; Souchon, I. Breakdown pathways during oral processing of different breads: Impact of crumb and crust structures. *Food Funct.* **2016**, *7*, 1446–1457. [CrossRef]
17. Atudorei, D.; Atudorei, O.; Codină, G. Dough Rheological Properties, Microstructure and Bread Quality of Wheat-Germinated Bean Composite Flour. *Foods* **2021**, *10*, 1542. [CrossRef] [PubMed]
18. Mironeasa, S.; Codină, G.G. Dough Rheological Behavior and Microstructure Characterization of Composite Dough with Wheat and Tomato Seed Flours. *Foods* **2019**, *8*, 626. [CrossRef]
19. Mancebo, C.M.; Martínez, M.M.; Merino, C.; De La Hera, E.; Gómez, M. Effect of oil and shortening in rice bread quality: Relationship between dough rheology and quality characteristics. *J. Texture Stud.* **2017**, *48*, 597–606. [CrossRef] [PubMed]
20. AACC International Method 54-30.02. Alveograph Method for Soft and Hard Wheat Flour. 1999. Available online: https://www.cerealsgrains.org/resources/Methods/Pages/54PhysicalDoughTests.aspx (accessed on 4 March 2022).
21. De Lamo, B.; Gómez, M. Bread Enrichment with Oilseeds. A Review. *Foods* **2018**, *7*, 191. [CrossRef] [PubMed]
22. Pojic, M.; Hadnadev, T.D.; Hadnaved, M.; Rakita, S.; Brlek, T. Bread supplimentation with hemp seed cake: A by-product of hemp oil processing. *J. Food Qual.* **2015**, *38*, 431–440. [CrossRef]
23. Day, L.; Golding, M. *Food Structure, Rheology, and Texture*; Elsevier BV: Amsterdam, The Netherlands, 2016; pp. 125–129.
24. Švec, I.; Hrušková, M. Crumb evaluation of bread with hemp products addition by means of image analysis. *Acta Univ. Agric. Silvic. Mendel. Brun.* **2013**, *61*, 1867–1872. [CrossRef]
25. Rusu, I.; Mureşan, C.C.; Mureşan, A.E.; Mureşan, V.; Pop, C.R.; Chiş, M.S.; Man, S.M.; Filip, M.R.; Onica, B.-M.; Alexa, E.; et al. Hemp (*Cannabis sativa* L.) Flour-Based Wheat Bread as Fortified Bakery Product. *Plants* **2021**, *10*, 1558. [CrossRef] [PubMed]
26. Apostol, L.; Belc, N.; Gaceu, L.; Oprea, O.B.; Popa, M.E. Sorghum Flour: A Valuable Ingredient for Bakery Industry? *Appl. Sci.* **2020**, *10*, 8597. [CrossRef]
27. Slade, L.; Levine, H.; Finley, J.W. Protein-water interactions: Water as a plasticizer of gluten and other protein polymers. In *Protein Quality and the Effects of Processing*; CRC Press: Boca Raton, FL, USA, 1989; Volume 9, p. 124.

28. Jarosław, K.; Lesław, J.; Witczak, M.; Ziobro, R. Effect of citrus fiber on the rheological properties of dough and quality of the gluten-free bread. *Appl. Sci.* **2020**, *10*, 6633.

29. Menteş, Ö.; Bakkalbaşşi, E.; Ercan, R. Effect of the Use of Ground Flaxseed on Quality and Chemical Composition of Bread. *Food Sci. Technol. Int.* **2008**, *14*, 299–306. [CrossRef]

30. Coelho, M.S.; Salas-Mellado, M.M. Effects of substituting chia (*Salvia hispanica* L.) flour or seeds for wheat flour on the quality of the bread. *LWT Food Sci. Technol.* **2015**, *60*, 729–736. [CrossRef]

31. Costantini, L.; Lukšič, L.; Molinari, R.; Kreft, I.; Bonafaccia, G.; Manzi, L.; Merendino, N. Development of gluten-free bread using tartary buckwheat and chia flour rich in flavonoids and omega-3 fatty acids as ingredients. *Food Chem.* **2014**, *165*, 232–240. [CrossRef]

32. Korus, J.; Witczak, M.; Ziobro, R.; Juszczak, L. Hemp (*Cannabis sativa* subsp. *sativa*) flour and protein preparation as natural nutrients and structure forming agents in starch based gluten-free bread. *LWT* **2017**, *84*, 143–150. [CrossRef]

33. Romero-Lopez, M.R.; Osorio-Diaz, P.; Bello-Perez, L.A.; Tovar, J.; Bernardino-Nicanor, A. Fiber Concentrate from Orange (*Citrus sinensis* L.) Bagase: Characterization and Application as Bakery Product Ingredient. *Int. J. Mol. Sci.* **2011**, *12*, 2174–2186. [CrossRef]

34. Melini, F.; Melini, V.; Luziatelli, F.; Ruzzi, M. Current and Forward-Looking Approaches to Technological and Nutritional Improvements of Gluten-Free Bread with Legume Flours: A Critical Review. *Compr. Rev. Food Sci. Food Saf.* **2017**, *16*, 1101–1122. [CrossRef] [PubMed]

35. Codină, G.G.; Mironeasa, S. Use of response surface methodology to investigate the effects of brown and golden flaxseed on wheat flour doughmicrostructure and rheological properties. *J. Food Sci. Technol.-Mysore* **2016**, *53*, 4149–4158. [CrossRef] [PubMed]

36. Coţovanu, I.; Mironeasa, S. Influence of Buckwheat Seed Fractions on Dough and Baking Performance of Wheat Bread. *Agronomy* **2022**, *12*, 137. [CrossRef]

37. Gao, J.; Ong, J.J.-X.; Henry, J.; Zhou, W. Physical breakdown of bread and its impact on texture perception: A dynamic perspective. *Food Qual. Prefer.* **2017**, *60*, 96–104. [CrossRef]

38. Purificación, G.S.; Igual, M.; Martínez-Monz, J. Physicochemical properties and consumer acceptance of bread enriched with alternative proteins. *Foods* **2020**, *9*, 933.

39. Agyare, K.; Addo, K.; Xiong, Y.; Akoh, C. Effect of structured lipid on alveograph characteristics, baking and textural qualities of soft wheat flour. *J. Cereal Sci.* **2005**, *42*, 309–316. [CrossRef]

40. Mao, Q.; Sun, Y.; Hou, J.; Yu, L.; Liu, Y.; Liu, C.; Xu, N. Relationships of Image Texture Properties with Chewing Activity and Mechanical Properties during Mastication of Bread. *Int. J. Food Eng.* **2016**, *12*, 311–321. [CrossRef]

41. Van Eck, A.; Hardeman, N.; Karatza, N.; Fogliano, V.; Scholten, E.; Stieger, M. Oral processing behavior and dynamic sensory perception of composite foods: Toppings assist saliva in bolus formation. *Food Qual. Prefer.* **2019**, *71*, 497–509. [CrossRef]

42. Gómez, A.; Ferrero, C.; Calvelo, A.; Añón, M.; Puppo, M. Effect of Mixing Time on Structural and Rheological Properties of Wheat Flour Dough for Breadmaking. *Int. J. Food Prop.* **2011**, *14*, 583–598. [CrossRef]

43. Beikzadeh, S.; Shojaee-Aliabadi, S.; Dadkhodazade, E.; Sheidaei, Z.; Abedi, A.-S.; Mirmoghtadaie, L.; Hosseini, S.M. Comparison of Properties of Breads Enriched with Omega-3 Oil Encapsulated in β-Glucan and Saccharomyces cerevisiae Yeast Cells. *Appl. Food Biotechnol.* **2019**, *1*, 11–20.

44. Puerta, P.; Laguna, L.; Villegas, B.; Rizo, A.; Fiszman, S.; Tarrega, A. Oral processing and dynamics of texture perception in commercial gluten-free breads. *Food Res. Int.* **2020**, *134*, 109233. [CrossRef] [PubMed]

45. Jourdren, S.; Saint-Eve, A.; Panouillé, M.; Lejeune, P.; Déléris, I.; Souchon, I. Respective impact of bread structure and oral processing on dynamic texture perceptions through statistical multiblock analysis. *Food Res. Int.* **2016**, *87*, 142–151. [CrossRef] [PubMed]

applied sciences

Article

Advance on the Capitalization of Grape Peels By-Product in Common Wheat Pasta

Mădălina Ungureanu-Iuga [1,2] and Silvia Mironeasa [1,*]

[1] Faculty of Food Engineering, "Ştefan cel Mare" University of Suceava, 13th University Street, 720229 Suceava, Romania; madalina.iuga@usm.ro
[2] Integrated Center for Research, Development and Innovation in Advanced Materials, Nanotechnologies, and Distributed Systems for Fabrication and Control (MANSiD), "Ştefan cel Mare" University of Suceava, 13th University Street, 720229 Suceava, Romania
* Correspondence: silviam@fia.usv.ro

Abstract: Capitalization of winery by-products has received high interest among scientists, producers and consumers concerned with healthy diet and environment protection. Grape peels are rich in fiber and polyphenols and can be used as ingredients in pasta matrix in order to increase the nutritional and functional value of such a staple food. The aim of this paper was to investigate the effects of grape peel flour added in various amounts (1–6%) to common wheat pasta dough viscoelasticity and texture and on pasta chemical composition, color, cooking behavior and texture, revealing at the same time the relations between characteristics. Grape peel flour induced the increase of the elastic (G′) and viscous (G″) moduli, dough hardness, springiness, cohesiveness, pasta crude ash, crude fat, crude fiber, total polyphenols and resistant starch contents, pasta water absorption, cooking loss and breaking force as the addition level was higher and compared to the control. On the other hand, dough resilience, pasta luminosity, chewiness and firmness decreased as the amount of grape peel flour raised. Significant correlations ($p < 0.05$) were obtained between the chemical composition and color parameters, while crude fiber, protein and fat were correlated with dough and pasta texture, total polyphenols with resistant starch content, cooking loss with crude fiber and dough textural parameters. The obtained results underlined the opportunity to use a valuable byproduct such as grape peels in novel pasta formulations, being helpful for processors to extend the product variety and to optimize the processes in order to better satisfy consumer's demand for functional foods.

Keywords: common wheat; vinification by-product; texture; dough rheology; physico-chemical properties

Citation: Ungureanu-Iuga, M.; Mironeasa, S. Advance on the Capitalization of Grape Peels By-Product in Common Wheat Pasta. *Appl. Sci.* **2021**, *11*, 11129. https://doi.org/10.3390/app112311129

Academic Editor: Claudio Medana

Received: 5 November 2021
Accepted: 22 November 2021
Published: 24 November 2021

Publisher's Note: MDPI stays neutral with regard to jurisdictional claims in published maps and institutional affiliations.

1. Introduction

Food processing trends are heading more and more to waste reduction and sustainable approaches in order to diminish the impact of these activities on the environment. Large amounts of by-products resulted can be used as alternative ingredients to create value-added food or extract bioactive compounds. One of the industries that generate byproducts with high functional properties is winemaking. Grape vine (*Vitis vinifera* L.) is cultivated all over the world, the total surface in 2020 summing up more than 7.3 million hectares, with a production of 230 mhL wine [1]. Grape pomace represents about 20–30% of the grape weight, with grape peel counting 63–75% of the total pomace [2]. Therefore, the disposal of such by-products greatly impacts the environment, the main uses being oriented towards animal feed or compost. Furthermore, some processors resort to the discarding of grape pomace on the soil without any pretreatment which will determine acidification due to the low pH of pomace and oxygen consumption in soil and ground waters caused by tannins and other compounds' presence [3,4].

Common wheat flour is used for pasta production mainly in countries where durum wheat is less accessible and more expensive [5]. Refined wheat flour presents a low nutritional value due to its proximate composition, grain parts like bran and endosperm are discarded during milling, depending on the extraction rate. Refined wheat flour has an extraction rate of 75% and thus presents low amounts of fiber, vitamins, minerals and bioactive compounds compared to whole wheat [6,7]. The lack of fiber in the human diet could be related to negative effects on the digestive system. Thus, the supplementation of wheat flour with various fiber-rich ingredients drew increasingly attention from researchers and producers.

Grape by-products give many health benefits due to the presence of bioactive compounds which are kept in grape pomace even after processing. Grape pomace has more than 60% dietary fiber [7,8], the main fraction being insoluble dietary fiber represented by cellulose and hemicelluloses. Nevertheless, soluble dietary fiber is also found in grape by-products and exerts many health benefits given by water-soluble polysaccharides such as β-glucan, pectin and gum [9]. It was demonstrated that grape pomace could act as a prebiotic in the colon due to its ability to ferment, acting as a substrate for the colon microbiota [7,10]. Grape by-products bioactive compounds include flavonoids such as anthocyanins, resveratrol, tannin and quercetin with antibacterial, anticancer, antioxidant and anti-inflammatory activities, hepatic and cardiovascular protection and chronic diseases prevention properties [11–15]. The proximate composition of grape pomace includes proteins, lipids and carbohydrates, the main micronutrients represented by vitamins and phenolic compounds that exert antioxidant activity against free-radicals [16,17]. The chemical composition of grape by-products depends on various factors such as cultivar, climate, ripening stage and soil [18].

In food products supplementation for functional value increase, the use of valuable bioactive compounds is an essential goal to satisfy the diversified, escalating market demand. There are some studies revealing the possibility to use wine grape by-products such as pomace, peels and seeds in bakery and pasta products, as it was summarized in our previous work [19]. According to the results presented by Marinelli et al. [20], grape marc incorporation in durum wheat pasta led to significant improvement of phenolics and anthocyanins contents and antioxidant activity compared to the control, higher bioaccessibility of polyphenols and lower glycemic index being obtained. Regarding cooked pasta texture, a decrease of firmness, adhesiveness and elasticity was reported, probably due to the gluten dilution effects [20]. Lou et al. [21] observed an increase in antioxidant activity and radical scavenging activities of biscuits enriched with grape pomace, while minor influence on the digestion rate was recorded. Furthermore, biscuits chewiness and hardness increased as the level of grape pomace was higher possibly as a result of the binding of polyphenols with protein or starch, determining a raised gluten matrix strength [21]. Textural properties of breadsticks were affected by grape pomace incorporation, revealing a decrease in hardness and breaking force proportional to the level added, while the antioxidant capacity and the fiber content increased, as stated by Rainero et al. [22]. The addition of grape peels, seeds and pomace flours in wheat dough caused the decrease of the elastic (G′) and viscous (G″) moduli, increasing at the same time tan δ values as the amount added raised [23]. The viscoelastic character of dough is related to its three-dimensional matrix given by the interaction between gluten proteins and starch or non-starch polysaccharides [24]. Balli et al. [25] studied the effect of 7% grape pomace addition in durum *tagliatelle* pasta and showed that pasta samples contained the same monoglycosylated and acetylated anthocyanins as found in grape pomace, while a good cooking resistance and texture after cooking were maintained. The incorporation of growing quantities of grape pomace flour in cakes led to the gradual increase of ash, lipid, proteins, fibers, free phenolics, anthocyanins, total polyphenol content and antioxidant activity, while the luminosity (*L**) and the red nuance (*a**) were reduced [4]. The most abundant phenolics identified in the enriched cakes were catechin, gallic acid, quercitin, protocatechuic acid, kaempferol and apigenin [4].

Our previous research [26,27] underlined the effects of grape peels on gluten free pasta made of corn flour and optimized the processing parameters for grape peels addition and hydrothermal treatment of wheat flour. Considering the factors mentioned above and the scarcity of papers regarding the influence of grape peel flour incorporation in common wheat pasta, this study aimed to clarify the contribution of this by-product on the rheological and textural behavior of dough, on pasta chemical properties, color parameters, uncooked pasta breaking force, cooking behavior and cooked pasta texture. The positive effects of grape peel flour on final product functionality were evidenced by fiber, total polyphenols and resistant starch determination. Compared to our previous research, this study aimed to evaluate dough rheology in terms of elastic and viscous moduli, dough texture regarding hardness, resilience, springiness and cohesiveness, pasta proximate composition, pasta color in terms of luminosity, red-green and blue-yellow nuances and pasta texture parameters such as elasticity, firmness and chewiness, and in order to complete the image of grape peel's influence on common wheat pasta. Furthermore, the correlations and the similarities between characteristics and samples were pointed out to better understand the interactions between wheat and grape peel flours components.

2. Materials and Methods

2.1. Materials Conditioning and Pasta Making

Common wheat (*Triticum aestivum*) flour (650 type) and grape peels from the Fetească Regală pomace were used in pasta making. This variety of grape peel was chosen in order to minimize the negative effects on color changes of the final product and taking into account that browner color is associated by the consumers with health, as stated in the literature [28].

Grape peel flour presented lower luminosity (L^* = 62.42 $\pm$ 0.07) compared to wheat flour (L^* = 89.49 $\pm$ 0.42), a red nuance (a^* = 3.24 $\pm$ 0.05) compared to the green nuance of wheat flour (a^* = $-5.04 \pm$ 0.04) and a higher yellow nuance (grape peel flour b^* = 25.32 $\pm$ 0.04, wheat flour b^* = 15.36 $\pm$ 0.12). Wheat flour proximate composition included 0.96 $\pm$ 0.02% crude fat, 12.39 $\pm$ 0.18% crude protein, 72.11 $\pm$ 0.08% carbohydrates, 0.52 $\pm$ 0.04% crude ash, 14.01 $\pm$ 0.16% moisture, 10.56 $\pm$ 0.44 mg GAE/100 g total polyphenols.

For grape peel flour processing, the fresh pomace was dehydrated in a convection oven at 50 °C for 18 h, then stems and seeds were manually removed. The resulted product, called grape peel, was ground and sieved to a particle size < 180 μm. Grape peel flour proximate composition comprised 2.30 $\pm$ 0.17% crude fat, 9.06 $\pm$ 0.06% crude proteins, 25.25 $\pm$ 0.05% crude fiber, 51.27 $\pm$ 0.20% carbohydrates, 4.11 $\pm$ 0.03% crude ash, 8.00 $\pm$ 0.08% moisture and 133.27 $\pm$ 1.44 mg GAE/100 g total polyphenols.

For composite flours making, wheat flour was mixed for 15 min in a Yucebas Y21 machine (Izmir, Turkey) with the appropriate amount of grape peel flour (1%, 2%, 3%, 4%, 5% or 6% replacing wheat flour).

Pasta dough was made by mixing the composite flour in a Kitchen Aid mixer (Whirlpool Corporation, Tulsa, OK, USA) with appropriate amounts of water calculated to achieve 40% moisture. After 15 min of resting at room temperature in closed glass containers, pasta dough was extruded by using a rigatoni mold of the Kitchen Aid machine. Modeled pasta was dried in agreement with the protocol described by Bergman et al. [29]: pasta were dried for 30 min in open air at room temperature, then the samples were placed in a convection oven for 60 min at 40 °C, then for 120 min at 80 °C and finally for 120 min at 40 °C (Figure 1).

2.2. Dough Visco-Elasticity Evaluation

For the rheological measurements, dough samples were previously laminated and left in closed containers for 30 min to allow internal strain elimination. The viscous (G') and elastic (G'') moduli were evaluated trough a frequency sweep test by means of a Thermo-HAAKE, MARS 40 (Karlsruhe, Germany) dynamic rheometer with parallel plates (40 mm diameter). G' and G'' variation with frequency in the range of 0.1–20 Hz was evaluated at a strain of 15 Pa which was in the linear viscoelastic region (LVR) previously tested.

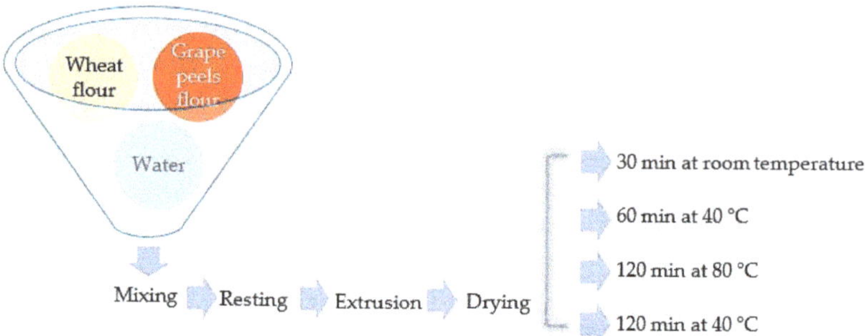

Figure 1. Pasta making process.

Pasta exterior diameter before drying varied between 1.15 and 1.20 cm and the length was comprised between 2.73 and 2.94 cm, while after drying the diameter varied from 1.00 to 1.10 cm and the length from 2.50 to 2.70 cm. Control and enriched pasta samples with 1 to 6% grape peel flour (GP) are presented in Figure 2.

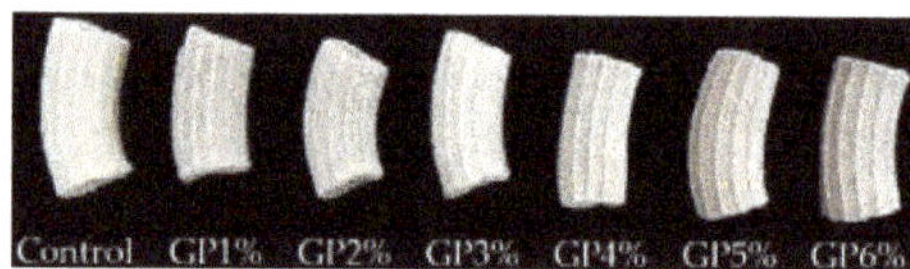

Figure 2. Pasta samples.

2.3. Texture of Dough Evaluation

Texture profile analysis of dough was made by double compression on balls of 50 g weight, at 50% height, with 5.0 mm/s test speed and a trigger force of 20 g, by using a Perten TVT-6700 texturometer (Perten Instruments, Hägersten, Sweden) [26]. Dough hardness, resilience, springiness and cohesiveness were recorded.

2.4. Chemical Compounds Determination

The chemical compounds in terms of moisture, crude ash, crude protein and crude fat from pasta samples were evaluated according to the Romanian and International standard methods: (SR EN ISO 712/2010), (SR ISO 2171/2002), (SR EN ISO 20483/2007) and (SR 91/2007), respectively [5]. The dietary fiber contents were estimated with a FOSS 6500 NIR (FOSS NIRSystems, Silver Springs, MD, USA) infrared spectrometer, the calibration being made using the off the shelf INGOT commercial calibrations (AUNIR, Towcester, UK). The carbohydrates were calculated by difference (Equation (1)).

$$Carbohydrates\ (\%) = 100 - (moisture + ash + protein + lipid + fiber) \tag{1}$$

The total polyphenols were determined by following Folin–Ciocalteu protocol [30]. For this purpose, grinded pasta samples were extracted with methanol 80% (v/v) and sonicated at 37 °C and 45 Hz for 40 min [31]. After filtering, Folin–Ciocalteu reagent (1N) and sodium carbonate 20% (v/v) were added in the diluted extract (1:4). After 40 min of resting away from light, the absorbance was read at 725 nm and the total polyphenols were determined by using a calibration curve (R^2 = 0.99) with gallic acid (GAE).

The resistant starch was evaluated by following the international AOAC 2017.16 protocol, by using Megazyme kit. The determination principle consists of pasta sample digestion with α-amylase and amyloglucosidase for 120 min, followed by second digestion only with amyloglucosidase and spectrophotometric glucose quantification (at 510 nm) by using glucose oxidase/peroxidase (GOPOD) reagent. All the chemical compounds were reported to product weight as it is.

2.5. Uncooked Pasta Color Evaluation

Uncooked pasta color in terms of luminosity (L^*), red or green nuance (a^*) and yellow or blue nuance (b^*) was evaluated by reflectance, in the CIE Lab system, on a Konica Minolta CR-400 (Tokyo, Japonia) colorimeter.

2.6. Pasta Cooking Behavior Determination

Pasta cooking loss and water absorption were determined gravimetrically according to the protocol described by Gimenez et al. [32], by boiling 10 g of pasta in 200 mL of water for the optimum cooking time.

2.7. Uncooked and Cooked Pasta Texture Evaluation

Uncooked pasta breaking force was measured on a Perten TVT-6700 texturometer (Perten Instruments, Hägersten, Sweden) equipped with an aluminum break rig set, at a test speed of 3 mm/s and a trigger force of 50 g [26].

Cooked pasta elasticity, chewiness and firmness were evaluated by means of double compression performed on one piece of sample, by using a Perten TVT-6700 equipment. A cylindric probe (35 mm diameter) was used for compression at 50% of sample height, at a test speed of 5.0 mm/s and a trigger force of 20 g [26].

2.8. Statistical Analysis

All of the determinations presented above were done at least in triplicate. The effect of grape peel flour on dough and pasta characteristics was evaluated by means of one-way ANOVA and Tukey test ($p < 0.05$), performed on XLSTAT for Excel 2021 version (Addinsoft, New York, NY, USA) software. The similarities and relations between characteristics were observed by applying Principal Component Analysis (PCA).

3. Results

3.1. Dough Visco-Elasticity

Wheat dough rheological properties in terms of elastic and viscous moduli were affected by grape peel flour incorporation. All the samples investigated presented a solid-like behavior since $G' > G''$, an increase of both moduli with frequency raise being obtained (Figure 3). The addition of grape peel flour led to increased G' and G'' values compared to the control and as the amount was higher.

3.2. Texture of Dough

Dough texture knowledge is important for the prediction of its behavior during mixing and handling and could be affected by the incorporation of fiber-rich ingredients. Dough hardness was significantly higher ($p < 0.05$) as the amount of grape peel flour increased and compared to the control (Table 1).

Springiness varied from 99.55 to 99.75×10^{-2}, the highest value being obtained for GP6%, while the control sample exhibited the lowest springiness. Cohesiveness values registered slight increases with grape peel flour level increase and compared to the control. On the other hand, a proportional decrease of dough resilience was observed as the addition level of grape peel flour raised.

3.3. Pasta Chemical Compounds

The nutritional and functional value of pasta products can be improved by using ingredients with high bioactive properties. Pasta crude ash values varied from 0.52% for the control to 0.74% for GP6% sample (Table 2). An increasing trend proportional to the addition level was also observed in the case of crude fiber (Figure 4), the differences being significant ($p < 0.05$).

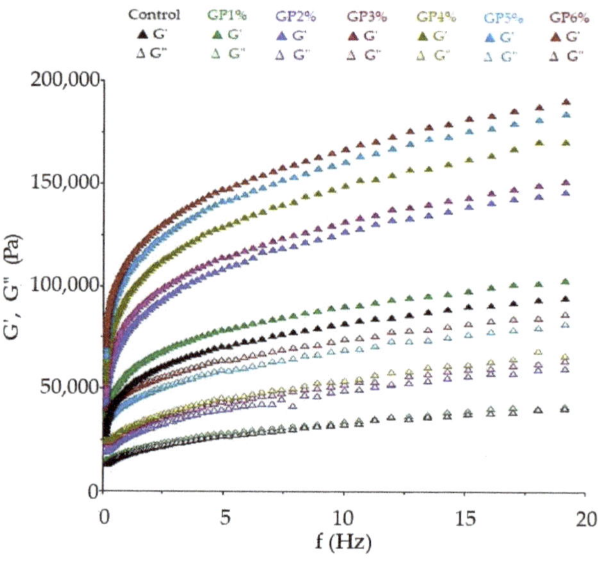

Figure 3. Effect of grape peel flour on dough elastic (G′) and viscous (G″) moduli.

Table 1. Effects of grape peel flour on dough texture.

Sample	Hardness (N)	Resilience (Adim.)	Springiness $\times\ 10^{-2}$ (Adim.)	Cohesiveness (Adim.)
Control	14.13 ± 0.62 [f]	1.22 ± 0.06 [abc]	99.55 ± 0.03 [b]	0.36 ± 0.02 [c]
GP1%	18.51 ± 0.18 [e]	1.28 ± 0.01 [a]	99.57 ± 0.08 [ab]	0.39 ± 0.01 [bc]
GP2%	24.71 ± 0.57 [d]	1.25 ± 0.05 [ab]	99.60 ± 0.07 [ab]	0.39 ± 0.01 [b]
GP3%	26.64 ± 0.36 [c]	1.22 ± 0.01 [abc]	99.62 ± 0.08 [ab]	0.40 ± 0.00 [b]
GP4%	28.21 ± 0.08 [c]	1.18 ± 0.04 [abc]	99.66 ± 0.11 [ab]	0.41 ± 0.01 [ab]
GP5%	32.11 ± 0.51 [b]	1.18 ± 0.02 [bc]	99.73 ± 0.00 [ab]	0.42 ± 0.01 [ab]
GP6%	38.77 ± 1.19 [a]	1.14 ± 0.01 [c]	99.75 ± 0.91 [a]	0.43 ± 0.01 [a]

Each experiment was carried out in triplicate and data was reported as means ± standard deviation. Means in the same column with different letters are significantly different ($p < 0.05$).

Table 2. Effects of grape peel flour on pasta chemical compounds content.

Sample	Moisture (%)	Crude Ash (%)	Crude Fat (%)	Crude Protein (%)	Carbohydrates (%)
Control	11.54 ± 0.03 [a]	0.52 ± 0.04 [d]	0.96 ± 0.02 [c]	12.39 ± 0.18 [a]	72.11 ± 0.08 [e]
GP1%	11.53 ± 0.03 [a]	0.56 ± 0.04 [cd]	0.97 ± 0.02 [bc]	12.36 ± 0.18 [a]	74.56 ± 0.12 [a]
GP2%	11.54 ± 0.02 [a]	0.59 ± 0.02 [bcd]	0.98 ± 0.01 [abc]	12.32 ± 0.17 [a]	74.21 ± 0.08 [ab]
GP3%	11.40 ± 0.17 [a]	0.63 ± 0.04 [abcd]	1.00 ± 0.04 [abc]	12.29 ± 0.15 [a]	74.09 ± 0.20 [b]
GP4%	11.42 ± 0.18 [a]	0.66 ± 0.06 [abc]	1.01 ± 0.06 [abc]	12.26 ± 0.17 [a]	73.60 ± 0.26 [c]
GP5%	11.48 ± 0.11 [a]	0.70 ± 0.04 [ab]	1.02 ± 0.02 [ab]	12.22 ± 0.14 [a]	73.27 ± 0.21 [cd]
GP6%	11.44 ± 0.16 [a]	0.74 ± 0.04 [a]	1.04 ± 0.02 [a]	12.19 ± 0.16 [a]	73.10 ± 0.07 [d]

Each experiment was carried out in triplicate and data was reported as means ± standard deviation. Means in the same column with different letters are significantly different ($p < 0.05$).

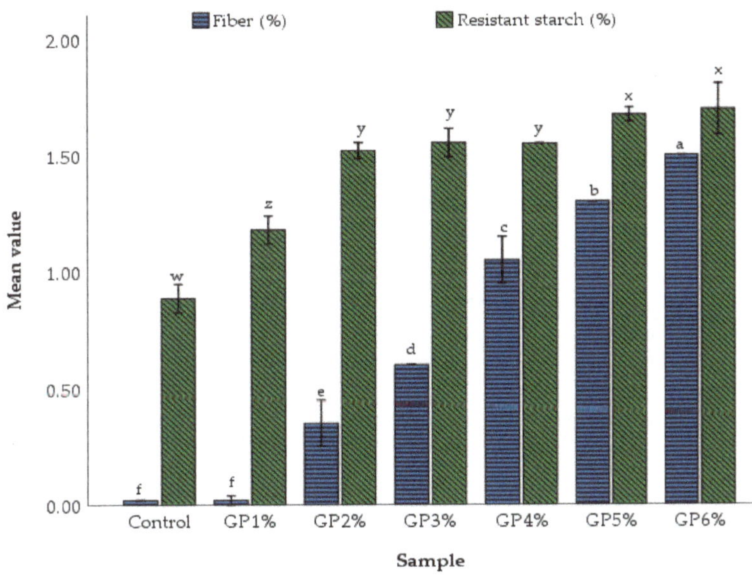

Figure 4. Effects of grape peel flour addition on fiber and resistant starch contents of pasta. Columns with different letters are significantly different ($p < 0.05$): a–f for fiber content and x–w for resistant starch variation.

Crude fat, protein and moisture of pasta did not exert significant differences ($p > 0.05$) among samples. The carbohydrates content decreased with grape peel flour level raise, the values being higher compared to the control, as it can be observed in Table 2.

The incorporation of grape peel flour in common wheat pasta caused significant increase of total polyphenols (Figure 5) and resistant starch content (Figure 4) compared to the control and with the raise of the amount added, the variations being from 9.50 to 14.12 mg GAE/100 g and from 0.89 to 1.70%, respectively.

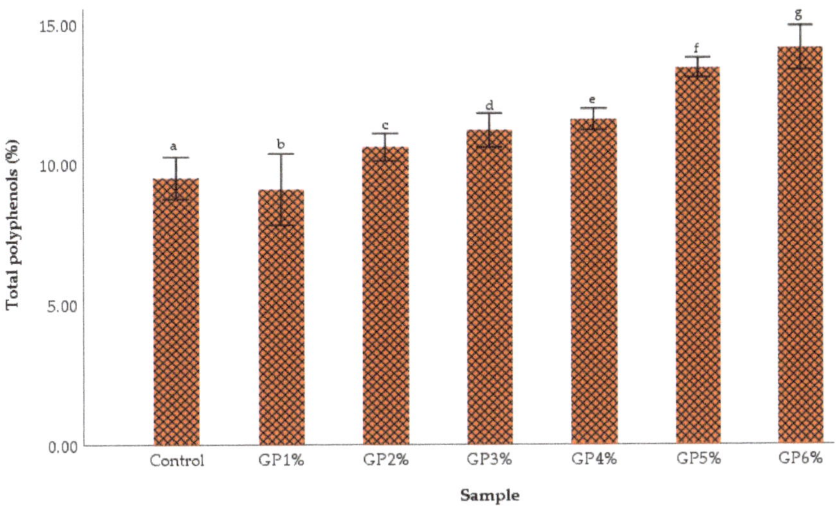

Figure 5. Effects of grape peel flour addition on pasta total polyphenols content. Columns with different letters are significantly different ($p < 0.05$).

3.4. Uncooked Pasta Color

Color parameters of the final product depend on the components of dough, being influenced at the same time by the drying methods applied. The luminosity (L^*) of pasta samples supplemented with grape peel flour was smaller compared to the control and decreased with the addition level increase (Table 3). The red nuance of the samples containing grape peel flour, given by the positive values of a^* parameter, showed significant ($p < 0.05$) intensification as the amount raised, while the control and GP1% exhibited tendence to green nuance (negative values of a^*).

Table 3. Effects of grape peel flour on pasta cooking behavior and color parameters.

Sample	Cooking Behavior		Color		
	Water Absorption (%)	Cooking Loss (%)	L^*	a^*	b^*
Control	161.14 ± 5.33 [a]	5.22 ± 0.12 [de]	72.54 ± 0.17 [a]	−2.31 ± 0.20 [e]	21.61 ± 0.24 [a]
GP1%	129.46 ± 3.55 [c]	4.55 ± 0.30 [e]	68.37 ± 0.09 [b]	−0.56 ± 0.02 [d]	21.45 ± 0.64 [a]
GP2%	139.38 ± 1.37 [bc]	5.43 ± 0.18 [cd]	65.97 ± 1.11 [c]	0.34 ± 0.06 [c]	20.94 ± 0.48 [a]
GP3%	145.76 ± 4.54 [abc]	5.72 ± 0.17 [cd]	62.88 ± 0.52 [d]	0.56 ± 0.02 [d]	20.57 + 0.49 [ab]
GP4%	147.02 + 12.62 [ab]	6.09 ± 0.31 [c]	58.32 ± 0.28 [e]	0.99 ± 0.07 [b]	19.62 ± 0.26 [bc]
GP5%	149.46 ± 1.08 [ab]	7.28 ± 0.36 [b]	55.20 ± 0.50 [f]	1.40 ± 0.04 [a]	18.73 ± 0.23 [cd]
GP6%	155.91 ± 5.04 [ab]	7.99 ± 0.14 [a]	52.57 ± 0.31 [g]	1.45 ± 0.26 [a]	18.48 ± 0.23 [d]

Each experiment was carried out in triplicate and data was reported as means ± standard deviation. Means in the same column with different letters are significantly different ($p < 0.05$).

The yellow nuance of pasta suggested by the positive values of b^* parameter presented a reduction trend as the level of grape peel flour was higher, the values ranging between 21.61 for the control sample and 18.48 for GP6% (Table 3).

3.5. Pasta Cooking Behavior

Pasta cooking behavior gives important quality characteristics which are given by the structure of dough and interactions between ingredients. Water absorption during boiling of pasta resulted in a proportional increase as the level of grape peel flour was higher, the obtained values being smaller compared to the control (Table 3). The amount of soluble solids lost during cooking (cooking loss) showed significant ($p < 0.05$) increases as the addition level of grape peel flour raised.

3.6. Uncooked and Cooked Pasta Texture

Uncooked pasta breaking force could give valuable information about its resistance to handling and transport. The supplementation of common wheat pasta with grape peel flour led to the increase in breaking force proportional to the amount added, the values being higher compared to the control, except for GP1% which did not show significant difference (Table 4).

Table 4. Effects of grape peel flour on uncooked and cooked pasta texture.

Sample	Uncooked Pasta		Cooked Pasta	
	Breaking Force (N)	Elasticity $\times$ 10^{-2} (adim.)	Chewiness (N)	Firmness (N)
Control	41.26 ± 1.24 [cd]	99.83 ± 0.07 [a]	48.79 ± 2.19 [a]	74.05 ± 0.91 [a]
GP1%	41.00 ± 1.01 [d]	99.87 ± 0.01 [a]	39.57 ± 0.09 [b]	62.67 ± 1.85 [b]
GP2%	42.73 ± 1.95 [cd]	99.87 ± 0.00 [a]	38.98 ± 1.65 [bc]	58.64 ± 0.26 [bc]
GP3%	44.32 ± 1.37 [c]	99.91 ± 0.08 [a]	36.29 ± 1.37 [bcd]	57.40 ± 2.89 [bc]
GP4%	49.02 ± 0.15 [b]	99.92 ± 0.07 [a]	35.51 ± 0.99 [cd]	56.43 ± 2.80 [cd]
GP5%	55.76 ± 0.54 [a]	99.94 ± 0.06 [a]	35.20 ± 1.17 [d]	54.04 ± 1.83 [cd]
GP6%	58.46 ± 0.17 [a]	99.96 ± 0.07 [a]	34.11 ± 1.01 [d]	51.67 ± 2.09 [d]

Each experiment was carried out in triplicate and data was reported as means ± standard deviation. Means in the same column with different letters are significantly different ($p < 0.05$).

Cooked pasta elasticity did not differ significantly ($p < 0.05$) among samples, a slight increase with the addition level of grape peel flour being noticed (Table 4). The chewiness property ranged between 48.79 N for the control and 34.11 N for GP6%, a decreasing trend being obtained as the amount was higher. Similar reduction tendency was observed in the case of firmness values which varied from 74.05 to 51.67 N.

3.7. Relations between Characteristics

Pearson's correlations coefficients (Table 5) were obtained to evaluate the relations between the studied characteristics. Significant very strong correlations ($p < 0.05$) were obtained between cooking loss and crude ash ($r = 0.92$), crude fat ($r = 0.92$), crude protein ($r = -0.92$), crude fiber ($r = 0.95$). Color parameters (L^*, a^* and b^*) were strongly correlated with the chemical composition of pasta ($p < 0.05$), except with moisture and carbohydrates, positive correlations being observed between L^* and crude protein ($r = 0.99$), a^* and crude ash ($r = 0.92$), crude fat ($r = 0.92$), crude fiber ($r = 0.86$), total polyphenols ($r = 0.82$) and resistant starch ($r = 0.99$), b^* and crude protein ($r = 0.98$), while negative correlations were obtained between L^* and crude ash ($r = -0.99$), crude fiber ($r = -0.99$), crude fat ($r = -0.99$), total polyphenols ($r = -0.95$) and resistant starch ($r = -0.89$), between a^* and crude protein ($r = -0.92$) and between b^* and crude ash ($r = -0.98$), crude fat ($r = -0.98$), crude fiber ($r = -0.99$), total polyphenols ($r = -0.98$) and resistant starch ($r = -0.82$) content. A strong positive correlation significant at $p < 0.05$ was observed between total polyphenols content and resistant starch ($r = 0.83$). The total polyphenols and starch content were significantly correlated ($p < 0.05$) with the proximate composition, except moisture and carbohydrates.

Dough textural parameters (hardness, resilience, springiness and cohesiveness), uncooked pasta breaking force and cooked pasta textural characteristics (elasticity, chewiness and firmness) were strongly correlated with pasta chemical composition in terms of crude fat, protein, ash and fiber and with cooking loss (Table 5). Furthermore, pasta total polyphenols showed significant correlation ($p < 0.05$) with dough, uncooked and cooked pasta texture, except with pasta resilience. The resistant starch content was correlated at $p < 0.05$ significance level with dough hardness ($r = 0.92$), springiness ($r = 0.83$) and cohesiveness ($r = 0.92$), with pasta elasticity ($r = 0.89$), chewiness ($r = -0.94$) and firmness ($r = -0.98$). Dough hardness was significantly correlated ($p < 0.05$) with all uncooked and cooked pasta textural parameters.

The similarities and dissimilarities between characteristics and samples were evaluated by means of Principal Component Analysis (PCA), the results being presented in Figure 6. The first principal component (PC1) explained 79.26% of data variance, the second principal component explained 16.12% of the variance, both summing 95.38% of the total variance. PC1 was associated with cooking, loss, pasta breaking force, total polyphenols, crude fiber, springiness, crude fat, crude ash, dough hardness, pasta elasticity, dough cohesiveness, resistant starch, color parameters, pasta firmness and chewiness, crude protein, while PC2 was associated with water absorption and carbohydrates content.

The moisture content of pasta exerts less influence on data variation, as it is suggested by its position closer to the origin of the graphic (Figure 6). Pasta samples which contained grape peel flour added were in opposition to the control which was associated with firmness and chewiness parameters. It can be observed that pasta with the highest grape peel flour levels (GP5%, GP6%) were associated with the fiber and total polyphenols contents, as expected.

Table 5. Correlations between variables.

Variables	WA	CL	L^*	a^*	b^*	Moisture	Crude Ash	Crude Fat	Crude Protein	Crude Fiber	Carbohydr.	TP	RS	Hardness	Resilience	Springiness	Cohesiveness	Breaking f.	Elasticity	Chewiness	Firmness
WA	1.00	0.53	−0.21	−0.11	−0.34	−0.23	0.24	0.24	−0.24	0.36	−0.94	0.42	−0.06	0.20	−0.70	0.30	0.02	0.41	0.17	0.25	0.16
CL		1.00	−0.91	0.72	−0.96	−0.50	0.92	0.92	−0.92	0.95	−0.36	0.99	0.73	0.91	−0.93	0.96	0.85	0.98	0.88	−0.62	−0.71
L^*			1.00	−0.92	0.98	0.66	−0.99	−0.99	0.99	−0.99	0.01	−0.95	−0.89	−0.97	0.83	−0.98	−0.98	−0.95	−0.98	0.87	0.90
a^*				1.00	−0.85	−0.63	0.92	0.92	−0.92	0.86	0.34	0.82	0.99	0.92	−0.59	0.86	0.95	0.77	0.92	−0.98	−0.99
b^*					1.00	0.58	−0.98	−0.98	0.98	−0.99	0.17	−0.98	−0.82	−0.94	0.89	−0.99	−0.93	−0.98	−0.95	0.77	0.82
Moisture						1.00	−0.67	−0.67	0.67	−0.65	−0.05	−0.57	−0.62	−0.62	0.64	−0.56	−0.58	−0.50	−0.74	0.66	0.60
Crude ash							1.00	0.99	−0.99	0.99	−0.02	0.97	0.90	0.98	−0.84	0.98	0.97	0.95	0.99	−0.86	−0.91
Crude fat								1.00	−0.99	0.99	−0.02	0.97	0.90	0.98	−0.84	0.98	0.97	0.95	0.99	−0.86	−0.91
Crude proteins									1.00	−0.99	0.02	−0.97	−0.90	−0.98	0.84	−0.98	−0.97	−0.95	−0.99	0.86	0.91
Crude fiber										1.00	−0.17	0.97	0.84	0.95	−0.91	0.98	0.93	0.97	0.96	−0.78	−0.83
Carbohydrates											1.00	−0.22	0.32	0.04	0.53	−0.13	0.17	−0.27	0.06	−0.48	−0.40
TP												1.00	0.83	0.95	−0.90	0.98	0.90	0.97	0.93	−0.73	−0.80
RS													1.00	0.92	−0.58	0.83	0.92	0.74	0.89	−0.94	−0.98
Hardness														1.00	−0.79	0.95	0.97	0.91	0.97	−0.87	−0.93
Resilience															1.00	−0.86	−0.71	−0.90	−0.81	0.49	0.55
Springiness																1.00	0.95	0.99	0.95	−0.78	−0.84
Cohesiveness																	1.00	0.90	0.97	−0.93	−0.96
Breaking force																		1.00	0.91	−0.68	−0.75
Elasticity																			1.00	−0.90	−0.91
Chewiness																				1.00	0.98
Firmness																					1.00

WA—water absorption, CL—cooking loss, TP—total polyphenols, RS—resistant starch. Values in bold are different from 0 with a significance level of 95%.

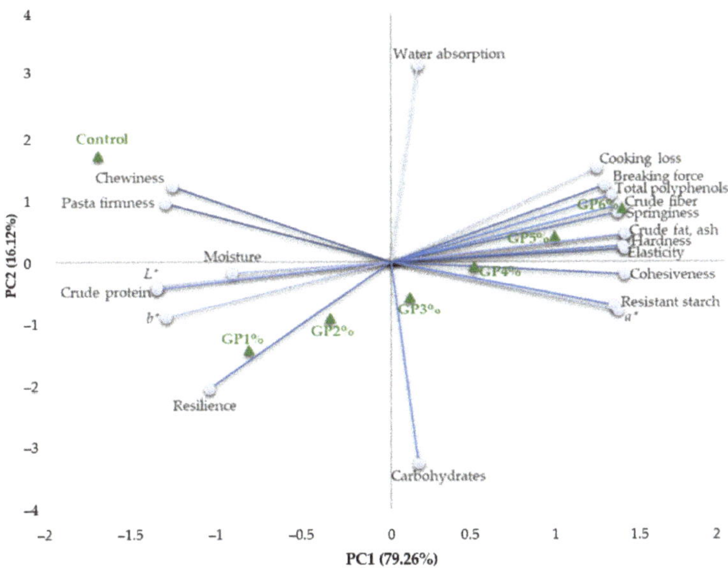

Figure 6. Principal Component Analysis (PCA) bi-plot.

4. Discussion

Wheat pasta dough and final product properties were proved to be significantly affected by grape peel flour inclusion. The dynamic rheological properties could provide information about the physical properties of dough, but also about the adhesiveness, kneading behavior, elasticity and hardness [23]. The elastic and viscous moduli of dough increased gradually as the addition level was higher, a possible cause being represented by the presence of hydroxyl groups from phenolic compounds from grape peels [33]. Protein–polyphenol interactions could change protein structure, resulting in different quality and functional characteristics of pasta. Some phenolic compounds could have been contributed to the formation of new complexes with gluten proteins, strengthening dough structure, as stated by Zhang et al. [34]. These authors demonstrated that the addition of tannic acid in wheat dough caused dough strengthening and elasticity increase, showing also the raise of sulfhydryl groups and the depletion of free amino groups in dough [34]. They explained these results by the formation of linkages between the tannic acid and gluten proteins, underlying that not only the disulfide bonds are responsible for dough strength, but also the complexes formed by polyphenols with flour proteins [34]. Thus, these results, along with those obtained by Meral and Dogan [33] for wheat dough with grape seeds, confirmed our findings. Chen et al. [35] revealed that the incorporation of 1% grape seeds flour in wheat noodles resulted in positive effects of final product quality and protein stability, obtaining at the same time an increase of free sulfhydryl groups which were proved to reduce dough strength. Thus, it can be concluded that the effects of grape peel flour on dough may not be assigned to the development of disulfide bonds, but possibly to the formation of non-covalent bonds such as hydrogen bonds, hydrophobic bonds and ionic bonds [35]. G′ and G″ increase was also supported by the raise of dough hardness proportional to the addition of grape peel flour level increase, significant correlation ($p < 0.05$) being obtained with the chemical composition of pasta, including proteins, fibers and polyphenols content. Dough resilience decreased as the amount of grape peel flour raised, while cohesiveness showed and opposite trend. Dough textural parameters are strongly influenced by ingredients particle size, as previously demonstrated [7,36,37]. The use of small particle size of grape peel flour (<180 μm) could have been contributed to these changes in dough texture since these particles have greater contact surface and consequently could absorb more water.

Furthermore, the chemical composition of the added ingredient had a major impact on dough texture which is given by the interactions between proteins, starch, lipids, fibers and sugars. Grape peel sugars could have a great contribution to dough cohesiveness increase since they could form bridges between starch and protein molecules [38]. Resilience was found to be directly influenced by protein quality, presence of soluble fiber and water availability [39]. Thus, the competition of gluten proteins with grape peel particles for water could have been a possible explanation of dough resilience decrease.

Pasta proximate composition was influenced by grape peel flour incorporation, significant changes in crude ash, crude fat and carbohydrates being observed. The intake of nutrients of grape peel flour determined the increase of final product ash and fat, while the protein content showed a slight decrease. Similar reduction in protein content and increase in ash was reported for durum wheat pasta enriched with apple flour which could be due to the dilution effect of wheat protein [40]. The carbohydrates content decrease was in line with the results reported by Bender et al. [41] which showed that the addition of grape skin lowered muffins carbohydrates values. One of the most important contribution of grape peel flour is to the raise of pasta fiber content, total polyphenols content and resistant starch. It is well known that grape peels are rich sources of soluble and insoluble dietary fibers and polyphenols such as flavonols, hydroxybenzoic, hydroxycinnamic acids, flavanols, tyrosol and condensed tannins [42,43]. The strong correlation ($r = 0.83$, $p < 0.05$) obtained between polyphenols and resistant starch content support the hypothesis of starch digestibility changes caused by polyphenols presence. Rocchetti et al. [44] demonstrated that polyphenols play an important role in the rate and the extent of starch digestion. Resistant starch formation could be promoted by the presence of polyphenols [45], the interactions between starch and polyphenols resulting in insoluble complexes development [46]. Furthermore, the phenolic compounds presence can raise resistant starch content due to the appearance of non-covalent phenolic/starch interactions and/or to the inhibition of digestive enzymes [44,45].

Total polyphenols and fiber content were correlated to pasta cooking loss increase ($r = 0.99$ and $r = 0.95$ respectively at $p < 0.05$). Similar raise was reported by Bustos et al. [47] for pasta supplemented with berry flour, being probably related to the gluten dilution effect and to the pectin presence which can contribute to the leaching of components in the boiling water. Grape peel flour gradually increased pasta water absorption which could be due to the affinity of the ingredient added for water [19]. Dietary fiber comprises cellulose, hemi-cellulose, pectin and others, pectin and galactomannan being proved to have higher water-holding capacity compared to cellulose [48]. Grape peels contain pectin [49] which can explain the increase in water absorption of pasta due to the formation of hydrogen linkages between water and hydroxyl groups of these polysaccharides. Pasta color was significantly influenced ($p < 0.05$) by grape peel flour incorporation, leading to lower luminosity and yellow nuance and higher red nuance as the level increased, similar findings being reported by Sant'Anna et al. [50] for pasta enriched with grape marc. These changes in color properties could be due to the presence of natural pigments in the ingredient added.

Uncooked pasta breaking force was significantly correlated ($p < 0.05$) to the chemical composition and to the texture of dough. The increasing trend of pasta breaking force could be related to the presence of polyphenols which may exert a strengthening effect, fact suggested also by the high correlation obtained ($r = 97$, $p < 0.05$). Similar tendence and correlation were reported by Šporin et al. [51] when grape pomace was incorporated in wheat dough. Cooked pasta chewiness and firmness decreased as the amount of grape peel flour was higher and compared to the control, while elasticity slightly increased. Similar raise for pasta elasticity was reported by Pasqualone et al. [52] for pasta supplemented with tomato flour. Chen et al. [35] reported a gradual decreasing trend of wheat gluten noodles when more than 1% grape seeds were added. Pasta firmness is determined by starch grains hydration during boiling and by the embedding of gelatinized starch grains in a network of partially denatured protein [53]. Polyphenols from grape peel can interact with starch, disturbing starch hydration and gelatinization and changing protein-starch matrix properties which led to hardness reduction [54].

5. Conclusions

Grape peel as a by-product resulted from wine processing are important sources of fibers and polyphenols that can give value added to the food products in which they can be incorporated. Pasta is a suitable matrix in which such ingredients could be incorporated to potentiate the nutritional and functional value. Grape peel flour addition in common wheat pasta resulted in higher visco-elastic moduli proportional to the level increase. Dough texture was significantly influenced by grape peel flour, leading to hardness, springiness and cohesiveness increase. The chemical composition of pasta showed significant increases of crude ash, crude protein and crude fat, while the protein content decreased as the addition level of grape peel flour raised. The purpose of this study of increasing the functionality of wheat pasta was achieved and confirmed by the significant higher fiber, total polyphenols and resistant starch proportional to the level raise. The gluten dilution effect was observed by the higher amount of soluble solids leached from pasta in the boiling water, with water absorption registering an increasing trend. Grape peel flour exhibited positive effects on pasta breaking force and elasticity by growing their value proportional to the level added, while cooked pasta chewiness and firmness decreased. Pasta luminosity and the yellow nuance were reduced as the amount of grape peel flour was higher, and the red nuance intensified due to the intake of pigments. The variations of textural and color parameters of dough and pasta were related to the chemical composition of the samples, significant correlations ($p < 0.05$) being observed. Thus, an addition level of about 5% grape peel flour would give significant pasta functional value improvement, while keeping acceptable quality parameters. The obtained results can contribute to the knowledge of grape peel's effects on dough and pasta properties, making up a source of inspiration for processors interested in functional products development.

Author Contributions: Conceptualization, M.U.-I. and S.M.; methodology, M.U.-I. and S.M.; software, M.U.-I. and S.M.; validation, M.U.-I. and S.M.; formal analysis, M.U.-I. and S.M.; investigation, M.U.-I. and S.M.; resources, M.U.-I. and S.M.; data curation, M.U.-I. and S.M.; writing—original draft preparation, M.U.-I. and S.M.; writing—review and editing, M.U.-I. and S.M.; visualization, M.U.-I. and S.M.; supervision, M.U.-I. and S.M.; project administration, M.U.-I. and S.M.; funding acquisition, S.M. All authors have read and agreed to the published version of the manuscript.

Funding: This work was supported by Romania National Council for Higher Education Funding, CNFIS, project number CNFIS-FDI-2021-0357.

Institutional Review Board Statement: Not applicable.

Informed Consent Statement: Not applicable.

Data Availability Statement: Not applicable.

Conflicts of Interest: The authors declare no conflict of interest.

References

1. OIV. State of the world vitivinicultural sector in 2019. *Int. Organ. Vine Wine* **2020**, 1–15.
2. Banu, C. *Food Industry Engineer's Manual*; Technical Publisher: Bucharest, Romania, 1999.
3. Dwyer, K.; Hosseinian, F.; Rod, M. The Market Potential of Grape Waste Alternatives Kyle. *J. Food Res.* **2014**, *3*, 91–106. [CrossRef]
4. Nakov, G.; Brandolini, A.; Hidalgo, A.; Ivanova, N.; Stamatovska, V.; Dimov, I. Effect of grape pomace powder addition on chemical, nutritional and technological properties of cakes. *LWT* **2020**, *134*, 109950. [CrossRef]
5. Iuga, M.; Mironeasa, S. Use of Grape Peels By-Product for Wheat Pasta Manufacturing. *Plants* **2021**, *10*, 926. [CrossRef] [PubMed]
6. Dewettinck, K.; Van Bockstaele, F.; Kühne, B.; Van de Walle, D.; Courtens, T.M.; Gellynck, X. Nutritional value of bread: Influence of processing, food interaction and consumer perception. *J. Cereal Sci.* **2008**, *48*, 243–257. [CrossRef]
7. Mironeasa, S.; Iuga, M.; Zaharia, D.; Mironeasa, C. Rheological Analysis of Wheat Flour Dough as Influenced by Grape Peels of Different Particle Sizes and Addition Levels. *Food Bioprocess. Technol.* **2019**, *12*, 228–245. [CrossRef]
8. Zhu, F.; Du, B.; Zheng, L.; Li, J. Advance on the bioactivity and potential applications of dietary fibre from grape pomace. *Food Chem.* **2015**, *186*, 207–212. [CrossRef]

9. Mildner-Szkudlarz, S.; Zawirska-Wojtasiak, R.; Szwengiel, A.; Pacyński, M. Use of grape by-product as a source of dietary fibre and phenolic compounds in sourdough mixed rye bread. *Int. J. Food Sci. Technol.* **2011**, *46*, 1485–1493. [CrossRef]
10. Yu, J.; Ahmedna, M. Functional components of grape pomace: Their composition, biological properties and potential applications. *Int. J. Food Sci. Technol.* **2013**, *48*, 221–237. [CrossRef]
11. Tang, G.Y.; Zhao, C.N.; Liu, Q.; Feng, X.L.; Xu, X.Y.; Cao, S.Y.; Meng, X.; Li, S.; Gan, R.Y.; Li, H. Bin Potential of grape wastes as a natural source of bioactive compounds. *Molecules* **2018**, *23*, 2598. [CrossRef]
12. Hosu, A.; Cristea, V.M.; Cimpoiu, C. Analysis of total phenolic, flavonoids, anthocyanins and tannins content in Romanian red wines: Prediction of antioxidant activities and classification of wines using artificial neural networks. *Food Chem.* **2014**, *150*, 113–118. [CrossRef] [PubMed]
13. Cádiz-Gurrea, M.D.L.L.; Borrás-Linares, I.; Lozano-Sánchez, J.; Joven, J.; Fernández-Arroyo, S.; Segura-Carretero, A. Cocoa and grape seed byproducts as a source of antioxidant and anti-inflammatory proanthocyanidins. *Int. J. Mol. Sci.* **2017**, *18*, 376. [CrossRef] [PubMed]
14. Ferri, M.; Rondini, G.; Calabretta, M.M.; Michelini, E.; Vallini, V.; Fava, F.; Roda, A.; Minnucci, G.; Tassoni, A. White grape pomace extracts, obtained by a sequential enzymatic plus ethanol-based extraction, exert antioxidant, anti-tyrosinase and anti-inflammatory activities. *New Biotechnol.* **2017**, *39*, 51–58. [CrossRef]
15. Khatkar, B.S.; David Schofield, J. Dynamic rheology of wheat flour dough. I. Non-linear viscoelastic behaviour. *J. Sci. Food Agric.* **2002**, *82*, 827–829. [CrossRef]
16. Taşeri, L.; Aktaş, M.; Şevik, S.; Gülcü, M.; Uysal Seçkin, G.; Aktekeli, B. Determination of drying kinetics and quality parameters of grape pomace dried with a heat pump dryer. *Food Chem.* **2018**, *260*, 152–159. [CrossRef]
17. Monteiro, G.C.; Minatel, I.O.; Junior, A.P.; Gomez-Gomez, H.A.; de Camargo, J.P.C.; Diamante, M.S.; Pereira Basílio, L.S.; Tecchio, M.A.; Pereira Lima, G.P. Bioactive compounds and antioxidant capacity of grape pomace flours. *LWT* **2021**, *135*, 110053. [CrossRef]
18. Peixoto, C.M.; Dias, M.I.; Alves, M.J.; Calhelha, R.C.; Barros, L.; Pinho, S.P.; Ferreira, I.C.F.R. Grape pomace as a source of phenolic compounds and diverse bioactive properties. *Food Chem.* **2018**, *253*, 132–138. [CrossRef]
19. Iuga, M.; Mironeasa, S. Potential of grape byproducts as functional ingredients in baked goods and pasta. *Compr. Rev. Food Sci. Food Saf.* **2020**, *19*, 2473–2505. [CrossRef] [PubMed]
20. Marinelli, V.; Padalino, L.; Conte, A.; Del Nobile, M.A.; Briviba, K. Red grape marc flour as food ingredient in durum wheat spaghetti: Nutritional evaluation and bioaccessibility of bioactive compounds. *Food Sci. Technol. Res.* **2018**, *24*, 1093–1100. [CrossRef]
21. Lou, W.; Zhou, H.; Li, B.; Nataliya, G. Rheological, pasting and sensory properties of biscuits supplemented with grape pomace powder. *Food Sci. Technol.* **2021**, *2061*, 1–10. [CrossRef]
22. Rainero, G.; Bianchi, F.; Rizzi, C.; Cervini, M.; Giuberti, G.; Simonato, B. Breadsticks fortification with red grape pomace: Effect on nutritional, technological, and sensory properties. *J. Sci. Food Agric.* **2021**. [CrossRef]
23. Lou, W.; Li, B.; Nataliya, G. The influence of Cabernet Sauvignon wine grape pomace powder addition on the rheological and microstructural properties of wheat dough. *CyTA J. Food* **2021**, *19*, 751–761. [CrossRef]
24. Yang, S.; Jeong, S.; Lee, S. Elucidation of rheological properties and baking performance of frozen doughs under different thawing conditions. *J. Food Eng.* **2020**, *284*, 110084. [CrossRef]
25. Balli, D.; Cecchi, L.; Innocenti, M.; Bellumori, M.; Mulinacci, N. Food by-products valorisation: Grape pomace and olive pomace (pâté) as sources of phenolic compounds and fiber for enrichment of tagliatelle pasta. *Food Chem.* **2021**, *355*, 129642. [CrossRef] [PubMed]
26. Ungureanu-Iuga, M.; Dimian, M.; Mironeasa, S. Development and quality evaluation of gluten-free pasta with grape peels and whey powders. *LWT* **2020**, *130*, 109714. [CrossRef]
27. Iuga, M.; Mironeasa, S. Simultaneous optimization of wheat heat moisture treatment and grape peels addition for pasta making. *LWT Food Sci. Technol.* **2021**, *150*, 112011. [CrossRef]
28. Walker, R.; Tseng, A.; Cavender, G.; Ross, A.; Zhao, Y. Physicochemical, Nutritional, and Sensory Qualities of Wine Grape Pomace Fortified Baked Goods. *J. Food Sci.* **2014**, *79*, S1811–S1822. [CrossRef] [PubMed]
29. Bergman, C.; Gualberto, D.; Weber, C. Development of a high-temperature-dried soft wheat pasta supplemented with cowpea (*Vigna unguiculata* (L.) Walp)—Cooking quality, color, and sensory evaluation. *Cereal Chem.* **1994**, *71*, 523–527.
30. FAO/IAEA. *Quantification of Tannins in Tree Foliage. A Laboratory Manual for the FAO/IAEA Co-Ordinated Research Project on Use of Nuclear and Related Techniques to Develop Simple Tannin Assays for Predicting and Improving the Safety and Efficiency of Feeding Rumina*; IAEA: Vienna, Austria, 2000.
31. Melilli, M.G.; Pagliaro, A.; Scandurra, S.; Gentile, C.; Di Stefano, V. Omega-3 rich foods: Durum wheat spaghetti fortified with Portulaca oleracea. *Food Biosci.* **2020**, *37*, 100730. [CrossRef]
32. Giménez, M.A.; González, R.J.; Wagner, J.; Torres, R.; Lobo, M.O.; Samman, N.C. Effect of extrusion conditions on physicochemical and sensorial properties of corn-broad beans (*Vicia faba*) spaghetti type pasta. *Food Chem.* **2013**, *136*, 538–545. [CrossRef]
33. Meral, R.; Dogan, I.S. Grape seed as a functional food ingredient in bread-making. *Int. J. Food Sci. Nutr.* **2013**, *64*, 372–379. [CrossRef]
34. Zhang, L.; Cheng, L.; Jiang, L.; Wang, Y.; Yang, G.; He, G. Effects of tannic acid on gluten protein structure, dough properties and bread quality of Chinese wheat. *J. Sci. Food Agric.* **2010**, *90*, 2462–2468. [CrossRef]

35. Chen, S.X.; Ni, Z.J.; Thakur, K.; Wang, S.; Zhang, J.G.; Shang, Y.F.; Wei, Z.J. Effect of grape seed power on the structural and physicochemical properties of wheat gluten in noodle preparation system. *Food Chem.* **2021**, *355*, 129500. [CrossRef] [PubMed]
36. Kurek, M.; Wyrwisz, J.; Piwińska, M.; Wierzbicka, A. The effect of oat fibre powder particle size on the physical properties of wheat bread rolls. *Food Technol. Biotechnol.* **2016**, *54*, 45–51. [CrossRef] [PubMed]
37. Tsatsaragkou, K.; Kara, T.; Ritzoulis, C.; Mandala, I.; Rosell, C.M. Improving Carob Flour Performance for Making Gluten-Free Breads by Particle Size Fractionation and Jet Milling. *Food Bioprocess. Technol.* **2017**, *10*, 831–841. [CrossRef]
38. Chevallier, S.; Colonna, P.; Buléon, A.; Della Valle, G. Physicochemical behaviors of sugars, lipids, and gluten in short dough and biscuit. *J. Agric. Food Chem.* **2000**, *48*, 1322–1326. [CrossRef] [PubMed]
39. Mudgil, D.; Barak, S.; Khatkar, B.S. Optimization of textural properties of noodles with soluble fiber, dough mixing time and different water levels. *J. Cereal Sci.* **2016**, *69*, 104–110. [CrossRef]
40. Espinosa-Solis, V.; Zamudio-Flores, P.B.; Tirado-Gallegos, J.M.; Ramírez-Mancinas, S.; Olivas-Orozco, G.I.; Espino-Díaz, M.; Hernández-González, M.; García-Cano, V.G.; Sánchez-Ortíz, O.; Buenrostro-Figueroa, J.J.; et al. Evaluation of cooking quality, nutritional and texture characteristics of pasta added with oat bran and apple flour. *Foods* **2019**, *8*, 299. [CrossRef]
41. Bender, A.B.B.; Speroni, C.S.; Salvador, P.R.; Loureiro, B.B.; Lovatto, N.M.; Goulart, F.R.; Lovatto, M.T.; Miranda, M.Z.; Silva, L.P.; Penna, N.G. Grape Pomace Skins and the Effects of Its Inclusion in the Technological Properties of Muffins. *J. Culin. Sci. Technol.* **2017**, *15*, 143–157. [CrossRef]
42. Gomes, T.M.; Toaldo, I.M.; da Silva Haas, I.C.; Burin, V.M.; Caliari, V.; Luna, A.S.; de Gois, J.S.; Bordignon-Luiz, M.T. Differential contribution of grape peel, pulp, and seed to bioaccessibility of micronutrients and major polyphenolic compounds of red and white grapes through simulated human digestion. *J. Funct. Foods* **2019**, *52*, 699–708. [CrossRef]
43. Deng, Q.; Penner, M.H.; Zhao, Y. Chemical composition of dietary fiber and polyphenols of five different varieties of wine grape pomace skins. *Food Res. Int.* **2011**, *44*, 2712–2720. [CrossRef]
44. Rocchetti, G.; Giuberti, G.; Busconi, M.; Marocco, A.; Trevisan, M.; Lucini, L. Pigmented sorghum polyphenols as potential inhibitors of starch digestibility: An in vitro study combining starch digestion and untargeted metabolomics. *Food Chem.* **2020**, *312*, 126077. [CrossRef] [PubMed]
45. Barros, F.; Awika, J.M.; Rooney, L.W. Interaction of tannins and other sorghum phenolic compounds with starch and effects on in vitro starch digestibility. *J. Agric. Food Chem.* **2012**, *60*, 11609–11617. [CrossRef] [PubMed]
46. Moraes, É.A.; Marineli, R.D.S.; Lenquiste, S.A.; Steel, C.J.; De Menezes, C.B.; Queiroz, V.A.V.; Maróstica Júnior, M.R. Sorghum flour fractions: Correlations among polysaccharides, phenolic compounds, antioxidant activity and glycemic index. *Food Chem.* **2015**, *180*, 116–123. [CrossRef] [PubMed]
47. Bustos, M.C.; Paesani, C.; Quiroga, F.; León, A.E. Technological and sensorial quality of berry-enriched pasta. *Cereal Chem.* **2019**, *96*, 967–976. [CrossRef]
48. Ajila, C.M.; Leelavathi, K.; Prasada Rao, U.J.S. Improvement of dietary fiber content and antioxidant properties in soft dough biscuits with the incorporation of mango peel powder. *J. Cereal Sci.* **2008**, *48*, 319–326. [CrossRef]
49. Spinei, M.; Oroian, M. The potential of grape pomace varieties as a dietary source of pectic substances. *Foods* **2021**, *10*, 867. [CrossRef]
50. Sant'Anna, V.; Christiano, F.D.P.; Marczak, L.D.F.; Tessaro, I.C.; Thys, R.C.S. The effect of the incorporation of grape marc powder in fettuccini pasta properties. *LWT Food Sci. Technol.* **2014**, *58*, 497–501. [CrossRef]
51. Šporin, M.; Avbelj, M.; Kovač, B.; Možina, S.S. Quality characteristics of wheat flour dough and bread containing grape pomace flour. *Food Sci. Technol. Int.* **2018**, *24*, 251–263. [CrossRef] [PubMed]
52. Pasqualone, A.; Gambacorta, G.; Summo, C.; Caponio, F.; Di Miceli, G.; Flagella, Z.; Marrese, P.P.; Piro, G.; Perrotta, C.; De Bellis, L.; et al. Functional, textural and sensory properties of dry pasta supplemented with lyophilized tomato matrix or with durum wheat bran extracts produced by supercritical carbon dioxide or ultrasound. *Food Chem.* **2016**, *213*, 545–553. [CrossRef]
53. Wójtowicz, A.; Mościcki, L. Effect of wheat bran addition and screw speed on microstructure and textural characteristics of common wheat precooked pasta-like products. *Polish J. Food Nutr. Sci.* **2011**, *61*, 101–107. [CrossRef]
54. Zhu, Y.; Feng, B.; He, S.; Su, Z.; Zheng, G. Resveratrol combined with total flavones of hawthorn alleviate the endothelial cells injury after coronary bypass graft surgery. *Phytomedicine* **2018**, *40*, 20–26. [CrossRef] [PubMed]

applied sciences

Article

The Quantification of Fatty Acids, Color, and Textural Properties of Locally Produced Bakery Margarine

Sergiu Pădureț

Faculty of Food Engineering, Stefan cel Mare University of Suceava, 13 University Str., 720229 Suceava, Romania; sergiu.paduret@fia.usv.ro

Abstract: This study focused on the analysis of bakery margarine samples divided into three groups according to physical-chemical analyses of their fat and water content. A Gas Chromatograph-Mass Spectrometer (GC-MS) was used for the evaluation of fatty acids, and from 37 fatty acids studied, only 18 were quantified. The highest concentration was occupied by the long-chain saturated fatty acids category (C14:0–C20:0), ranging between 85.61 µg/mg and 127.30 µg/mg. The dominant fatty acid was palmitic acid for all margarine samples. The texture parameters (hardness, mechanical work of plastic deformation, and fracturability) analyzed in this study with three different penetrometers and a puncture test showed that bakery margarine is a hard plastic material with a pronounced fracturability. The margarine's fracturability varied from 0.35 N to 8.23 N. The highest values were measured using the 10 mm diameter spherical penetrometer. Of the outside and inside evaluated color parameters, only the b* color parameter indicated an influence on the principal component analysis samples' projections; its values are also positively correlated with polyunsaturated fatty acids (PN).

Keywords: bakery margarine; fatty acids; fracturability; color; PCA

Citation: Pădureț, S. The Quantification of Fatty Acids, Color, and Textural Properties of Locally Produced Bakery Margarine. *Appl. Sci.* **2022**, *12*, 1731. https://doi.org/10.3390/app12031731

Academic Editor: Francisco Artés-Hernández

Received: 7 December 2021
Accepted: 3 February 2022
Published: 8 February 2022

Publisher's Note: MDPI stays neutral with regard to jurisdictional claims in published maps and institutional affiliations.

1. Introduction

Food emulsion represents a mixture of two non-miscible phases, with one of the phases being dispersed as small spherical drops in the other one [1].

In food technologies, emulsions take place partially or totally in the structures of various natural (milk, cream, butter) and processed (ice cream, mayonnaise, margarine, cake batters) food products. Certain foods are already emulsified in some phases of production; emulsion science has an important role in quality assurance and improving food quality or production techniques benefiting from emulsion principles [2].

Margarine was produced for the first time in 1869 by Hippolyte Mège Mouriès as a spreadable butter substitute and is a much cheaper water-in-oil (W/O) emulsion than butter. In addition to water, the aqueous phase also contains preservatives and salt, while the fatty phase represents a mixture of liquid oil and crystalline fat that confers texture and consistency to the margarine [3,4]. The solid or semisolid structure of margarine is attained by the fat crystal aggregate matrix, in which small drops of water are entrapped. The most commonly used emulsifiers are represented by lecithin, mono-and diglycerides of fatty acids, diacylglycerol, and distilled monoacylglycerol, to which are added antioxidants, dyes, vitamins, and flavorings [5].

According to the Codex Alimentarius [6], margarine is defined as a food in the form of a plastic or fluid emulsion that is mainly composed of a minimum of 80% m/m fat phase and a maximum of 16% m/m aqueous phase. In addition to this type of margarine, there are also variants with a lower fat content (less than 80% fat), whose role is to reduce caloric intake and reduce heart disease risk.

Since its invention, margarine has been modified and improved, resulting in the development of the range of products now available on the market that can be used for spreading, baking, or cooking. The physical characteristics of margarine are influenced largely by the physical properties of oil triacylglycerols, the total solid fat content, the fat's polymorphic state, and also the distribution of these solid fats [7]. The properties of triacylglycerols that are of great importance for margarine include the number of carbons in the fatty acid chain, the degree of unsaturation, and the main fatty acid [3].

Generally, to obtain margarine, vegetable oil is subject to technological operations to become semisolid or solid, depending on the desired final functionalities. Hydrogenation, interesterification (enzymatic and chemical), fractionation, and blending are the main technological operations used by the margarine industry to harden oils to desirable degrees [8].

Initially, the non-selective hydrogenation process led to products with a high concentration of saturated fatty acids (the most abundant fatty acid was stearic acid) and disagreeable textural characteristics, but the selective hydrogenation technique reduced the fatty acids' saturation. Thus, the most abundant fatty acid was oleic acid, which contributes to the softness and spreadability of margarine. Selective hydrogenation has the disadvantage of producing trans-fatty acids, which are known to have a negative effect on human health [9,10].

Recently, the European Union regulated the content of trans fats (other than trans-fat naturally occurring in the fat of animal sources) in food intended for the final consumer and food intended for supply to retail, at a limit of a maximum of 2% [11].

Depending on fat hardness and the melting point, margarine can be classified into hard and medium plastic margarine used for cooking or baking (bakery margarine) and medium plastic and soft margarine, known as table margarine (spreadable margarine) [3].

Both the textural and rheological properties of emulsions and margarine, respectively, are very important for consumers and also for industry. The rheological properties are important practical properties during processing, whereas in the final product, the textural properties represent the significant factor of mouthfeel. Good margarine should not present with oil separation, discoloration, hardening, sandiness, graininess, or water separation [12].

Taking into account the above, when it comes to margarine, its quality is influenced by the type, quantity, and mixture of oils and also by the technology applied. Therefore, the purpose of this study was to investigate the chemical composition and the fatty acid concentration of locally produced margarine used by bakery manufacturers in close correlation with textural properties like hardness, fracturability, and plasticity measured with different penetrometers; additionally, color parameters were evaluated.

2. Materials and Methods

The materials used in this study were represented by locally (Romania) produced margarine (n = 9) used by bakery manufacturers. The margarine was from the same producer and, based on the fat/water content, was classified into 3 categories (M1, M2, M3). The margarine samples were stored in refrigerated conditions at 8–10 °C until further analysis.

2.1. Physicochemical and Color Analysis

The fat content of margarine samples was conducted according to Toma et al. [13] using the Soxhlet method and petroleum ether as extraction solvent. The results were expressed as percentages by mass. The moisture content was evaluated by heating 5 g of margarine sample at 103 °C $\pm$ 2 °C until a constant weight was achieved [14,15].

Both outside and inside color parameters of margarine samples were evaluated using a ChromaMeter CR-400 from Konica Minolta (Konica Minolta, Japan) based on the CIE L*a*b* (Commission Internationale de l'Eclairage) uniform color space method and using the C illuminant [16]. The measured and calculated parameters were: L*—brightness, a*—red-green parameters, b*—yellow–blue parameters, h^0—hue angles, WI—whiteness indexes, C*—color intensities, YI—yellowness indexes, and ΔE*—color differences [17].

2.2. Fatty Acid Composition Analysis

The fat was extracted from the margarine samples at 50 °C, and the resulting fat phase was filtered through filter paper with anhydrous Na_2SO_4 [18]. The fatty acid methyl ester (FAME) preparation assumed the solubilization of margarine's fat in n-hexane. Then, 0.2 mL methanolic potassium hydroxide (2 mol/L) was added as a transesterification agent, following the method described by Jirarattanarangsri [19]. The standard used contained a FAME mix of 37 components (FAME Mix, Restek, Bellefonte, PA, USA), and the identification of margarine FAME was made by comparing their retention time with those of the FAME mix standard. Additionally, the resulting mass spectra were confronted with the ones from the GC-MS database (NIST MSSearch 2.0). The GC-MS (GC MS QP 2010 Plus, Shimadzu, Kyoto, Japan) following the method described by Oroian et al. [20], using a SUPELCOWAX10 capillary column (60 m length, 0.25 mm in diameter, with 0.25 μm film thickness, Supelco Inc., Bellefonte, PA, USA) and helium gas with a flow rate of 0.8 mL/min and a split ratio of 1/24; the injection volume was set at 0.001 mL.

2.3. Textural Evaluation

The texture parameters of margarine samples were evaluated with a Mark 10-ESM 301 (Mark 10 Corporation, Copiague, NY, USA) texturometer using three penetrometers of different geometries (120° conical penetrometer—4 mm penetration, 5 mm diameter spherical penetrometer—2 mm penetration, and 10 mm diameter spherical penetrometer—4 mm penetration). The MESUREgauge software was used for data collection at a reading rate of 5 points per second, and the curves were represented as force (expressed in Newtons) versus travel (expressed in millimeters). The cubic (side of 30 mm) margarine samples were tested at a speed of 10 mm/min. The measured texture parameters were: hardness, plasticity, and fracturability. Hardness (H) was calculated as the maximum force recorded at the end of the penetration. Plasticity (P) was calculated as the positive area/work under the penetration curve and was expressed as (N·mm or mJ). The margarine fracturability (F) was quantified as the height of the peaks of the force-penetration curves and expressed in Newtons [21,22].

All reagents used for physicochemical and fatty acid composition analysis were of analytical grade (Sigma Aldrich, Darmstadt, Germany).

2.4. Statistical Analysis

The results were expressed as the mean of three measurements, and the difference between the analyzed margarine samples was studied by one-way ANOVA; variance analysis was executed with STATGRAPHICS CENTURION XVI software (Trial Version). The OriginPro software was used to perform principal component analysis, while the Pearson correlation matrix was obtained by SPSS 13.0 software (Chicago, IL, USA).

3. Results and Discussion

Margarine is commonly consumed in various diets because it is cheaper than butter, contains no cholesterol, and many consumers consider that margarine has a lower atherogenic index than butter [23]. In Romania, according to the Population Consumption Availability report from recent years (2017–2019), the consumption per capita of margarine varies between 3.4–4.1 kg per year. This is a much larger amount than butter, which has an annual average consumption per capita of 0.7–0.8 kg [24,25].

3.1. Physicochemical and Color Analysis

Figure 1 shows the average fat and moisture results of margarine samples, and it can be noticed that fat content ranged between 58.96% and 76.16%, while moisture content ranged between 22.93% and 39.66%. The M1 margarine samples presented a lower amount of fat, while the moisture content had high values. In contrast, the M3 samples presented a high fat content and a low moisture content. The M2 margarine samples had an average fat content of 69.63% and a moisture content of 29.50%. The determined protein content was insignificant.

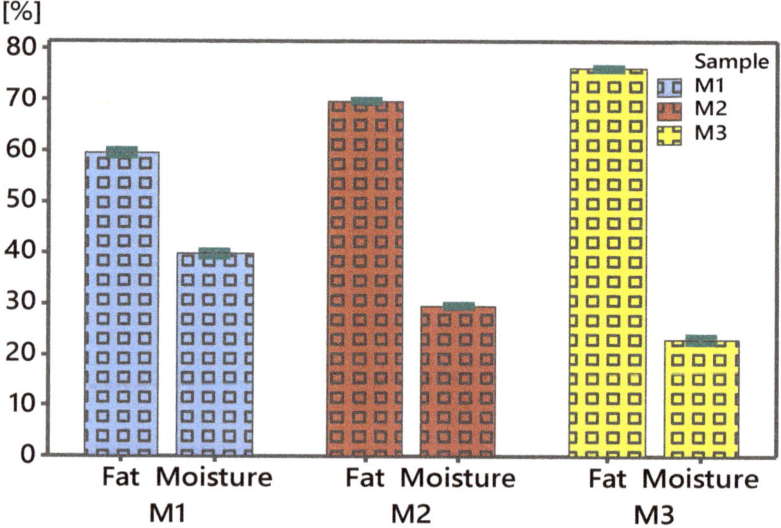

Figure 1. Chemical composition of margarine samples.

For consumers, the color of food represents a major quality attribute, being related to food freshness, nutritional value, and sensory properties. Thus, it is critically important to consumers. The instrumental or physical color evaluation represents a rapid and objective non-destructive method and uses analytical tools that offer rigorous data for quality control [26].

The measured and calculated color parameters of analyzed margarine samples are shown in Table 1. The color properties such as L* (brightness or luminosity), a* (+red/−green), and b* (+yellow/−blue) were the measured properties, while the h^0 (hue angle or tone), WI (whiteness), C* (color intensities), YI (yellowness), and ΔE* (color differences) were the calculated color parameters for both inside and outside color evaluation. The samples' brightness ranged from 95.67 to 91.54, with the highest value being recorded for the M1 inside evaluation. The ANOVA statistical analysis divided the luminosity results into five groups. For M2 samples, there was no difference between inside and outside measurements, with the results belonging to the same statistical group (c). The values of the a* color parameter were in the negative region for all analyzed margarine samples, revealing that this color parameter is more green. Between the outside measurements of the M1 and M2 samples, the ANOVA analysis showed no differences. In the case of margarine samples, the positive b* color parameter represents the degree of yellow, and as can be seen, the b* results varied from 35.74 to 22.22; the lowest value was measured for outside M1 samples, while the highest value was recorded for outside M3 samples. A higher value of the b* color parameter means more yellowness in the margarine sample. Additionally, the outside results presented higher values than the inside ones; the ANOVA highlighted these differences at a level of $p < 0.001$, dividing the measurements into six statistical groups. Furthermore, the measured color parameters were in the same range as the results discussed by Smetana et al. [12].

Table 1. The CIE L*a*b* color parameters of margarine samples.

Sample	Outside Color Parameters—Mean (SD)						
	L*	a*	b*	C	h^0	YI	WI
M1	95.41 [b] (0.16)	−6.46 [e] (0.07)	22.22 [d] (0.16)	23.14 [d] (0.17)	106.21 [b] (0.07)	33.27 [d] (0.30)	76.40 [b] (0.20)
M2	94.99 [c] (0.16)	−6.51 [e] (0.05)	23.94 [c] (0.53)	24.81 [c] (0.52)	105.21 [d] (0.19)	36.01 [c] (0.85)	74.68 [c] (0.54)
M3	91.81 [d] (0.02)	−5.82 [b] (0.07)	35.74 [a] (0.55)	36.21 [a] (0.55)	99.25 [f] (0.02)	55.61 [a] (0.87)	62.87 [e] (0.54)
	Inside Color Parameters—Mean (SD)						
M1	95.67 [a] (0.07)	−6.39 [d] (0.01)	21.68 [e] (0.05)	22.60 [e] (0.05)	106.42 [a] (0.04)	32.37 [e] (0.11)	76.98 [a] (0.06)
M2	95.14 [c] (0.06)	−6.21 [c] (0.01)	22.50 [d] (0.16)	23.34 [d] (0.15)	105.43 [c] (0.09)	33.79 [d] (0.26)	76.15 [b] (0.16)
M3	91.54 [e] (0.25)	−5.51 [a] (0.05)	32.33 [b] (0.22)	32.79 [b] (0.21)	99.67 [c] (0.16)	50.45 [b] (0.49)	66.13 [d] (0.27)
F	1002	312.78	1905	1773	4937	1913	1785
p	$p < 0.001$	$p < 0.001$	$p < 0.001$	$p < 0.001$	$p < 0.001$	$p < 0.001$	$p < 0.001$

SD—standard deviation. Different lowercase letters (a–f) between columns (L*, a*, b*, C, h^0, YI, WI) indicate a significant difference of mean values ($p < 0.05$) tested with one-way ANOVA.

The chroma values (C*) represent the color saturation, and in this study, the chroma results ranged between 36.21 and 22.60. Higher values were measured for outside color evaluations. The whiteness (WI) and yellowness (YI) indexes of margarine, butter, and spreadable fats can be considered a critical color parameter. The whiteness indexes indicate the degree of sample whiteness and mathematically combine three color parameters (brightness, red-green, and yellow-blue) in a single color parameter, whereas the yellowness indexes indicate the degree of sample yellowness, combining two color parameters (brightness and yellow-blue) in a single term [17]. The highest yellowness index was registered for M3 outside measurements (55.61 [a]), while the highest whiteness index was registered for M1 inside measurements (76.98 [a]). The hue angle values (h^0) varies between 0 and 360 and describes the main spectral component, like red for 0° or 360°, yellow for 90°, green for 180°, or blue for angles of 270° [27]. The margarine hue angle values were higher than 99.25 and lower than 106.42, presenting a yellow main spectral component with some green interferences and also presenting variations in the margarine's chemical composition. Regarding the chemical composition, it can be observed that the margarine samples with a high fat content (M3 samples) also present high values for b*, C*, and yellowness index color parameters.

The total color difference represents the magnitude of color variation between any two samples [28]; the human perception of total color difference is influenced by the observed color and eye sensitivity. If the total color difference is smaller than 1, the differences cannot be distinguished by the human eye. If the total color difference is between 1 and 3, minor color differences can be distinguished. If the total color difference is greater than 3, the color differences can be easily perceived by the human eye [29].

Table 2 shows the results of margarine color differences calculated between samples and between outside and inside color measurements. As we can see, the highest color differences were obtained for M1 inside and M3 outside margarine samples (14.59), followed by M1 outside and M3 outside color differences (14.00). Smaller color differences were calculated for M1 inside and M1 outside (0.60) and for M2 inside and M1 outside (0.46); these samples showed close color parameters. It can be observed that the color difference is especially greater between the samples with the minimum and maximum fat content; the fat content is an important factor in assessing color differences of margarine samples. In addition, the results of total color differences were also assessed by ANOVA statistical analysis, resulting in a significant difference at a level of $p < 0.05$ between the margarine samples. The evaluated margarine color parameters are of great importance in the classification of the analyzed samples. The total color differences present different values influenced by the fat content of the compared samples.

Table 2. The CIE L*a*b* color differences of margarine samples.

Sample		Outside			Inside		
		M1	**M2**	**M3**	**M1**	**M2**	**M3**
M1		-	1.77	14.00	0.60	0.46	10.86
M2	outside		-	12.23	2.36	1.47	9.12
M3				-	14.59	13.65	3.43
M1					-	0.99	11.45
M2	inside					-	10.48
M3							-

3.2. Fatty Acids Analysis

Table 3 presents the concentration of the fatty acids of analyzed margarine samples, and as can be seen from 37 fatty acids investigated, only 18 were quantified. The 18 quantified fatty acids were grouped depending on the chain length and according to the degree of unsaturation into 4 groups as follows: short- and middle-chain saturated, long-chain saturated, monounsaturated, and polyunsaturated. Of the 4 categories mentioned, the highest amount was the long-chain saturated fatty acids category (C14:0–C20:0), ranging between 85.61 µg/mg for M1 samples and 127.30 µg/mg for M3 samples, while the short and middle-chain saturated fatty acids (C8:0–C12:0) were present in a small amount (from 12.95 µg/mg to 19.22 µg/mg). The monounsaturated fatty acid methyl esters (C14:1–C20:1) also represent an important category of the chemical composition, varying from 53.50 µg/mg (M1) to 65.64 µg/mg (M3). The polyunsaturated fatty acid methyl esters (C18:2–C18:3) of margarine samples ranged between 31.16 µg/mg and 46.10 µg/mg. All analyzed margarine samples contained a significant quantity of lauric, myristic, palmitic, and stearic saturated fatty acids and considerable quantities of oleic and linoleic unsaturated fatty acids.

From the saturated fatty acids quantified, it can be seen that the dominant fatty acid was palmitic acid for all margarine samples; the M1 margarine samples presented a lower concentration of 60.01 µg/mg, whereas the highest concentration of 94.76 µg/mg was recorded for M3 margarine samples. Similar results were also reported by Anwar et al. [30] and Karabulut & Turan [31] on both margarines and shortenings. According to Kandhro et al. [32], the palmitic acid results designate the greater influence of palm oil in the margarine production process. Furthermore, the most abundant monounsaturated fatty acid was oleic acid, ranging from 50.28 µg/mg for M1 samples to 55.89 µg/mg for M2 and 62.81 µg/mg for M3 samples. From the polyunsaturated category, only linoleic, γ-linolenic, and linolenic acids were identified; significant concentrations of linoleic acid were present.

Of the three groups of samples analyzed, the M3 group presented the highest values of fatty acid concentration, whereas the M1 group presented lower levels of fatty acid concentration; Pearson correlation highlighted the positive connection between a sample's fat content and the concentration of fatty acids at a level of $p < 0.05$ (r = 0.986).

Taking into account that the maragarines analyzed contain between 53.79% and 56.73% saturated fatty acids (SFAs) and between 43.26% and 46.20% unsaturated fatty acids (UFAs), and considering the World Health Organization recommendation, the ratio of UFAs and SFAs was also calculated. The results for margarine samples varied from 0.76 to 0.85, which are smaller than 1.6; the World Health Organization recommends a ratio of UFAs and SFAs higher than 1.6 [33].

Table 3. Fatty acids concentration (µg/mg) of margarine samples.

Name	Abbreviation	RT ± 0.5 min	Margarine Samples—Mean (SD)		
			M1	**M2**	**M3**
Caprylic	C8:0	10.82	1.68 (0.12)	1.81 (0.11)	2.40 (0.01)
Capric	C10:0	14.25	1.54 (0.11)	1.69 (0.12)	2.44 (0.01)
Lauric	C12:0	17.63	9.72 (0.72)	10.72 (0.05)	14.37 (0.22)
Short- and middle-chain saturated			12.95	14.23	19.22
Myristic	C14:0	20.48	6.12 (0.20)	7.21 (0.46)	10.18 (0.30)
Pentadecanoic	C15:0	22.31	0.25 (0.02)	0.34 (0.01)	0.46 (0.02)
Palmitic	C16:0	23.96	60.01 (2.53)	74.48 (2.45)	94.76 (1.82)
Heptadecanoic	C17:0	25.77	1.87 (0.23)	0.54 (0.03)	0.68 (0.01)
Stearic	C18:0	27.91	17.34 (0.56)	16.66 (0.57)	20.78 (0.01)
Arachidic	C20:0	30.82	0.00	0.95 (0.11)	0.41 (0.01)
Long-chain saturated			85.61	100.20	127.30
Tetradecenoic	C14:1	21.42	0.40 (0.02)	0.41 (0.01)	0.71 (0.01)
cis-10-pentadecenoic	C15:1	23.00	0.28 (0.03)	0.19 (0.02)	0.27 (0.01)
Palmitoleic	C16:1	24.51	1.19 (0.22)	0.92 (0.12)	0.97 (0.02)
cis-10 Heptadecanoic	C17:1	26.43	1.27 (0.08)	0.91 (0.03)	0.84 (0.02)
Oleic	C18:1 cis (n9)	28.56	50.28 (1.89)	55.89 (2.53)	62.813 (1.78)
cis-11 Eicosenoic	C20:1 (n9)	33.66	0.06 (0.01)	0.03 (0.01)	0.02 (0.01)
Monounsaturated			53.509	58.37	65.64
Linoleic	C18:2 cis (n6)	29.85	30.97 (1.23)	34.15 (1.98)	45.86 (2.45)
γ-Linolenic	C18:3 (n3)	31.84	0.00	0.28 (0.02)	0.12 (0.02)
Linolenic	C18:3 (n6)	34.61	0.19 (0.01)	0.10 (0.01)	0.12 (0.01)
Polyunsaturated			31.16	34.55	46.10
Unsaturated/saturated fatty acids			0.85	0.81	0.76
Atherogenicity index			1.11	1.22	1.35

SD—standard deviation.

Regarding the fatty acids with atherogenic potential, it is well known that lauric acid (12:0), myristic acid (14:0), and palmitic acid (16:0) have this effect. The quantity of those acids found in the analyzed margarine samples ranged from 75.86 µg/mg for M1 samples to 119.33 µg/mg for M3 samples; the obtained results were much smaller than those reported by Vučić et al. [23] for Serbian margarines, but closer to those reported by Alonso et al. [34] for Spanish margarines. In contrast to the ratio of UFAs and SFAs, the atherogenicity index is a more complex indicator that evaluates the potential effects of fatty acids on the incidence of pathogenic phenomena. The atherogenicity index (IA) has been calculated as follows: IA = [C12:0 + (4 × C14:0) + C16:0]/UFA. For the analyzed margarine samples, this index ranged between 1.11 for M1 group samples and 1.22 for M2 and 1.35 for M3 margarine samples. The obtained results were greater than those reported for soft margarines (0.42), [35] but close to the ones of hard margarines (1.17–1.67) [23]. Regarding the atherogenicity index of margarine compared to butter (1.78–1.85), we can observe that the former is 27% to 36% smaller, being influenced also by the fat content of the sample [36]. Therefore, the consumption of low-atherogenicity index foods can decrease the total cholesterol of the blood.

3.3. Textural Evaluation

Texture properties represent one of the most important attributes of food products [37], and when it comes to margarine, butter, or shortening, the main textural parameter is hardness (firmness) [38]. The most commonly used instrumental assays include major deformations such as puncture, extrusion, and compression, which lead to the destruction of the material's structure under analysis. Given the fact that the analyzed samples fall into the plastics materials category, another very important parameter is the mechanical work of plastic deformation (or plasticity).

The evolution of force (N) versus deformation (mm) of margarine samples determined using the 10 mm diameter spherical penetrometer is displayed in Figure 2. The profile of the margarine test curves shows that it is a material with a pronounced fracturability. The margarine's fracturability varied from 0.35 N to 8.23 N; the highest values were measured using the 10 mm diameter spherical penetrometer, and from the point of view of the analyzed samples, the M2 margarine group presented the highest values of fracturability.

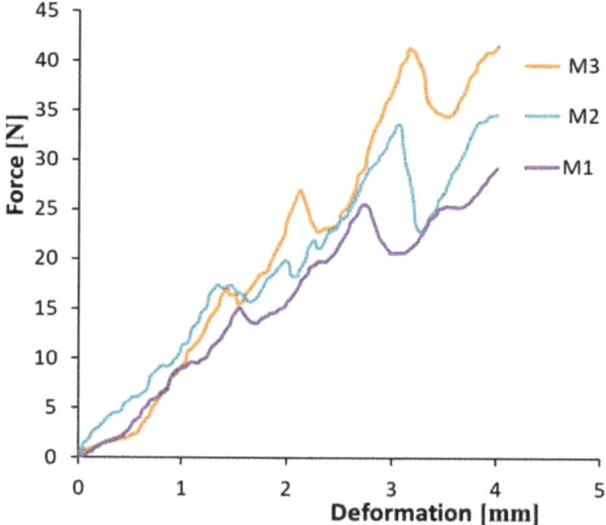

Figure 2. Force versus deformation curves of margarine samples.

The textural parameters of margarine samples analyzed in this study using three different penetrometers and a puncture test are displayed in Table 4. As in the case of butter and corresponding to ISO 16305 [39], the margarine hardness was also expressed in newtons (N) and ranged between 4.76 N and 52.28 N. The 5 mm diameter spherical penetrometer hardness measurements showed the smallest values, varying from 8 N for M3 samples to 4.76 N for M1 samples; by comparison, the 10 mm diameter spherical penetrometer and conical penetrometer presented close values varying from 41.56 N to 29.18 N and from 52.28 N to 19.02 N, respectively. The mechanical work of plastic deformation (plasticity) was calculated as the area under the deformation curve and ranged between 6.12 and 82.84 mJ (millijoules), with the greatest value being recorded for M3 margarine samples. In contrast, the lowest plasticity was measured for M1 samples. Of the three probes used to test the margarine samples, the 10 mm diameter spherical penetrometer presented the highest values (82.84–63.48 mJ).

Table 4. The textural parameters of margarine samples.

Probe	PS_5			PS_{10}			PC		
Sample	H (N)	P (mJ)	F (N)	H (N)	P (mJ)	F (N)	H (N)	P (mJ)	F (N)
M1	4.76 (0.40)	6.12 (0.62)	0.68 (0.02)	29.18 (1.85)	63.48 (2.25)	4.25 (0.92)	19.02 (2.78)	38.93 (3.54)	1.36 (0.28)
M2	7.26 (0.25)	7.76 (0.32)	0.35 (0.10)	34.66 (1.22)	73.37 (1.87)	8.23 (1.20)	30.06 (2.42)	53.55 (5.10)	2.66 (0.54)
M3	8 (0.30)	7.79 (0.43)	0.74 (0.12)	41.56 (1.89)	82.84 (2.02)	5.39 (0.33)	52.28 (3.25)	70.76 (4.83)	0.42 (0.08)

SD—standard deviation, PS_5—5 mm diameter spherical penetrometer, PS_{10}—10 mm diameter spherical penetrometer, PC—120° conical penetrometer, H—hardness, P—plasticity, F—fracturability, SD—standard deviation.

Both the hardness and the plasticity values of the margarine samples were influenced by their chemical composition, especially by the fat content and by the saturated and unsaturated fatty acids. Pearson correlation showed a positive influence ($p < 0.05$) of palmitic fatty acid content on the cone ($p < 0.05$; r = 0.995), the 10 mm spherical hardness ($p < 0.01$; r = 0.998) and on plasticity values ($p < 0.05$; r = 0.994; r = 0.999). Another consideration is that the hardness and plasticity of the 10 mm spherical penetrometer and cone penetrometer are positively correlated ($p < 0.01$; r = 0.992; r = 0.997). Furthermore, the content of cis-10 heptadecanoic and cis-11 eicosenoic fatty acids have a negative influence ($p < 0.05$; r = −0.998; r= 0.991; r = −0.990) on margarine plasticity and hardness evaluated with 5 and 10 mm spherical penetrometers.

In addition, to highlight the differences between analyzed margarine samples, emphasize the significant information obtained from the evaluated parameters, and diminish the amount of the variables, a PCA analysis (principal component analysis) was performed. The PCA results are represented by the biplot (score and loading) from Figure 3. PCA analysis was accomplished on physicochemical properties, fatty acid groups, measured color properties, and texture parameters of analyzed margarine samples. The two principal components, PC1 and PC2 cover 98.37% of data variation: the first component, PC1, describes 96.34% of the variation, while the second component, PC2, describes 2.03% of the variation. According to these PCA scores, it can be observed that the margarine samples are distributed into diverse quadrants according to chemical composition and evaluated textural parameters. The first component (PC1) separates the M1 and M2 margarine group samples from M3 margarine samples based on monounsaturated fatty acids (MN), 10 mm spherical penetrometer hardness, cone plasticity, and the long-chain saturated fatty acid (LCS) content. The second component (PC2) separated the M2 group from the other samples based on the cone and 10 mm spherical fracturability values, with a smaller contribution having the 5 mm hardness and plasticity. From the loadings plot, it can be observed that the projection of M3 margarine samples was significantly influenced by the long-chain saturated fatty acids (LCS) and by the cone hardness values and polyunsaturated fatty acids (PN), which have a smaller influence. Of the outside and inside measured color parameters, only the b* color parameter indicates an influence in the M3 sample distribution; its values are also positively correlated with polyunsaturated fatty acids (PN). Regarding the projection of the M1 samples, it can be observed that moisture content has a strong influence on their positioning in the PCA diagram.

The high value achieved in the dispersal of data variation highlights the utility of PCA in margarine classification using chemical composition, color parameters, and textural properties.

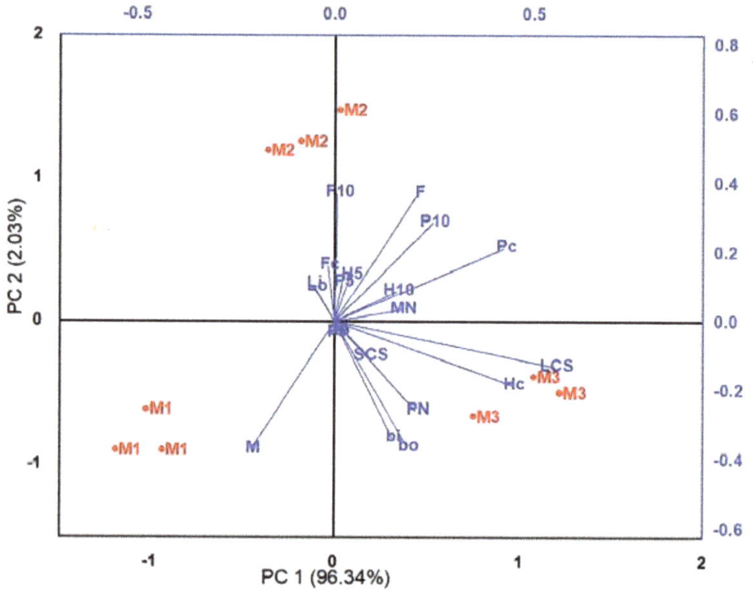

Figure 3. Principal component analysis (score and loading) of margarine samples: Lo, ao, bo-outside color parameters, Li, ai, bi-inside color parameters, F-fat content, M–moisture content, SCS-short- and middle-chain saturated fatty acids, LCS-long-chain saturated fatty acids, MN-monounsaturated fatty acids, PN-polyunsaturated fatty acids, Hc-cone penetrometer hardness, Pc-cone penetrometer plasticity, Fc-cone fracturability, H5-5 mm spherical penetrometer hardness, P5-5 mm spherical penetrometer plasticity, F5-5 mm spherical fracturability, H10-10 mm spherical penetrometer hardness, P10-10 mm spherical penetrometer plasticity, F10-10 mm spherical fracturability.

4. Conclusions

In this study on bakery margarines with different chemical compositions, it was observed that from the saturated fatty acids quantified, the dominant fatty acid was palmitic acid, whereas the most abundant unsaturated fatty acid was oleic. The atherogenicity index of margarine samples, calculated based on fatty acid concentration, was smaller than that of butter but close to hard margarine, being also influenced by the fat content of the sample. Regarding the penetrometer's geometry used for margarine texture evaluation, the cone and 10 mm diameter spherical penetrometer registered high values for both hardness and plasticity, which have been influenced by the sample's chemical composition, especially by the fat content and the saturated and unsaturated fatty acids. Bakery margarine is a plastic material with a pronounced fracturability, and the 10 mm diameter spherical penetrometer is recommended for evaluating this texture parameter. In terms of color parameters, both the outside and inside b* color parameter influence the differentiation of the analyzed samples. The ANOVA highlighted these differences at a level of $p < 0.001$, dividing the measurements into six statistical groups, whereas the PCA analysis indicates an influence in the fattest margarine samples' projection.

Funding: This research was funded by "Stefan cel Mare" University of Suceava.

Institutional Review Board Statement: Not applicable.

Informed Consent Statement: Not applicable.

Data Availability Statement: Not applicable.

Conflicts of Interest: The author declares no conflict of interest.

References

1. McClements, D.J. *Food Emulsions: Principles, Practices, and Techniques*, 3rd ed.; CRC Press: New York, NY, USA, 2015; pp. 1–26.
2. Serdaroglu, M.; Öztürk, B.; Kara, A. An Overview of Food Emulsions: Description, Classification and Recent Potential Applications. *Turk. J. Agric. Food Sci. Technol.* **2015**, *3*, 430–438. [CrossRef]
3. Miskandar, M.S.; Man, Y.C.; Yusoff, M.S.A.; Rahman, R.A. Quality of margarine: Fats selection and processing parameters. *Asia Pac. J. Clin. Nutr.* **2005**, *14*, 387–395. [PubMed]
4. Makeri, M.; Sahri, M.M.; Ghazali, H.M.; Ahmad, K.; Muhammad, K. Polymorphism, textural and crystallization properties of winged bean (Psophocarpus tetragonolobus, D.C) oil-based trans-fatty acids free ternary margarine blends. *LWT* **2018**, *100*, 158–166. [CrossRef]
5. Bongers, P.; Almeida-Rivera, C. Dynamic modelling of the margarine production process. In *Computer Aided Chemical Engineering*; Elsevier: Amsterdam, The Netherlands, 2011; pp. 1301–1305. [CrossRef]
6. FAO; WHO. Codex Alimentarius. In *Codex Standards for Margarine, Fats, Oils and Related Products*, 2nd ed.; Bernan Assoc.: Rome, Italy, 2001; Volume 8, pp. 1–5.
7. Andersen, A.J.C.; Williams, P.N. *Margarine*, 2nd ed.; Elsevier: Amsterdam, The Netherlands, 1965; pp. 5–27.
8. Puprasit, K.; Wongsawaeng, D.; Ngaosuwan, K.; Kiatkittipong, W.; Assabumrungrat, S. Non-thermal dielectric barrier discharge plasma hydrogenation for production of margarine with low trans-fatty acid formation. *Innov. Food Sci. Emerg. Technol.* **2020**, *66*, 102511. [CrossRef]
9. Fard, N.E.; Bahmaei, M.; Eshratabadi, P. Comparison of physicochemical characteristics of margarine and butter in Iranian market during storage. *J. Pharm. Health Sci.* **2016**, *4*, 181–191.
10. Gebauer, S.K.; Chardigny, J.M.; Jakobsen, M.U.; Lamarche, B.; Lock, A.L.; Proctor, S.D.; Baer, D.J. Effects of ruminant trans fatty acids on cardiovascular disease and cancer: A comprehensive review of epidemiological, clinical, and mechanistic studies. *Adv. Nutr.* **2011**, *2*, 332–354. [CrossRef]
11. Commission Regulation (EU) 2019/649 of the European Parliament and of the Council as Regards Trans-Fat, Other than Trans-Fat Naturally Occurring in Fat of Animal Origin. 2019. Available online: https://eur-lex.europa.eu/eli/reg/2019/649/oj (accessed on 6 December 2021).
12. Smetana, S.; Leonhardt, L.; Kauppi, S.M.; Pajic, A.; Heinz, V. Insect margarine: Processing, sustainability and design. *J. Clean. Prod.* **2020**, *264*, 121670. [CrossRef]
13. Toma, M.A.; Amin, M.R.; Alim, M.A. Evaluation of Margarine Quality Prepared from Sun-flower and Coconut Oil. *Asian Food Sci. J.* **2020**, *15*, 20–28. [CrossRef]
14. ISO. Animal and vegetable fats and oils. Determination of moisture content and volatile matter. *Polish Standard Method PN-EN ISO* **1998**, *665*, 1–9.
15. Chougui, N.; Djerroud, N.; Naraoui, F.; Hadjal, S.; Aliane, K.; Zeroual, B.; Larbat, R. Physi-cochemical properties and storage stability of margarine containing Opuntia ficus-indica peel extract as antioxidant. *Food Chem.* **2015**, *173*, 382–390. [CrossRef]
16. Zimbru, R.O.; Pădureţ, S.; Amariei, S. Physicochemical and color evaluation of confectionery mousses. *Food Environ. Saf. J.* **2020**, *19*, 228–236.
17. Milovanovic, B.; Djekic, I.; Miocinovic, J.; Djordjevic, V.; Lorenzo, J.M.; Barba, F.J.; Mörlein, D.; Tomasevic, I. What Is the Color of Milk and Dairy Products and How Is It Measured? *Foods* **2020**, *9*, 1629. [CrossRef] [PubMed]
18. Torres, D.; Casal, S.; Oliveira, M.P. Fatty acid composition of Portuguese spreadable fats with emphasis on trans isomers. *Eur. Food Res. Technol.* **2002**, *214*, 108–111. [CrossRef]
19. Jirarattanarangsri, W. Trans fatty acid content in a selection of margarines sourced from a local market in Thailand. *Food Appl. Biosci. J.* **2019**, *7*, 33–41.
20. Oroian, M.; Ursachi, F.; Dranca, F. Ultrasound-Assisted Extraction of Polyphenols from Crude Pollen. *Antioxidants* **2020**, *9*, 322. [CrossRef]
21. Rebeaud, S.G.; Jaylet, A.; Cotter, P.-Y.; Camps, C.; Christen, D. A Multi-Parameter Approach for Apricot Texture Analysis. *Agriculture* **2019**, *9*, 73. [CrossRef]
22. Liu, Y.X.; Cao, M.J.; Liu, G.M. Texture analyzers for food quality evaluation. In *Evaluation Technologies for Food Quality*; Zhong, J., Wang, X., Eds.; Woodhead: Kidlington, UK, 2019; pp. 441–463.
23. Vucic, V.; Arsic, A.; Petrovic, S.; Milanovic, S.; Gurinovic, M.; Glibetic, M. Trans fatty acid content in Serbian margarines: Urgent need for legislative changes and consumer information. *Food Chem.* **2015**, *185*, 437–440. [CrossRef]
24. Brodeală, A.M.; Munteanu, M. Population Consumption Availability 2017–2019. *Natl. Inst. Stat.* **2020**, *1*, 1–23.
25. Food Market in Romania. 2016. Available online: https://www.dutchromaniannetwork.nl/wp-content/uploads/2017/01/Food-Report-Romania-2016.pdf (accessed on 3 November 2021).
26. Bowen, A.J.; Blake, A.; Tureček, J. Development and validation of a color evaluation process for sweet potato preference characterization. *J. Sens. Stud.* **2019**, *34*, e12524. [CrossRef]
27. Bermúdez-Aguirre, D.; Mawson, R.; Versteeg, K.; Barbosa-Cánovas, G.V. Composition properties, physicochemical characteristics and shelf life of whole milk after thermal and thermo-sonication treatments. *J. Food Qual.* **2009**, *32*, 283–302. [CrossRef]
28. Fernandez-Avila, C.; Gutierrez-Merida, C.; Trujillo, A.J. Physicochemical and sensory char-acteristics of a UHT milk-based product enriched with conjugated linoleic acid emulsified by Ultra-High Pressure Homogenization. *Innov. Food Sci. Emerg. Technol.* **2017**, *39*, 275–283. [CrossRef]

29. Quintanilla, P.; Beltrán, M.; Molina, A.; Escriche, I.; Molina, M. Characteristics of ripened Tronchón cheese from raw goat milk containing legally admissible amounts of antibiotics. *J. Dairy Sci.* **2019**, *102*, 2941–2953. [CrossRef] [PubMed]

30. Anwar, F.; Bhanger, M.; Iqbal, S.; Sultana, B. Fatty acid composition of different margarines and butters from Pakistan with special emphasis on trans unsaturated contents. *J. Food Qual.* **2006**, *29*, 87–96. [CrossRef]

31. Karabulut, I.; Turan, S. Some properties of margarines and shortenings marketed in Turkey. *J. Food Compos. Anal.* **2006**, *19*, 55–58. [CrossRef]

32. Kandhro, A.; Sherazi, S.; Mahesar, S.; Bhanger, M.; Talpur, M.Y.; Rauf, A. GC-MS quantification of fatty acid profile including trans FA in the locally manufactured margarines of Pakistan. *Food Chem.* **2008**, *109*, 207–211. [CrossRef] [PubMed]

33. Who, J.; Consultation, F.E. Diet, nutrition and the prevention of chronic diseases. *World Health Organ Tech. Rep. Ser.* **2003**, *916*, 1–8.

34. Alonso, L.; Fraga, M.J.; Juárez, M. Determination of trans fatty acids and fatty acid profiles in margarines marketed in spain. *J. Am. Oil Chem. Soc.* **2000**, *77*, 131–136. [CrossRef]

35. Pérez-Llamas, F.; Larqué, E.; Garaulet, M.; Zamora, S.; Tebar, F.J. Fatty acid composition and nutritional relevance of most widely consumed margarines in Spain. *Grasas Aceites* **2003**, *54*, 65–70. [CrossRef]

36. Pǎdureţ, S. The Effect of Fat Content and Fatty Acids Composition on Color and Textural Properties of Butter. *Molecules* **2021**, *26*, 4565. [CrossRef]

37. Nishinari, K.; Fang, Y. Perception and measurement of food texture: Solid foods. *J. Texture Stud.* **2018**, *49*, 160–201. [CrossRef]

38. Ayuningtias, A.; Nurhasanah, S.; Nurhadi, B.; Subroto, E. Textural Characteristic of Margarine Enriched with Pectin Fiber by Using Blending Method. *Int. J. Adv. Sci Eng. Inf. Tech.* **2013**, *3*, 268–272. [CrossRef]

39. *ISO 16305*; Butter—Determination of Firmness. ISO: Geneva, Switzerland, 2005; pp. 1–18.

Article

Plant-Based Milks: Alternatives to the Manufacture and Characterization of Ice Cream

Ana Leahu *, Sorina Ropciuc and Cristina Ghinea

Faculty of Food Engineering, Stefan cel Mare University of Suceava, 720229 Suceava, Romania; sorina.ropciuc@fia.usv.ro (S.R.); cristina.ghinea@fia.usv.ro (C.G.)
* Correspondence: analeahu@fia.usv.ro

Abstract: This study investigated the potential use of dietary fibers (psyllium and pectin fibers added in different proportions of 0–10%) to improve the rheological, textural, and sensory characteristics of vegetable ice cream using vegetable milk (almond and hemp milk). Hemp milk was obtained from the peeled seeds of the industrial hemp plant, which includes varieties of *Cannabis sativa*, which have a low content of the psychotropic substance tetrahydrocannabinol (THC) and are grown for food. The rheological characteristics of the mix and ice cream were determined by using the Haake Mars rheometer. Compared with the control sample, the viscosities of the mix in all samples analyzed were enhanced with the addition of dietary fibers, due to the occurrence of interactions and stabilizations. The viscoelastic modules G′ G″ were determined on ice cream samples at a temperature of $-10\ °C$. The elastic and viscous modulus showed high values with the increase of the addition of 6% dietary fibers. The textural characteristics were assessed by the shear strength of a layer of ice cream at a temperature of $-4\ °C$. Hardness, firmness, and adhesiveness were influenced by the size of their ice crystals, the fat content, and the percentage of dietary fibers added. The sensory analysis of the ice cream showed higher overall scores for the almond milk ice cream, because the sweet taste was appreciated with a maximum score, while the hemp milk ice cream was evaluated for flavor and taste.

Keywords: vegetable ice cream; dietary fibers; sensorial properties; rheological properties

Citation: Leahu, A.; Ropciuc, S.; Ghinea, C. Plant-Based Milks: Alternatives to the Manufacture and Characterization of Ice Cream. *Appl. Sci.* **2022**, *12*, 1754. https://doi.org/10.3390/app12031754

Academic Editor: Alessandra Durazzo

Received: 19 January 2022
Accepted: 5 February 2022
Published: 8 February 2022

Publisher's Note: MDPI stays neutral with regard to jurisdictional claims in published maps and institutional affiliations.

1. Introduction

In recent times, there has been a trend towards the development of innovative, readily available, reasonably priced, yet safe foods that offer consumers benefits such as improved health or the prevention of health complications associated with diet beyond basic nutritional functions [1]. It has been found that changes in the eating habits of health-conscious consumers, strongly influenced by the increasing incidence of lifestyle diseases, such as heart disorders and depression, have given rise to new functional foods and products on the market [2]. Ice cream is a very popular and delicious frozen dessert, made from dairy products such as milk, cream, and natural or artificial sweeteners, while in recent years, plant-based milk substitutes have begun to be used [3]. In this sense, dietary ice cream is a new product on the Romanian market and is increasingly preferred, due to a different taste, along with being enriched with various delicious flavors. This dietary ice cream is chosen especially by vegetarians, raw vegans, people with lactose intolerance or those who generally prefer a healthy lifestyle. According to Eurostat [4], ice cream consumption in Romania is the lowest in Europe, respectively 1.6 kg/capita, and ice cream is considered to be a refreshing product, in the hot season as a dessert. Ice cream is a. physico-chemical complex characterized by hardness and melting properties, usually contains at least 10% fat, stabilizers, and sweeteners. The use of plant-based milk substitutes from soy (*Glycine max*), cashew (*Anacardium occidentale*), hazelnut (*Corylus*), coconut (*Cocos nucifera*), hemp (*Cannabis sativa*), or almonds (*Prunus dulcis*) and dietary fiber allow for final products with the texture, hardness, and firmness characteristics of milk ice cream [5]. Plant-based milk is free of cholesterol and has unsaturated fats, vitamins, minerals, and antioxidants, which is why they are considered functional foods and nutraceuticals [6].

Almonds (*Prunus dulcis* Mill. DA Webb) have a high content of fatty acids, lipids, amino acids, proteins, carbohydrates (including dietary fiber), vitamins, and minerals which contribute to improving artery health, reducing high blood pressure, has positive effects in improving diabetes and metabolic syndrome [7]. In recent years, almond milk beverage is promoted as a healthy alternative to cow's milk have potential prebiotic properties due to the presence of bioactive compounds such as vitamins, especially vitamin E which cannot be synthesized by the body, flavonoids, and polyphenols [8,9].

In addition, consumers choose plant-based milk alternatives instead of milk due to its numerous positive health effects on the human body, a desire for a healthy lifestyle, and environmental awareness. The solution to this problem may be to drink almond drink instead of cow's milk, as it contains a vitamin E content of 6.33 mg 100 g^{-1}, which represents 42% of the recommended daily allowance of 15 mg [6,10].

The non-drug variety *Cannabis sativa* subsp. *sativa*, hemp (*Cannabis sativa* L.) seeds contain approx. 20–25% protein with a biological value similar to chicken egg white, as well as considerable amounts of vitamins, minerals (magnesium, copper, phosphorus, and calcium) and dietary fiber. Hemp milk has a high nutritional value with a low content of saturated fats, a good percentage of polyunsaturated fatty acids (PUFA) ω-3 and ω-6, and low allergenicity [9]

According to Szparaga et al. [11], hemp-based milk substitutes provide very low amounts of protein (3.23 g/100 g) and lipids (21.08 g/100 g) and can be an alternative to consumers who seek gluten-free food. The use of hemp seed milk has not yet been widely tested in food production. Hemp milk is unstable with a tendency to flocculate, being an oil-in-water emulsion, which is a challenge in the industry, as it leads to loss of quality and validity. The use of stabilizing substances can increase the cost of production [12]. Hemp seed milk can be a very valuable new ingredient for the food industry due to its potential as natural emulsions of polyunsaturated fatty acids (PUFAs), essential fatty acids (EFAs), and other fat-soluble bioactive compounds.

Dietary fibers include resistant starch and non-starch polysaccharides, oligosaccharides, and various lignified compounds that are not digested by enzymes of the gastrointestinal tract, but also have functional properties in food, for example, increase water retention, or contribute to emulsification or gel formation. Moreover, the addition of dietary fiber to foods can help change texture, avoid syneresis and increase the shelf life of foods [1]. Dietary fiber is prebiotic, improving the activity of beneficial intestinal bacteria, so they can prevent many gastrointestinal disorders [13]. Additionally, dietary fiber plays an important role in the treatment of diabetes, decreases the risk of coronary heart diseases, reduces the risk of cancer in the colon or rectum, and relieves symptoms caused by lactose intolerance [2]. A recent meta-analysis by McRorie et al. [14] showed that highly viscous soluble fiber (β-glucan, psyllium, and crude guar gum) can effectively reduce high serum cholesterol levels and improve glycemic control. Whole grains (especially the pericarp), vegetables, fruits, and nuts are one of the most important sources of dietary fiber [6]. Resistant starch present in food is considered dietary fiber that is non-digestible by enzymes in the small intestine of healthy people. Therefore, to produce vegetable milk ice cream with the desired textural properties and sensory attributes, it can be used a mixture of plant-based milk on almond and hemp and dietary fiber.

In recent decades, there have been significant changes in food patterns and food consumption. In association with this, there has been a growing demand for plant-based milk alternatives due to milk allergy, lactose intolerance but which can provide health benefits. Moreover, in the presented study, agave syrup was used as an ingredient to prepare a vegan ice-cream mix. Agave syrup (natural sweetener) is a product with functional properties, and the intake has beneficial properties for human health, such as high prebiotic capacity and a low glycemic index score.

The aim of this study was to investigate the impact of different concentrations (0–10%) of psyllium and pectin fibers of vegetable ice cream, which is comparable to ordinary ice cream in terms of rheological, textural, and sensorial parameters.

2. Materials and Methods

2.1. Materials

Almond seeds (Pronat SRL Romania), agave syrup, and vanilla were all purchased from the local market. High quality peeled hemp (Pronat SRL Romania) was purchased from local pharmacies. Other ingredients were psyllium and pectin fiber were purchased from Enzymes & Derivates SA Neamt, Romania.

2.2. Methods

2.2.1. Almond Milk Preparation

The almond milk was prepared according to the method previously described. The almond seeds (300 g) were soaked overnight at 4 °C in 900 mL of distilled water. For milk extraction, the almonds were mixed with distilled water (1:10 *g/v* ratio), ground in a blender and filtered. The almond milk obtained was stored for further analysis at 4 °C [15].

2.2.2. Hemp Milk Preparation

Hemp milk was freshly prepared in the laboratory. The dehulled seeds of hemp (300 g) were soaked for 10 h at 20 °C in 900 mL of distilled water. For milk extraction, the hemp seeds were mixed with distilled water (1:10 *g/v* ratio), ground in a Tefal laboratory easy soup (France) and filtered [11].

2.2.3. Ice Cream Preparation

In order to obtain the ice cream, the mixtures were prepared using the formula as shown in Table 1. The milk was heated to 80 °C, then homogenized and frozen in a 1.5-L ice cream maker. The steps followed for the ice cream manufacturing are illustrated in Figure 1, as described before by Leahu [16]. The ice cream was made in Ice Cream Maker Machines (Cuisinart.com). Finally, the samples were frozen and stored in a refrigerator at −22 °C for subsequent analysis. The ice cream production process referred to the description in previous studies [17–19].

Table 1. The content of components used in ice cream mix formulation.

Samples	% Dietary Fiber	Volume of Milk (mL)	% Almonds/Hemp Flour	% Agave Syrup	Vanilin %
A_Psy	0	100	3	5	1
A1_Psy	2	100	3	5	1
A2_Psy	4	100	3	5	1
A3_Psy	6	100	3	5	1
A4_Psy	8	100	3	5	1
A5_Psy	10	100	3	5	1
H_Pec	0	100	3	5	1
H1_Pec	2	100	3	5	1
H2_Pec	4	100	3	5	1
H3_Pec	6	100	3	5	1
H4_Pec	8	100	3	5	1
H5_Pec	10	100	3	5	1

A_Psy—ice cream with almond milk with 0% psyllium; A1_Psy—ice cream with almond milk with 2% psyllium; A2_Psy—ice cream with almond milk with 4% psyllium; A3_Psy—ice cream with almond milk with 6% psyllium; A4_Psy—ice cream with almond milk with 8% psyllium; A5_Psy—ice cream with almond milk with 10% psyllium; H_Pec—ice cream with hemp milk with 0% pectin; H1_Pec—ice cream with hemp milk if 2% pectin; H2_Pec—ice cream with hemp milk if 4% pectin; H3_Pec—ice cream with hemp milk if 6% pectin, H4_Pec—ice cream with hemp milk if 8% pectin; and H5_Pec—ice cream with hemp milk if 10% pectin.

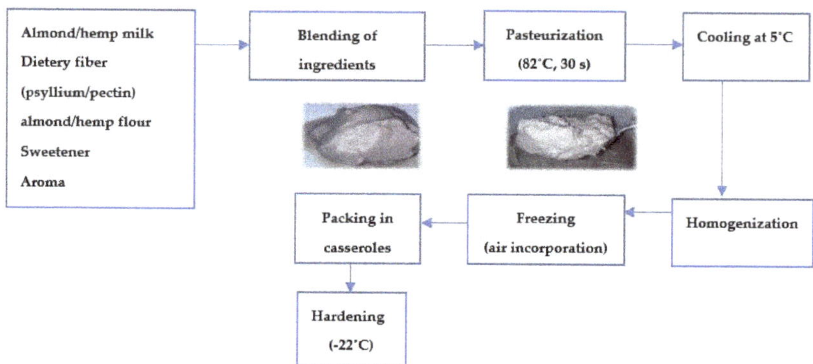

Figure 1. Manufacturing process of ice cream.

2.2.4. Ice Cream Analysis

Physico-Chemical Analyses of Ice Cream Mix

The titratable acidity was determined according to the AOAC 920.124 method [20]. The pH values were measured according to the AOAC 14.022 method using a digital pH meter HACH HQ 30d flexi [21]. The soluble solids percentage was determined by using an Abbé refractometer. Two drops of the ice cream mix were placed on the prism of the refractometer. The soluble solids content was read directly on the refractometer scale [21]. The fat content was determined by applying the chemical method of using an ice cream butyrometer (according to Roeder's weighing method). The protein content was determined by using the Kjeldahl chemical method [22].

Rheological Analysis of Ice Cream Mix

The dynamic rheometer, model Haake Mars 40, was used for determination of ice cream mix rheological characteristic, after cooling at 5 °C. Viscosity and vascoelastic modulus of the ice cream mix were determined over time, by using the plate/plate system with a diameter of 80 mm for the plate and 40 mm for the rotor. The temperature during the analysis remained constant. Vascoelastic modules were obtained by recording the frequency from 10^{-1} to 10 Hz at a temperature of 5 °C. Numerical values for viscosity and modulus G prime G s. were obtained by using Microsoft Excel [23].

Textural Analysis

A 50-g sample of ice cream was used to determine the texture. Ice cream tubs were taken out of the freezer and left at room temperature for two minutes. The ice cream samples were analyzed texturally by using the Perten Texturometer with 45° cone probe (P-CO45S), a trigger force of 4 g, a compression percentage of 30%, and a test speed of 2 mm/s. The hardness and adhesiveness of the ice cream samples were determined.

Sensory Analysis

The 12 ice cream samples were proposed for sensory analysis to a group of 30 members selected from the students of the Faculty of Food Engineering. The coded samples were offered to the tasters and the sensory evaluation was performed by using the preference test. The preference or ranking test was carried out according to the ISO 8587:2006 standard.

It used a 9-point hedonic scale (from 1 = "Dislike Extremely" to 9 = "Like Extremely", with 5 = "Neither Like nor Dislike"). The investigated attributes included flavor, taste, appearance, consistency, and overall acceptability [24].

Statistical Analysis

The experiments were assayed in triplicates ($n = 3$) and the results were expressed as mean $\pm$ S.E.M (standard mean error) values. For the comparison among the different formulations of the average values of all the compositional and physical–chemical properties, an analysis of variance (ANOVA) with a confidence interval of 95% ($p < 0.05$) using the Tukey test was carried out. The statistical analysis was carried out using XLSTAT statistical software, version 2021. The correlations between the quantitative variables and the qualitative (sensory characteristics) variables were interpreted by analyzing the main components using the Matlab 2021b program.

3. Results and Discussion

Given the importance of adding fiber in improving the quality characteristics of ice cream, in this research the viscosity, hardness, firmness, adhesiveness, physicochemical, and sensorial properties of the ice cream formulations were studied. Five formulations were prepared with almond milk and psyllium fiber, while the other five were prepared with hemp milk and pectin.

3.1. Physicochemical Analyses of Cream Mix

In Table 2, the results obtained after physicochemical analysis of ice cream samples are presented. The titratable acidity is maintained in the range of 0.17–0.20 g/100 g lactic acid for all samples of the almond cream mixture (including those with the addition of psyllium fiber). Hemp milk with added pectin has higher titratable acidity values as the percentage of added pectin increases. Compared to the control sample (H0_Pec) which shows the value of 0.26 g/100 g lactic acid, the addition of 10% pectin leads to an increase in acidity to the value of 0.80 g/100 g lactic acid. From Table 2, it can be observed that the pH of almond milk ice cream and addition of psyllium fibers samples did not vary, while the pH of hemp milk ice cream samples with pectin decreased with the addition of a higher concentrations of pectin [25].

Table 2. Values of the physicochemical parameters.

Sample	Titratable Acidity, g Lactic Acid/100 g of Total Solids	pH	Soluble Solids, °Brix	Fat, %	Protein, %
A0_Psy	0.16 ± 0.02 [h]	6.2 ± 0.02 [a]	23.12 ± 0.01 [b]	2.67 ± 0.03 [c]	1.09 ± 0.03 [f]
A1_Psy	0.17 ± 0.01 [g]	6.1 ± 0.01 [a]	23.26 ± 0.02 [a]	2.58 ± 0.03 [d]	1.12 ± 0.03 [f]
A2_Psy	0.17 ± 0.01 [g]	6.1 ± 0.02 [a]	23.46 ± 0.01 [a]	2.5 ± 0.04 [e]	1.32 ± 0.02 [e]
A3_Psy	0.17 ± 0.01 [g]	6.1 ± 0.01 [a]	23.16 ± 0.01 [b]	2.57 ± 0.04 [d]	1.46 ± 0.03 [e]
A4_Psy	0.17 ± 0.01 [g]	6.1 ± 0.03 [a]	23.13 ± 0.03 [b]	2.61 ± 0.02 [c]	1.68 ± 0.03 [d]
A5_Psy	0.17 ± 0.02 [g]	6.1 ± 0.02 [a]	23.13 ± 0.04 [b]	2.53 ± 0.01 [e]	1.75 ± 0.02 [d]
H0_Pec	0.26 ± 0.03 [f]	5.3 ± 0.02 [b]	21.25 ± 0.03 [d]	3.24 ± 0.01 [b]	2.04 ± 0.02 [c]
H1_Pec	0.28 ± 0.02 [e]	5.1 ± 0.03 [b]	21.39 ± 0.04 [d]	3.20 ± 0.02 [b]	2.32 ± 0.02 [c]
H2_Pec	0.34 ± 0.01 [d]	4.9 ± 0.02 [c]	21.54 ± 0.04 [c]	3.41 ± 0.04 [a]	2.57 ± 0.01 [b]
H3_Pec	0.48 ± 0.02 [c]	4.7 ± 0.01 [c]	21.56 ± 0.03 [c]	3.5 ± 0.04 [a]	2.61 ± 0.02 [b]
H4_Pec	0.63 ± 0.04 [b]	4.2 ± 0.01 [d]	21.33 ± 0.03 [d]	3.5 ± 0.04 [a]	2.79 ± 0.01 [a]
H5_Pec	0.80 ± 0.03 [a]	3.8 ± 0.01 [e]	21.47 ± 0.03 [c]	3.24 ± 0.03 [b]	2.66 ± 0.03 [b]

[a–h] Mean values in the same column (corresponding to the same letter) indicate statistically differ significantly ($p < 0.05$).

The pH values of the almond milk cream mix samples are around 6, while pH values of the hemp cream mix samples are in the range of 5.3 for the sample without the addition of pectin and 3.8 for the sample with 10% pectin (Table 2).

The refractometric soluble solids (Table 2) showed values between 23.12–23.46 °Brix to the mixture of almond milk ice cream and addition of psyllium fibers for the control sample and the sample with 4% psyllium. Hemp milk samples had soluble solids values in the range of 21.25 °Brix in the control sample and 21.56 °Brix in the sample with the addition of 6% pectin.

The fat content (Table 2) determined for the almond milk ice cream mix was 2.5–2.6%, being correlated with the fat of the almond milk. Hemp milk samples had a fat content relative to milk fat, which was around 3.5%.

The results obtained for the protein content (Table 2) indicate values of 1.09% in the control sample with almond milk and a maximum of 1.75% in the sample with the addition of 10% psyllium fiber. Hemp milk ice cream mix had higher values for protein content due to the fact that hemp milk has a higher protein content, the control sample with hemp milk had a content of 2.04% and the sample with 8% pectin fibers had a higher protein content, 2.79%. According to Kozłowicz et al. [26] the protein content in the ice cream supplemented with Moldavian dragonhead bagasse (MDB) changed significantly ($p < 0.05$) from 10.16 g per 100 g (w/w) for the 1% MDB sample to 12.07 g per 100 g (w/w) for ice cream with 3% MDB addition.

The analysis of the correspondences between the quantitative variables obtained after physico-chemical parameters determination is represented in Figure 2. The analysis of the correspondence places the results at a very short distance between them. This arrangement indicates a strong association between the chemical characteristics of pH and titratable acidity—they practically overlap. The variables that express protein and fat are distanced from the values obtained for pH and titratable acidity. No statistically significant changes in the pH and titratable acidity of plant-based milks ice cream were observed regardless of the milk used. The pH value correlated with the titratable acidity of the product and is not influenced by the protein content, but such properties can directly affect the quality and ultimately the acceptability of the product by the consumer. The fat content of the ice cream does not correlate with the pH value. Also, it was desired to develop a recipe based on maintaining a constant content total solids and sugar.

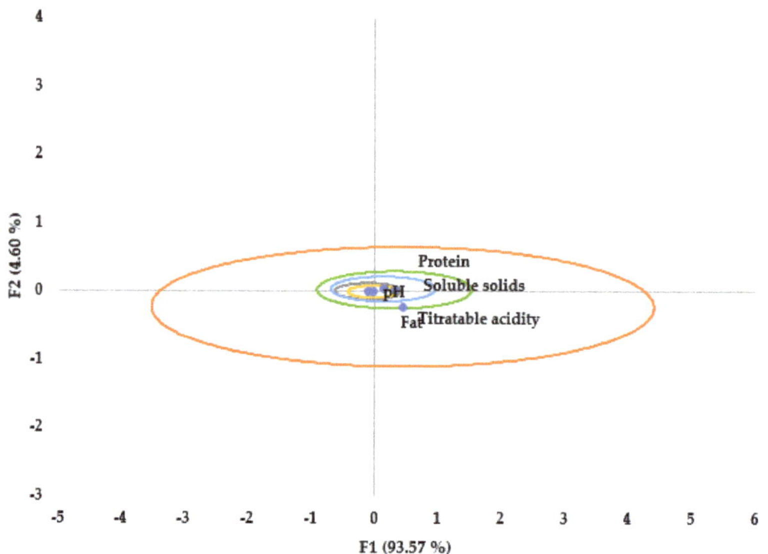

Figure 2. Correspondence analysis between quantitative variables.

3.2. Rheological Analysis of Ice Cream Mix

The average values of the rheological characteristics of the apparent viscosity and the viscoelastic modules made with different milk and with additions of psyllium/pectin are presented in Figures 3–8.

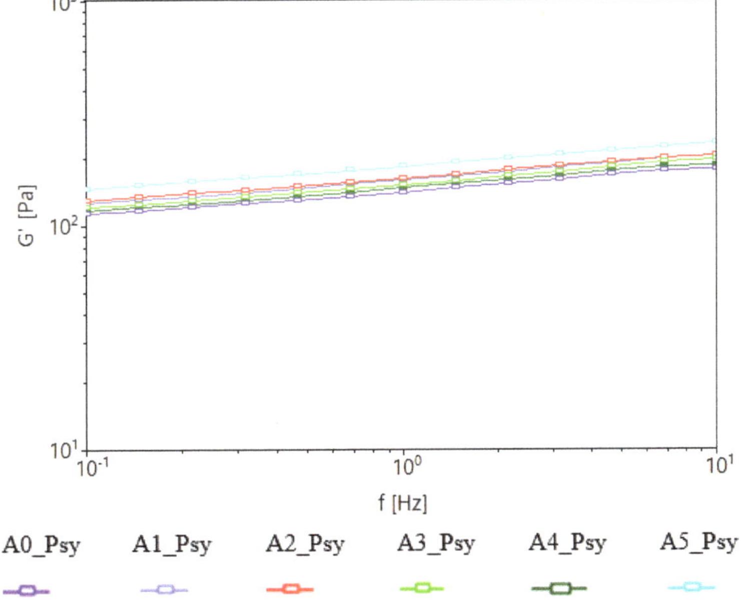

Figure 3. The elastic modulus curves to the ice cream mix with almond milk and psyllium.

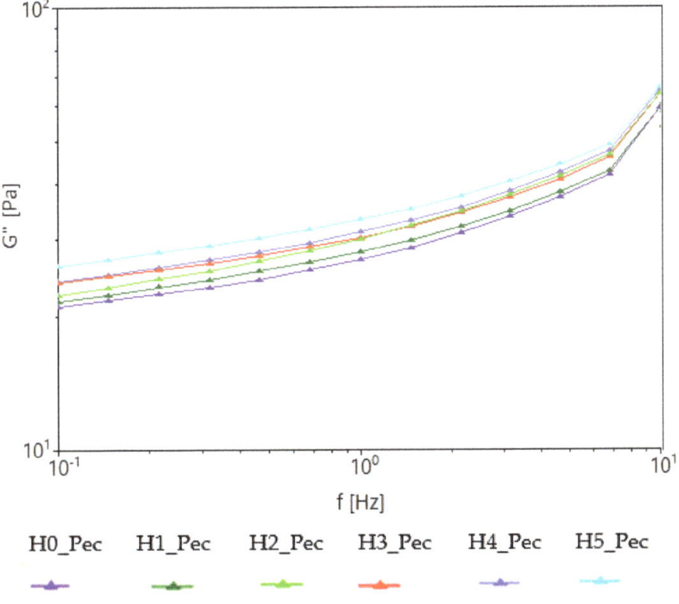

Figure 4. The viscous modulus curves to the mix of ice cream with almond milk and psyllium.

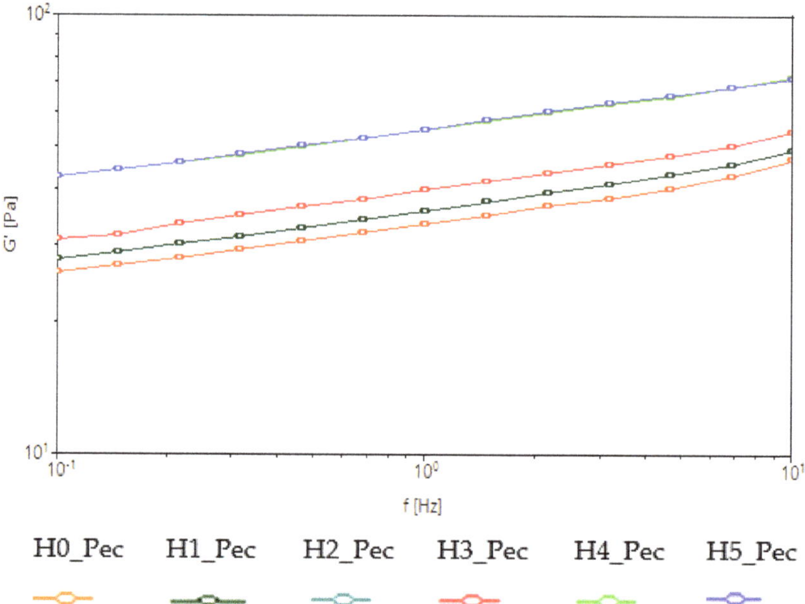

Figure 5. The elastic modulus curves to the ice cream mix with hemp milk and pectin.

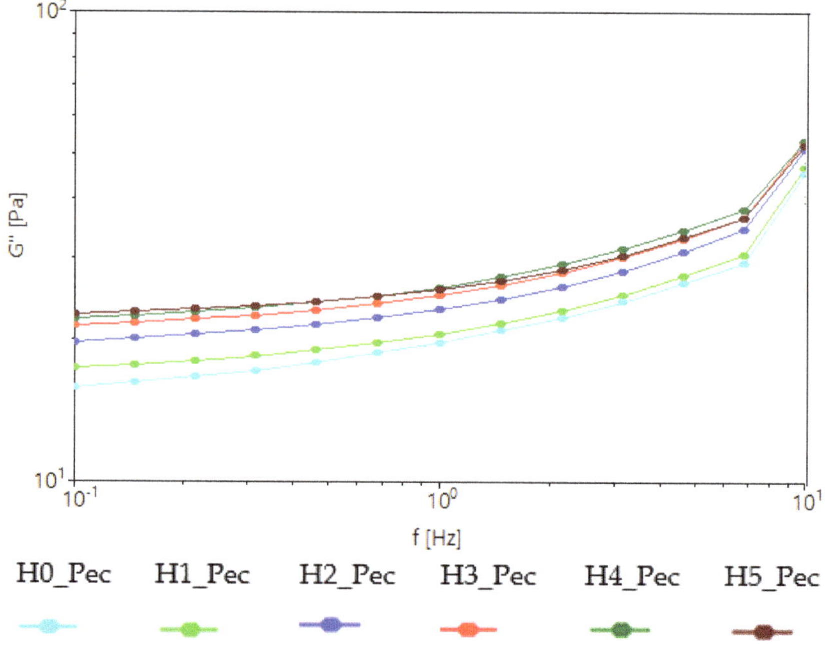

Figure 6. The viscous modulus curves to the mix of ice cream with hemp milk and pectin.

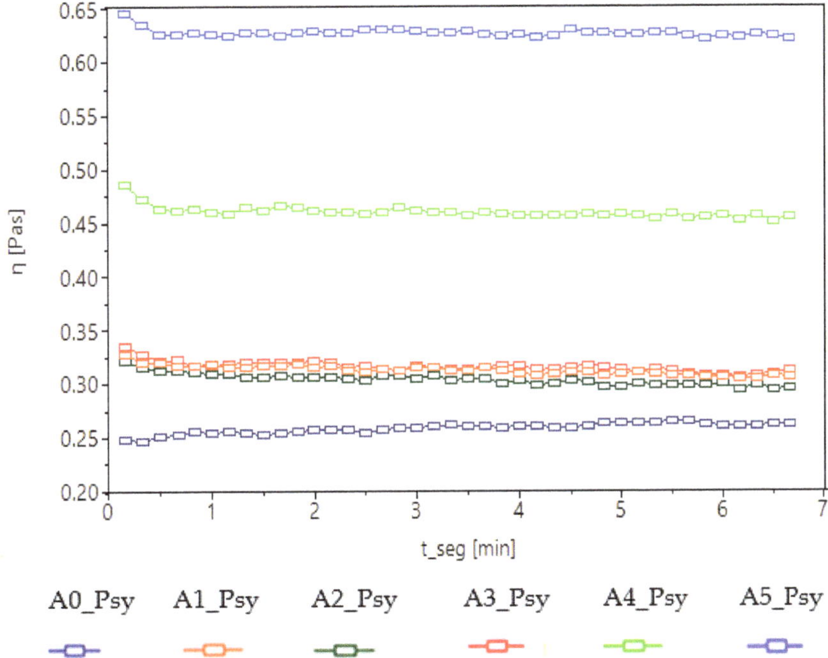

Figure 7. Viscosity curves for the ice cream mix with almond milk the addition of psyllium.

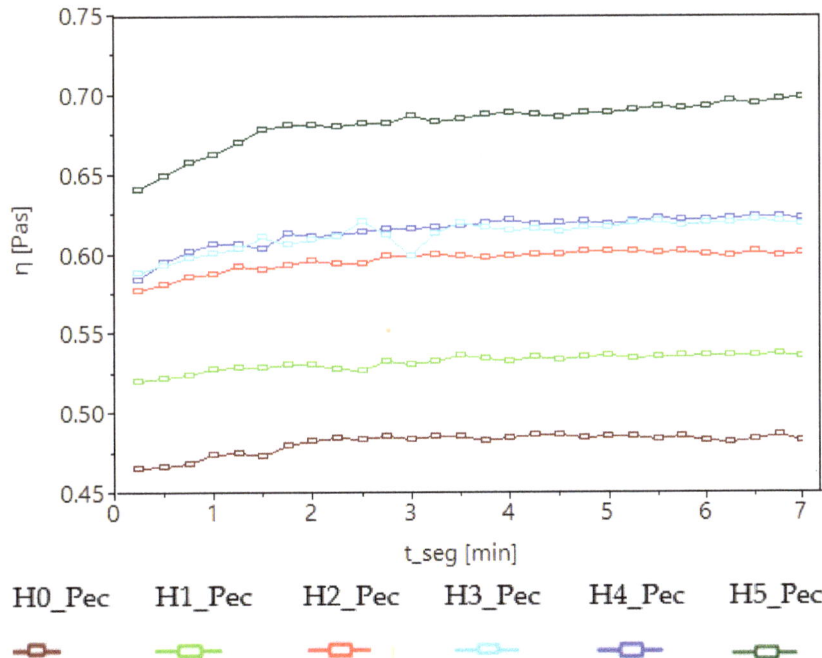

Figure 8. Viscosity curves for the ice cream mix with hemp milk the addition of pectin.

The curves that express the results of the rheological behavior of the ice cream mix with vegetable milk (Figures 3–6) demonstrate that the addition of psyllium and pectin fibers leads to a significant increase in the viscous and elastic mode [27]. The rheological properties of the almond (Figure 3) ice cream mixes showed that an increase in viscoelastic parameters (G') when the psyllium fiber concentrations increased comparatively with hemp ice cream with pectin fiber (Figure 4). Viscosity curves have the same tendency to increase as the percentage of psyllium added to the mix increases (Figure 5). By comparison, the elastic modulus of pectin-added ice cream mixes increases significantly at 8–10% pectin added (Figure 5).

The viscous modulus of the ice cream mix with hemp milk and pectin fiber shows the same tendency to increase the viscosity with the percentage of pectin (Figure 6). The increase in the viscosity and elasticity of the mix influences the foaming capacity of the mix, the retention of air in the structure of the ice cream and the stability of the emulsion will have a positive effect on the melting resistance of the ice cream [27,28]. The idea that psyllium and pectin fibers form a stable gel structure with high viscosity and pseudoplastic flow behavior can be supported. According to the results obtained by Soukoulis et al. [1] and Valera et al. [25], all samples showed high elastic and viscous behavior in the mixtures obtained with vegetable milk and fiber additives with the role of hydrocolloids [1,25]. The viscosity curves shown in Figures 7 and 8 describe the increase in the viscosity of the ice cream mix as the amount of psyllium and pectin fibers increases. Compared to the control samples (A0_Psy and H0_Pec), the mix has a high viscosity due to the addition of psyllium fiber and pectin, thus increasing the flow resistance and the thickening effect gives the mix stability to the shear action. The higher thickening effect is observed with the addition of psyllium fiber compared to pectin fiber (Figure 7).

The Pearson correlations that are established between the chemical properties of the mixture obtained with almond milk or hemp milk are presented in Table 3. The correlation coefficient between viscosity and fat (r = 0.710), respectively protein (r = 0.753) indicates that the presence of proteins and fat in vegetable milk helps the stability of the mix and the formation of gel binding mechanisms.

These results are justifiable according to Makinen et al. [28]; Elsamani et al. [29] Shinyoung et al. [30]; Adapa et al. [31] which demonstrated that a high protein and fat content of the ice cream mixture with vegetable milk led to an increase in viscosity and a good elasticity of the ice cream mix.

Table 3. Correlation matrix (Pearson (n)) between chemical variables and rheological variables.

Variables	G″ Pa	G′ Pa	Viscosity	Titratable Acidity	pH	Soluble Solids	Fat	Protein
G″ Pa	1	**0.990**	0.001	0.346	−0.365	−0.409	0.446	0.238
G′ Pa	**0.990**	1	0.079	0.455	−0.460	−0.487	0.517	0.341
viscosity	0.001	0.079	1	0.453	**−0.697**	**−0.748**	**0.710**	**0.792**
Titratable acidity	0.346	0.455	0.453	1	**−0.871**	**−0.722**	**0.753**	**0.845**
pH	−0.365	−0.460	**−0.697**	**−0.871**	1	**0.942**	**−0.970**	**−0.964**
Soluble solids	−0.409	−0.487	**−0.748**	**−0.722**	**0.942**	1	**−0.960**	**−0.904**
Fat	0.446	0.517	**0.710**	**0.753**	**−0.970**	**−0.960**	1	**0.924**
Protein	0.238	0.341	**0.792**	**0.845**	**−0.964**	**−0.904**	**0.924**	1

3.3. Textural Properties

The hardness and adhesiveness of the vegetable milk ice cream are represented in Figures 9 and 10. It can be stated that the hardness value increases with the addition of psyllium and pectin fibers. Hemp milk ice cream samples have a higher fat content which leads to a lower increase in hardness compared to almond milk ice cream samples. Changing the type and amount of protein and fat in vegetable milk for the preparation of various ice cream samples have a significant effect on the textural properties of ice cream, which influences the formation of ice cream crystals and increases the hardness and consistency of ice cream. The error limits in Figures 9 and 10 represents the standard deviations.

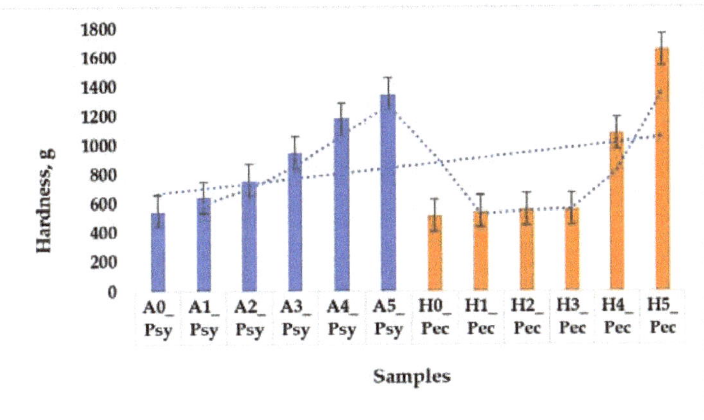

Figure 9. Graphical representation of the results describing the hardness of the ice cream samples.

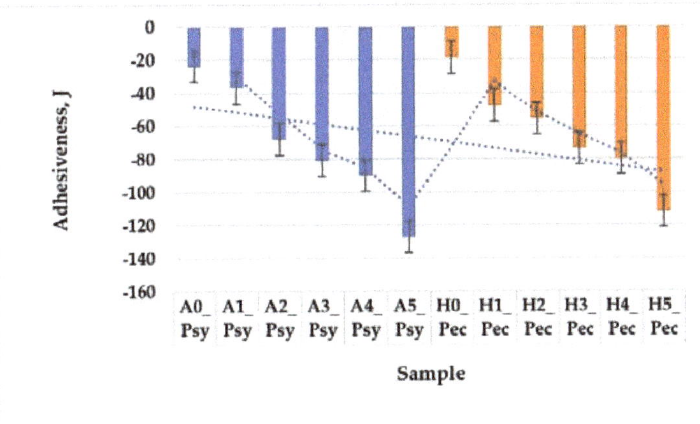

Figure 10. Graphical representation of the results describing the adhesiveness of the ice cream samples.

3.4. Sensory Properties Results

Descriptive sensory panelists evaluated the almond milk and psyllium fiber ice cream samples and also the ice cream samples with hemp milk and pectin fiber. In general, they gave higher overall scores for the almond milk ice cream because the sweet taste was appreciated with maximum score and the hemp milk ice cream was rated for flavor and taste (Table 4). Compared to almond milk ice cream, hemp milk ice cream was characterized by specific sensory properties: high-intensity, slightly unpleasant aromas that were considered to be specific to hemp milk. The consistency and appearance of ice cream with hemp milk and pectin was appreciated as dense, smooth, without ice needles, which allows for the development of taste and consumption in ice cream. Samples of almond milk and psyllium fiber ice cream were less appreciated by sensory experts due to their high consistency and adhesiveness. The panelists reported that the aroma scores of the hemp milk ice cream samples with 8% and 10% pectin fibers were much improved compared to the control samples, while those formulated with 2% and 4% were evaluated with a lower score. The panelists criticized these samples for having a strong hemp aroma. The research conducted in this study may be useful to vegetable ice cream producers [32–34]. This study is the first to define the sensory characteristics of vegetable ice cream with almond milk and hemp with the addition of psyllium fiber and pectin. The descriptive attributes developed can be used for the development of new varieties of vegetable milk ice cream [18,19,35].

Table 4. Average sensory acceptance results on a 9-point hedonic scale.

Formulation	Attributes of Sensory Acceptance				
	Flavor	Taste	Appearance	Consistency	Overall Acceptability
A0_Psy	6.1 ± 0.17 [d]	6.0 ± 0.23 [d]	7.2 ± 0.02 [b]	5.2 ± 0.34 [f]	6.3 ± 0.02 [d]
A1_Psy	6.5 ± 0.08 [b]	6.3 ± 0.18 [c]	7.3 ± 0.26 [b]	6.1 ± 0.44 [e]	6.9 ± 0.03 [c]
A2_Psy	6.3 ± 0.62 [c]	6.3 ± 0.12 [c]	7.2 ± 0.02 [b]	7.3 ± 0.18 [d]	7.0 ± 0.03 [c]
A3_Psy	6.3 ± 0.24 [c]	6.4 ± 0.40 [c]	7.2 ± 0.03 [b]	8.0 ± 0.28 [c]	8.0 ± 0.01 [b]
A4_Psy	6.5 ± 0.31 [b]	6.4 ± 0.35 [c]	7.4 ± 0.04 [b]	8.7 ± 0.22 [b]	8.0 ± 0.02 [b]
A5_Psy	6.7 ± 0.12 [b]	6.8 ± 0.27 [b]	7.8 ± 0.04 [a]	9.0 ± 0.23 [a]	8.2 ± 0.01 [b]
H0_Pec	5.3 ± 0.22 [e]	6.2 ± 0.22 [d]	5.4 ± 0.07 [e]	6.1 ± 0.20 [e]	5.3 ± 0.02 [e]
H1_Pec	5.9 ± 0.31 [d]	5.6 ± 0.31 [e]	5.6 ± 0.08 [e]	6.5 ± 0.16 [d]	6.2 ± 0.01 [d]
H2_Pec	5.6 ± 0.09 [e]	6.0 ± 0.42 [d]	5.8 ± 0.00 [d]	6.9 ± 0.17 [d]	6.7 ± 0.03 [c]
H3_Pec	6.0 ± 0.27 [d]	6.4 ± 0.12 [c]	6.2 ± 0.00 [d]	8.4 ± 0.21 [c]	6.9 ± 0.01 [c]
H4_Pec	6.2 ± 0.36 [c]	7.0 ± 0.24 [b]	6.8 ± 0.04 [c]	8.7 ± 0.17 [b]	7.0 ± 0.00 [c]
H5_Pec	7.2 ± 0.14 [a]	7.8 ± 0.16 [a]	7.9 ± 0.06 [a]	9.0 ± 0.00 [a]	8.8 ± 0.01 [a]

Means that do not share a letter ([a–f]) are significantly different ($p \leq 0.05$).

The analysis of the main components (PCA) was applied to the average values of the textural and rheological attributes to express the correlation ratio between them. The PCA plot (Figure 11) explains the variability of the results obtained by chemical, rheological, textural, and sensory analyses. Axis F1, indicates the positive correlation for almond milk ice cream and on axis F2 are grouped the ice cream samples according to the attributes that characterize them. There is a significant positive association between the control formulations (H0, H1, H2, and H3 in the samples obtained from hemp milk and differentiates in samples H4 and H5 to which pectin was added in proportion of 8 and 10%. For this grouping of samples H4 and H5, it is also noticeable in the sensory analysis; these being appreciated with a high score. The separation between the two F2 axis groups is well differentiated because the results obtained in the determination of chemical, rheological, textural, and sensory characteristics separated the two samples (H5_Pec and H6_Pec) from the rest of the group. The ice cream samples that were formulated with almond milk and psyllium fiber are grouped on the F1 axis according to the positive characteristics. The control sample A0_Psy is significantly different from the rest of the samples, we can express a negative correlation between the control sample and the other almond milk samples.

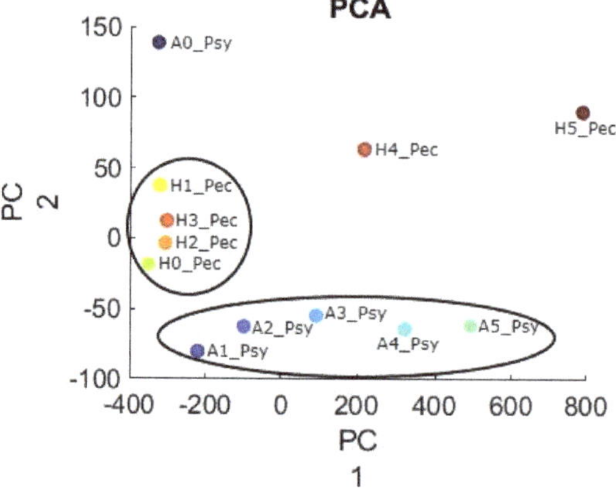

Figure 11. The PCA plot of correlation configuration of samples and characteristic attributes.

4. Conclusions

Given the results obtained from the chemical, rheological, textural, and sensory determinations, it can be concluded that a vegetable milk ice can be formulated by completely replacing the conventional stabilizer with psyllium or pectin fibers. From a technological point of view, this modification of the recipe can be allowed without hindering the technological process and without significantly changing the recipe. Ice cream with added psyllium can be obtained with a maximum of 6% added fiber and pectin fiber can be obtained with a percentage of at least 8% added in order to obtain a specific consistency and well-appreciated sensory characteristics. There is an increasing trend for the consumption of vegetable milk ice cream and through this study good results were obtained so that these formulations can be made. The increasing share of pectin added to hemp milk ice cream improved the physical properties of the ice cream, increasing the viscosity and consistency, and the appearance of the ice cream was appreciated as dense and without ice needles. It can be concluded that the almond milk ice cream was more appreciated by the panelists from a sensory point of view, while the hemp milk ice cream is more advantageous from a physical-chemical and rheological points of view due to the addition of pectin fibers. It is known that the choice of ice cream is determined primarily by taste preferences. Thus, the taste preferences for different types ice cream (almond and hemp milk) with great characteristics, which can be considered as possible applications of vegan ice cream, were investigated. Although a conflict between taste preferences and health objectives can be detected, such an approach can be treated as the main objective for a potential producer.

Author Contributions: Author Contributions: Conceptualization, A.L. and S.R.; methodology, A.L.; software, S.R.; validation, A.L., S.R. and C.G.; formal analysis, A.L. and S.R.; investigation, A.L. and S.R.; resources, A.L. and S.R.; data curation, S.R. and C.G.; writing—original draft preparation, A.L. and S.R.; writing—review and editing, A.L. and S.R.; visualization, A.L. and S.R.; supervision, C.G.; funding acquisition, A.L. All authors have read and agreed to the published version of the manuscript.

Funding: This research was funded by "Stefan cel Mare" University of Suceava.

Institutional Review Board Statement: Not applicable.

Informed Consent Statement: Not applicable.

Data Availability Statement: Not applicable.

Acknowledgments: This research was supported by "Stefan cel Mare" University of Suceava.

Conflicts of Interest: The authors declare no conflict of interest.

References

1. Soukoulis, C.; Lebesi, D.; Tzia, C. Enrichment of ice cream with dietary fibre: Effects on rheological properties, ice crystallization and glass transition phenomena. *Food Chem.* **2009**, *115*, 665–671. [CrossRef]
2. Rana, J.; Paul, J. Consumer behavior and purchase intention for organic food: A review and research agenda. *J. Retail. Consum. Serv.* **2017**, *38*, 157–165. [CrossRef]
3. Tsai, S.-Y.; Tsay, G.J.; Li, C.-Y.; Hung, Y.-T.; Lin, C.-P. Assessment of Melting Kinetics of Sugar-Reduced Silver Ear Mushroom Ice Cream under Various Additive Models. *Appl. Sci.* **2020**, *10*, 2664. [CrossRef]
4. Eurostat. Available online: https://ec.europa.eu/eurostat/web/products-eurostat-news/-/ddn-20210803-1 (accessed on 18 January 2022).
5. Velotto, S.; Parafati, L.; Ariano, A.; Palmeri, R.; Pesce, F.; Planeta, D.; Alfeo, V.; Todaro, A. Use of stevia and chia seeds for the formulation of traditional and vegan artisanal ice cream. *Int. J. Gastron. Food Sci.* **2021**, *26*, 100441. [CrossRef]
6. Aydar, E.F.; Tutuncu, S.; Ozcelik, B. Plant-based milk substitutes: Bioactive compounds, conventional and novel processes, bioavailability studies, and health effects. *J. Funct. Foods* **2020**, *70*, 103975. [CrossRef]
7. Smeriglio, A.; Mandalari, G.; Bisignano, C.; Filocamo, A.; Barreca, D.; Bellocco, E.; Trombetta, D. Polyphenolic content and biological properties of Avola almond (*Prunus dulcis* Mill. DA Webb) skin and its industrial byproducts. *Ind. Crops Prod.* **2016**, *83*, 283–293. [CrossRef]
8. Nissen, L.; di Carlo, E.; Gianotti, A. Prebiotic potential of hemp blended drinks fermented by probiotics. *Food Res. Int.* **2020**, *131*, 109029. [CrossRef]
9. Sethi, S.; Tyagi, S.K.; Anurag, R.K. Plant-based milk alternatives an emerging segment of functional beverages: A review. *J. Food Sci. Technol.* **2016**, *53*, 3408–3423. [CrossRef] [PubMed]

10. Chalupa-Krebzdak, S.; Long, C.J.; Bohrer, B.M. Nutrient density and nutritional value of milk and plant-based milk alternatives. *Int. Dairy J.* **2018**, *87*, 84–92. [CrossRef]
11. Szparaga, A.; Tabor, S.; Kocira, S.; Czerwińska, E.; Kuboń, M.; Płóciennik, B.; Findura, P. Survivability of probiotic bacteria in model systems of non-fermented and fermented coconut and hemp milks. *Sustainability* **2019**, *11*, 6093. [CrossRef]
12. Paul, A.A.; Kumar, S.; Kumar, V.; Sharma, R. Milk Analog: Plant based alternatives to conventional milk, production, potential and health concerns. *Crit. Rev. Food Sci. Nutr.* **2020**, *60*, 3005–3023. [CrossRef]
13. Pool-Zobel, B.L. Inulin-type fructans and reduction in colon cancer risk: Review of experimental and human data. *Br. J. Nutr.* **2005**, *93*, S73–S90. [CrossRef] [PubMed]
14. McRorie, J.W., Jr.; McKeown, N.M. Understanding the physics of functional fibers in the gastrointestinal tract: An ev-idence-based approach to resolving enduring misconceptions about insoluble and soluble fiber. *J. Acad. Nutr. Diet.* **2017**, *117*, 251–264. [CrossRef]
15. Manzoor, M.F.; Siddique, R.; Hussain, A.; Ahmad, N.; Rehman, A.; Siddeeg, A.; Yahya, M.A. Thermosonication effect on bioactive compounds, enzymes activity, particle size, microbial load, and sensory properties of almond (*Prunus dulcis*) milk. *Ultrason. Sonochem.* **2021**, *78*, 105705. [CrossRef] [PubMed]
16. Leahu, A.; Ghinea, C.; Damian, C. The influence of inulin and psyllium addition to ice-cream and its effects on the sensorial properties. *Food Environ. Saf.* **2019**, *17*, 363–371.
17. Yu, B.; Zeng, X.; Wang, L.; Regenstein, J.M. Preparation of nanofibrillated cellulose from grapefruit peel and its application as fat substitute in ice cream. *Carbohydr. Polym.* **2021**, *254*, 117415. [CrossRef]
18. Espinoza, L.A.; Purriños, L.; Centeno, J.A.; Carballo, J. Chemical, microbial and sensory properties of a chestnut and milk ice cream with improved healthy characteristics. *Int. J. Food Prop.* **2020**, *2*, 2271–2294. [CrossRef]
19. Aboulfazli, F.; Salihin, A.; Misran, M. Effect of Vegetable Milks on the Physical and Rheological Properties of Ice Cream. *Food Sci. Technol. Res.* **2014**, *20*, 987–996. [CrossRef]
20. AOAC. *Official Methods of Analysis of the Association of Official Analytical Chemists*; Association of Official Analytical Chemists: Gaithersburg, MD, USA, 2000; Volume 9.
21. AOAC. Hydrogen-Ion Activity (pH). In *Official Methods of Analysis*; Association of Official Analytical Chemists: Gaithersburg, MD, USA, 1990.
22. AOAC International. 41.1.28A AOAC Official Method 996.06 Fat (Total, Saturated, and in Foods. Hydrolytic Extraction Gas Chromatographic Method Unsaturated). In *Official Methods of Analysis of AOAC International*; AOAC International: Arlington, TX, USA, 2001.
23. Karaca, O.B.; Güven, M.; Yasar, K.; Kaya, S.; Kahayoglu, T. The functional, rheological and sensory characteristics of ice creams with various fat replacers. *Int. J. Dairy Technol.* **2009**, *62*, 93–99. [CrossRef]
24. Herald, T.J.; Aramouni, F.M.; Mahmoud, A.-G. Comparison study of egg yolks and egg alternatives in French vanilla ice cream. *J. Texture Stud.* **2008**, *39*, 284–295. [CrossRef]
25. Varela, P.; Pintor, A.; Fiszman, S. How hydrocolloids affect the temporal oral perception of ice cream. *Food Hydrocoll.* **2014**, *36*, 220–228. [CrossRef]
26. Kozłowicz, K.; Nazarewicz, S.; Różyło, R.; Nastaj, M.; Parafiniuk, S.; Szmigielski, M.; Bieńczak, A.; Kozłowicz, N. The Use of Moldavian Dragonhead Bagasse in Shaping the Thermophysical and Physicochemical Properties of Ice Cream. *Appl. Sci.* **2021**, *11*, 8598. [CrossRef]
27. Hua, Z.; Jianle, C.; Junhui, L.; Chaoyang, W.; Xingqian, Y.; John, S.; Shiguo, C. Pectin from citrus canning wastewater as potential fat replacer in ice cream. *Molecules* **2018**, *23*, 925.
28. Makinen, O.E.; Wanhalinna, V.; Zannini, E.; Arend, T.E.K. Foods for special dietary needs: Non-dairy plantbased milk substitutes and fermented dairy type products. *Crit. Rev. Food Sci. Nutr.* **2016**, *56*, 339–349. [CrossRef] [PubMed]
29. Elsamani, M.O. Probiotics, organoleptic and physicochemical properties of vegetable milkbased bio-ice cream supplemented with skimmed milk powder. *Int. J. Nutr. Food Sci.* **2016**, *5*, 361–366. [CrossRef]
30. Shinyoung, K.; Stephen, L.M.; Juan, L.S.; Wesley, S.M.; Lurdes, S.W. Sensory and nutritional characteristics of concept frozen desserts made from underutilized sweetpotato roots. *Hort.Technol.* **2021**, *31*, 259–265.
31. Adapa, S.; Dingeldein, H.; Schmidt, K.A.; Herald, T.H. Rheological properties of ice cream mixes and frozen ice creams containing fat and fat replacers. *J. Dairy Sci.* **2000**, *83*, 2224–2229. [CrossRef]
32. Dervisoglu, M.; Yazici, F.; Ayedemir, O. The effect of soy protein concentrate addition on the physical, chemical, and sensory properties of strawberry flavored ice cream. *Eur. Food Res. Technol.* **2005**, *221*, 466–470. [CrossRef]
33. Muse, M.R.; Hartel, R.W. Ice cream structural elements that affect melting rate and hardness. *J. Dairy Sci.* **2004**, *87*, 1–10. [CrossRef]
34. Mahdian, E.; Karazhian, R. Effects of Fat Replacers and Stabilizers on Rheological, Physicochemical and Sensory Properties of Reduced-fat Ice Cream. *JAST* **2013**, *15*, 1163–1174.
35. David-Birman, T.; Romano, A.; Aga, A.; Pascoviche, D.; Davidovich-Pinhas, D.; Lesmes, U. Impact of silkworm pupae (*Bombyx mori*) powder on cream foaming, ice cream properties and palatability. *Innov. Food Sci. Emerg. Technol.* **2022**, *75*, 102874. [CrossRef]

Article

Study on Kinetics of Trans-Resveratrol, Total Phenolic Content, and Antioxidant Activity Increase in Vine Waste during Post-Pruning Storage

Alina Lenuța Crăciun * and Gheorghe Gutt

Faculty of Food Engineering, Stefan cel Mare University of Suceava, 720225 Suceava, Romania; g.gutt@fia.usv.ro
* Correspondence: alina.craciun@usm.ro

Abstract: There is increasing evidence surrounding the health benefits of E-resveratrol; this has triggered interest in stilbenoids in grapes, wine, and by-products. On the one hand, there is an enormous amount of underutilized vine waste, rich in bioactive substances during wine production. On the other hand, there is a growing demand for promising phytochemicals, for dietary and pharmaceutical purposes. Vine shoots are promising sources of stilbenoids; they have economic potential because they are sources of high-value phytochemicals. Recent studies have shown that, due to biosynthesis pathway genes, especially STS (forming trans-resveratrol), which is abundant during storage periods of vine shoots—trans-resveratrol accumulates up to 40-fold. The objective of this research was to determine the most economical part of vine waste to be exploited, and to study the kinetics of resveratrol increase in a 90-day period, to determine the optimal storage period to reach a maximum trans-resveratrol content. Total phenolic content (TPC) and antioxidant activity (AA) were studied to determine possible correlations. In Fetească Neagră vine shoot varieties stored at laboratory temperatures, trans-resveratrol content increased to a maximum (2712.86 mg/kg D.W.) at day 70, and then slightly decreased until day 90. TPC remained constant and there was a slight increase in AA. Vine shoots contained the largest amounts of trans-resveratrol (1658.22 mg/kg D.W.), followed by tendrils (169.92 mg/kg D.W.), and leaves (43.54 mg/kg D.W.).

Keywords: trans-resveratrol; post-pruning storage; total phenolic content; antioxidant activity

Citation: Crăciun, A.L.; Gutt, G. Study on Kinetics of Trans-Resveratrol, Total Phenolic Content, and Antioxidant Activity Increase in Vine Waste during Post-Pruning Storage. *Appl. Sci.* **2022**, *12*, 1450. https://doi.org/10.3390/app12031450

Academic Editor: Antony C. Calokerinos

Received: 24 November 2021
Accepted: 21 January 2022
Published: 29 January 2022

Publisher's Note: MDPI stays neutral with regard to jurisdictional claims in published maps and institutional affiliations.

1. Introduction

Viticulture is one of the most important agricultural activities in the world; in 2018 approximately 7.4 million hectares were cultivated globally, of which 4.3 million hectares were cultivated in Europe. Statistics place Spain in first place regarding the areas cultivated with grape vines (969,000 hectares); in second place is China (875,000 hectares); in third place is France (793,000 hectares) [1]. The average number of byproducts harvested from annual vine trimmings is 1.3 kg of wood/live log, thus "accumulating" to 2×10^7 tons of wood in the world. The only use for this waste is to grind it and use it as a fertilizer, by distributing it on vineyards [2].

Recent studies have shown that vine waste from annual trimmings accumulate via a wide range of stilbenoids, from resveratrol to complex stilbene oligomers [3]. Thereby, due to the fact that vine waste contains valuable bioactive substances, with potential applications in medicine and agriculture (due to multiple pharmacological and phytopathogenic properties), these can represent promising bioresources [4].

Stilbenoids are produced by plant secondary metabolisms and are a part of the non-flavonoid phenolic compound family. They are synthetized by plants as defense mechanisms in response to biotic and abiotic stresses. There is increased interest surrounding stilbenoids due to their potential effects on human health [5].

Stilbenes are a reduced group of phenylpropanoids characterized by a 1,2-diphenylethylene general structure. Plant stilbenes are derived from the general phenylpropanoid pathway [6]. Only a few plant species are capable of producing stilbenes, although all higher plants can synthesize esters of cinnamic acid derivatives and malonyl-CoA. The first enzymes of the phenylpropanoid pathway are phenylalanine ammonia lyase (PAL), cinnamate-4-hydroxylase (C4H), and 4-coumarate: CoA ligase (4CL). Stilbene synthase (STS) catalyzes, in a single reaction, the biosynthesis of stilbene's general structure from one CoA-ester of a cinnamic acid derivative (p-coumaroyl-CoA—most (or frequently) cinnamoyl-CoA) and three molecules of malonyl-CoA, and is only characteristic to plants that are able to produce stilbene [7]. First, the purified protein of the STS gene was from the stressed cell suspension cultures of the peanut (*Arachis hypogaea*) [8]. Subsequently, Scots pine (*Pinus sylvestris*) and grape (*Vitis vinifera*) were used to clone STS genes [9]. STS genes are found in many plant species as families of related genes. The vine was the only plant capable of producing stilbene, whose genome was sequenced until recently. It has been demonstrated in early studies that more than 20 STS genes are contained by the grapevine genome [10]. The large size of the multigene family was confirmed by an analysis of the first sketch of the grapevine genome, with an estimated 20–40 STS genes. A minimum of 20 different STS genes are expressed in the grape after infection with *Plasmopara viticola* [11].

Resveratrol is synthesized in grape vines using three molecules of malonyl-CoA and one molecule of coumaroyl-CoA, by the catalysis of stilbene synthase (EC 2.3.1.95; STS, malonyl-CoA:4-coumaroyl-CoA malonyltransferase). Most stilbenes synthesized by plants are derived from the basic unit trans-resveratrol (*trans*-3,5,4′-trihydroxystilbene) [12].

Generally, plants synthesize trans-resveratrol (3,5,4′-trihydroxystilbene) as a phytoalexin, naturally, in response to exogenous factors, such as ultraviolet (UV) irradiation, traumatic damage, pathogen infection, and other stresses. Only a few dietary sources have been found to contain this compound; the main representatives being peanuts, grapes, hops, strawberries, blueberries, and the products derived therefrom [13].

Several reviews and research studies have shown that trans-resveratrol, due to its pharmacological activities, is an important anti-carcinogenic and anti-aging/anti-inflammatory agent, with neuroprotective and antioxidant proprieties, and that it could protect, to some extent, against cardiac and metabolic disorders. Therefore, a growing demand for trans-resveratrol is expected in areas such as nutraceuticals, health, cosmetics, and food, due to its pharmacological importance [14].

The majority of resveratrol supplements available on the market are made of extracts obtained from the roots of Japanese knotweed (*Polygonum cuspidatum*) and the degree of purity can vary widely. Research studies have shown that emodin, a compound that could have laxative effects, is found in unpurified (or partially purified) resveratrol extracts [15]. The Japanese knotweed is an invasive plant that grows in heavily polluted environments and is contaminated with heavy metals. Additional concerns about the safety of resveratrol supplements have been raised, due to the possibility of the root tissue of Japanese knotweed to cellular uptake of these contaminants [16,17].

Houillé et al. studied the biosynthetic origins of trans-resveratrol accumulation in grape canes during postharvest storage and they assayed the structural genes PAL, C4H, 4CL, and STS (forming E-resveratrol) through a quantitative real-time PCR. They demonstrated that, due to transcript abundance of the structural genes, the resveratrol content increases during storage time. In the first four weeks of storage, the STS gene was induced; during the 6 weeks of storage, the three genes (PAL, 4CL, and C4H) of the general phenylpropanoid pathway were constitutively expressed. These results confirm that vine shoots during the storage time is still transcriptionally active [14]. The results can vary widely, despite recent advances in the determination of stilbenoid compositions of vine shoots [18]. This variability can be attributed to the various factors influencing the stilbenoid composition, especially the provenance and the variety of the grape cane [19]. Gyongyi et al., (2017) reported a resveratrol content ranging from 0.95 ± 0.08 to 3.94 ± 0.21 g/100 g D.W. for 10 samples of vine waste of different varieties extracted with 60% ethanol [20].

The aim of this study was to characterize trans-resveratrol accumulation in pruned vine shoots, and particularly to determine the optimal storage period for vine variety Fetească

Neagră, to identify which vine waste contains a higher amount of trans-resveratrol, to be exploited economically and to study possible correlations among the kinetics of resveratrol increase, total polyphenol content, and antioxidant activity of vine waste.

2. Materials and Methods

2.1. Vine Waste Samples

The samples of grape shoots, leaves, and tendrils of the grape variety Fetească Neagră were collected from the region of Cotnari, Iași (476,708,864;269,361,914) on 17 November 2020.

2.2. Reagent and Chemicals

The diethyl ether, trans-resveratrol standard (99% GC), sodium carbonate, and DPPH solution were purchased from Sigma-Aldrich Co (Burlington, MA, USA). Ethanol, methanol, and acetonitrile (LiChrosolv for HPLC) were obtained from Merck (Darmstadt, Germany). The aqueous solutions were prepared with twice distillated and demineralized water. The 0.22 μm PTFE membrane filters were purchased from Phenomenex (Torrance, CA, USA).

2.3. Standard Solutions

A stock solution of resveratrol with a concentration of 1 g/L was prepared. To guarantee a complete dissolution of the trans-resveratrol standard, it was initially dissolved in a small volume of methanol.

A set of standards was prepared with methanol, 99%, before the analysis, from the stock solution. To avoid degradation of the standard solutions, special care was taken, keeping them protected from exposure to light and air, and storing them in brown glass containers at $-20\,°C$.

2.4. Sample Collection and Preparation

Vine shoots collected on 17 November, 2020, were analyzed in turn—one each day for 90 days. The last sample was prepared on 14 February, 2021. Meanwhile, the other samples were kept at laboratory temperature (22 °C) to increase the concentrations of trans-resveratrol; optimal parameters were as recommended by the latest research in the field. Each sample was previously dried in the oven at 45 °C to constant weight, after which they were crushed with a grinder and dried for an additional 24 h at the same temperature. The resulting powder was macerated for 72 h with ethanol–diethyl ether solution at a 4:1 ratio, and then filtered through filter paper. The extracts obtained were partially purified before HPLC analysis. The extracts were first evaporated using a rotary evaporator near to dryness and then redissolved in diethyl ether, and the solution was washed three times with a 5% sodium bicarbonate solution in a separatory funnel. The layer of diethyl ether was recovered, evaporated on a rotary evaporator, and redissolved in methanol. The obtained extract was filtered through a 0.22 μm PTFE membrane filter prior to HPLC analysis.

Another set of samples was processed on the day of collection, as follows: the vine shoots were cut into pieces of different dimensions: 2, 3, 4, 5, and 10 cm, and pieces smaller than 1 cm, cut in sections. One sample was left uncut. The samples were left for 6 weeks at laboratory temperature (22 °C) to determine the maximum increase of trans-resveratrol, depending on sample processing. Leaves and tendrils were left for 6 weeks at laboratory temperature (22 °C). The samples were prepared in the same way as the samples describe above. The extracts were kept in a refrigerator at 4 °C until used in the analyses.

2.5. Analytical HPLC Procedure

An HPLC instrument (Shimadzu, Kyoto, Japan) coupled with an SPD-M-20A diode array detector was used to determine the trans-resveratrol content of the samples. The system was equipped with a Phenomenex Kinetex 2.6 μm Biphenyl 100 Å HPLC Column 150 × 4.6 mm and thermostated at 20 °C.

The mobile phases were in pure water (solvent A) and acetonitrile (solvent B). Elution was carried out after a modified method described previously by Marshall et al., (2012), with a binary gradient as follows: linear gradient from 0% to 10% B in 42 min, 10–40% B in 42.6 min, 40–90% B in 46.5 min. Total time of running was 49.5 min [21].

For the determination of trans-resveratrol, the detection wavelength was set at 306 nm and the flow rate of the solvent was 0.5 mL/min.

A high degree of linearity was obtained for standard calibration curves ($R^2 > 0.9996$). LC solution software version 1.21 (Shimadzu, Kyoto, Japan) was used to perform data collection and subsequent processing. The quantification was made based on the peak area, using the external standard method. The analyses were performed in duplicate.

2.6. Determination of Total Phenolic Content

Total phenolic content was determined by the Folin–Ciocalteu method proposed by Luque-Rodríguez et al., (2006) [22]. The Folin–Ciocalteu reagent was diluted with distilled water in a ratio of 1:10, and then 180 µL was added to 90 µL of the tested extract (10 mg/mL). A total of 730 µL of Na_2CO_3 (100 mM) was added, and then it was left in the dark for an hour for incubation. The absorbance of the samples was then measured with an Ocean Optics HR4000CG-UV-NIR spectrometer (SUA) at 765 nm. The calibration curve of gallic acid (0–1000 mg/L, $R^2 = 0.9856$) was used to calculate TPC, and expressed in terms of gallic acid equivalents (GAE).

2.7. Determination of the Antioxidant Activity by the DPPH Method

To determine the DPPH (1,1-diphenyl-2-picrylhydrazyl) radical scavenging activity of the extracts, we used a method proposed by Liu et al., (2015) and Brand-Williams et al., (1995), with some modifications [23,24]. Vine shoot extracts, at 10 mg/mL, were diluted in methanol (0.25 mL) after liquid–liquid extraction, and then the DPPH solution (60 µM) (9.75 mL) was added. The total volume of 10 mL was mixed properly and left to react for 60 min at 25 °C. The absorbance was measured at 515 nm on Ocean Optics QE65000 spectrometer (SUA), using absolute methanol as a blank control. The percentage of free radical scavenging by DPPH was calculated using the following formula:

$$\text{DPPH (\% inhibition)} = ((\text{absorbance of DPPH} - \text{absorbance of sample})/\text{absorbance of DPPH}) \times 100.$$ All determinations were performed in triplicate

2.8. Statistical Analysis

Statgraphics Centurion XIX software-trial version (Manugistics Corp., Rockville, MD, USA) was used to carried out ANOVA.

3. Results

3.1. Statistical Analysis of Variance

The tables presented bellow contain the mean values and standard deviation of resveratrol content, TPC and antioxidant activity for different samples of vine wastes. In the first table (Table 1) are shown the result for three types of vine wastes, leaves, shoots and tendrils. Table 2 presents the results for samples of vine shoots analyzed for 90 days in a row post-pruning. In Table ?? are summarized the results obtained after analyzing the samples of different length after 6 weeks of storage.

Table 1. Analysis of variance of resveratrol content, TPC, and antioxidant activity in leaves, tendrils, and shoots.

Parameters	Waste Type			F-Ratio
	Leaves	**Shoots**	**Tendrils**	
Resveratrol content, mg/kg D.W.	43.54 (±0.08)	1658.22 (±0.17)	169.92 (±0.03)	131,472,834.15 ***
TPC mg GAE/g D.W.	30.42 (±0.32)	29.36 (±0.41)	28.65 (±0.81)	50.43 ***
DPPH %	73.19 (±0.04)	52.72 (±0.22)	14.19 (±0.05)	95,811.06 ***

(Mean values ± standard deviation). ns not significant ($p > 0.05$), *** $p < 0.001$.

Table 2. Analysis of variance of resveratrol content, TPC, and antioxidant activity in vine shoots for 90 days.

Parameters	Storage Days																		F-Ratio
	5	10	15	20	25	30	35	40	45	50	55	60	65	70	75	80	85	90	
Resveratrol content, mg/kg D.W.	86.58 ($\pm$1.24)	163.56 ($\pm$2.31)	282.53 ($\pm$3.01)	386.44 ($\pm$5.12)	557.92 ($\pm$7.89)	923.10 ($\pm$13.05)	1190.79 ($\pm$16.84)	1236.28 ($\pm$17.48)	1349.59 ($\pm$19.09)	1401.91 ($\pm$19.24)	1510.93 ($\pm$18.62)	1524.73 ($\pm$19.40)	1563.39 ($\pm$22.24)	1605.61 ($\pm$22.70)	1753.50 ($\pm$24.80)	1532.31 ($\pm$21.67)	1335.31 ($\pm$18.89)	1135.43 ($\pm$16.05)	2044.65 ***
TPC mg GAE/g D.W.	35.86 ($\pm$0.53)	24.75 ($\pm$0.34)	26.45 ($\pm$0.37)	35.43 ($\pm$0.58)	28.16 ($\pm$0.45)	28.48 ($\pm$0.41)	23.53 ($\pm$0.34)	20.64 ($\pm$0.35)	25.86 ($\pm$0.43)	27.09 ($\pm$0.41)	35.38 ($\pm$0.56)	20.30 ($\pm$0.32)	20.45 ($\pm$0.35)	29.37 ($\pm$0.42)	21.97 (0.36)	23.24 ($\pm$0.37)	23.70 ($\pm$0.31)	29.18 ($\pm$0.43)	3380.74 ***
DPPH %	31.90 ($\pm$0.45)	39.84 ($\pm$0.56)	44.09 ($\pm$0.62)	35.16 ($\pm$0.50)	41.77 ($\pm$0.59)	35.81 ($\pm$0.51)	44.19 ($\pm$0.62)	58.06 ($\pm$0.82)	48.09 ($\pm$0.68)	58.22 ($\pm$0.82)	59.53 ($\pm$0.84)	56.34 ($\pm$0.80)	60.72 ($\pm$0.86)	50.85 ($\pm$0.72)	51.38 ($\pm$0.73)	48.44 ($\pm$0.69)	76.52 ($\pm$1.08)	35.97 ($\pm$0.51)	518.39 ***

(Mean values $\pm$ standard deviation). Ns not significant ($p > 0.05$), *** $p < 0.001$.

Table 3. Analysis of variance of resveratrol content, TPC, and antioxidant activity in vine shoot samples of different lengths.

Parameters	Shoot Length								F-Ratio
	<1 cm	1 cm	2 cm	3 cm	4 cm	5 cm	10 cm	Uncut	
Resveratrol content, mg/kg D.W.	1667.49 ($\pm$23.82)	231.6 ($\pm$3.31)	509.92 ($\pm$7.28)	525.285 ($\pm$7.50)	618.67 ($\pm$8.84)	999.03 ($\pm$14.27)	1406.09 ($\pm$20.09)	1641.64 ($\pm$23.45)	2619.29 ***
TPC mg GAE/g D.W.	22.03 ($\pm$0.31)	22.95 ($\pm$0.41)	22.14 ($\pm$0.32)	28.63 ($\pm$0.44)	35.55 ($\pm$0.52)	26.45 ($\pm$0.43)	24.75 ($\pm$0.41)	29.03 ($\pm$0.42)	2963.32 ***
DPPH %	78.45 ($\pm$1.12)	52.725 ($\pm$0.76)	36.74 ($\pm$0.52)	39.255 ($\pm$0.56)	22.515 ($\pm$0.32)	32.135 ($\pm$0.46)	50.64 ($\pm$0.72)	52.195 ($\pm$0.74)	1225.19 ***

(Mean values $\pm$ standard deviation). Ns not significant ($p > 0.05$), *** $p < 0.001$.

3.2. Kinetics of Resveratrol Increase during Post-Pruning Storage

Trans-resveratrol increase in vine shoots after pruning and storage at laboratory temperature (~20 °C) was analyzed for 90 days in a row and the results are displayed in Figure 1.

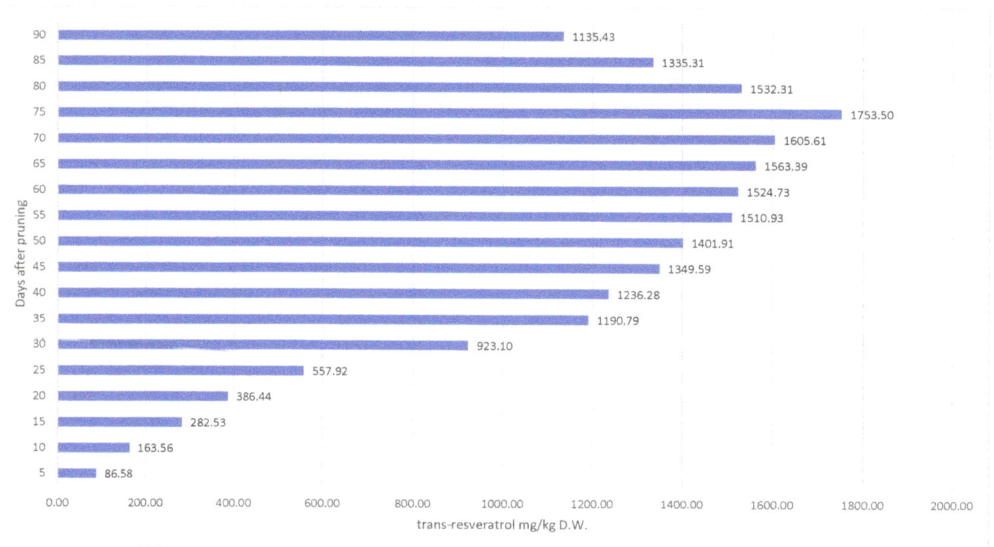

Figure 1. Trans-resveratrol increase in vine shoots for 90 days.

3.3. Total Phenolic Content

Total phenolic content of shoots after pruning and storage at laboratory temperature (~20 °C) was analyzed for 90 days in a row and the results are displayed in Figure 2.

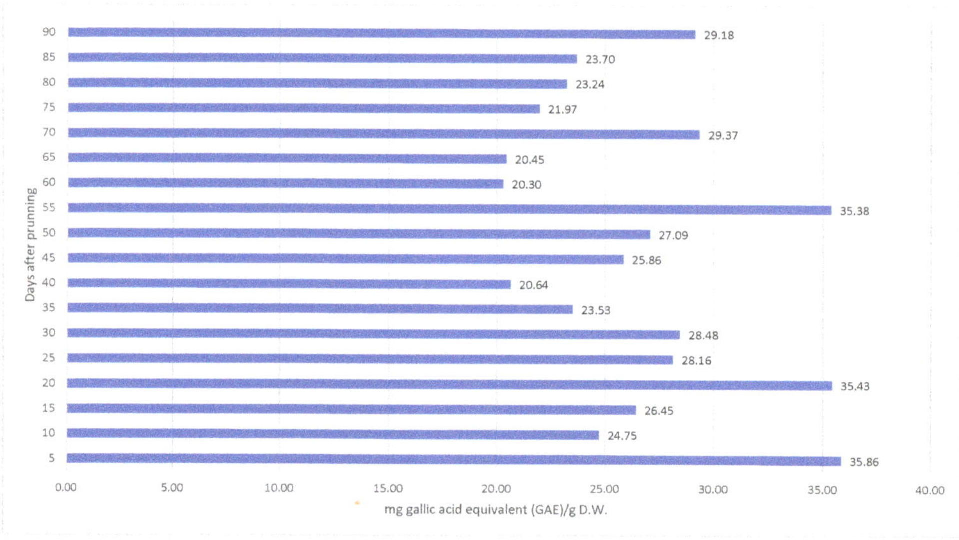

Figure 2. Total phenolic content on grape canes during the 90-day post-pruning.

3.4. Antioxidant Activity by the DPPH Method

Antioxidant activity in vine shoots after pruning and storage at laboratory temperature (~20 °C) was analyzed for 90 days in a row and the results are displayed in Figure 3.

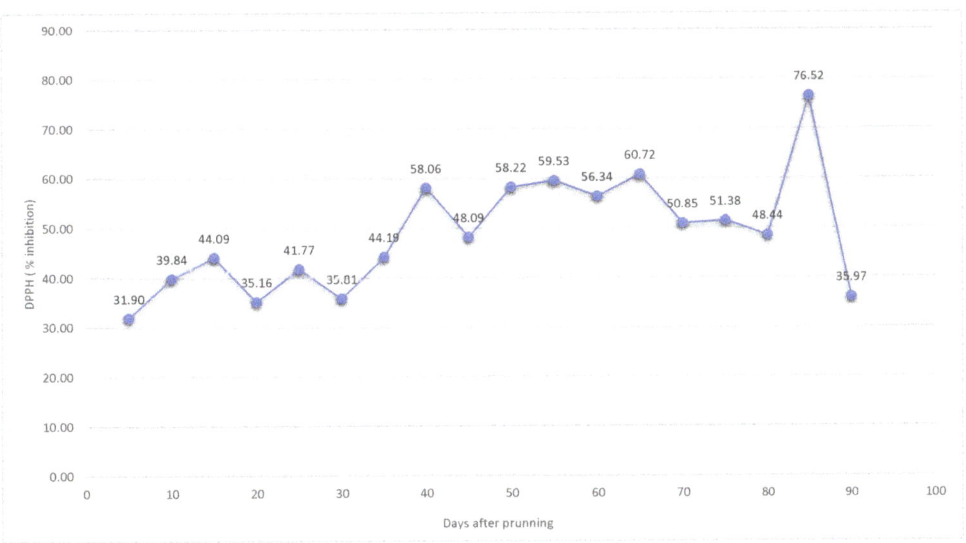

Figure 3. DPPH % inhibition in vine shoots during the 90-day post-pruning time storage.

4. Discussion

4.1. Analysis of Variance of Resveratrol Content, TPC, and Antioxidant Activity in Leaves, Tendrils, and Shoots

The cleaning of the vine generates considerable amounts of vegetable waste each year (tendrils, leaves, and shoots). This green waste could be an important source for high-value phytochemicals, as it (now) does not have any usage, and it could contribute to sustainability programs [25]. Trans-resveratrol is found in abundance in this vine waste, and its content can vary widely among the waste types [26].

Results for samples analyzed after 42 days are represented in Table 1. As it can be seen, the largest amount of resveratrol is contained in vine shoots (1658.22 mg/kg D.W.), followed by tendrils (169.92 mg/kg dry weight (D.W.) and leaves (43.54 mg/kg D.W.). Vine shoots contain up to 10 times more trans-resveratrol than tendrils and up to 40 times more than leaves. Although, besides vine shoots, this vineyard waste also contains trans-resveratrol—extraction and purification are not economically recommended.

These results agree with other studies, such as the study by Lachman et al., (2016), which determine the concentration of resveratrol on vine shoots, tendrils, and leaves. Vine shoots had the highest content of trans-resveratrol (12.5 mg/kg D.W.), followed by tendrils (0.51 mg/kg D.W.) and leaves (0.24 mg/kg D.W.) [26].

A study conducted on vines from the Bohemian and Moravian Region, Czech Republic, determined the resveratrol content in different parts of grapevines (leaves, rachis, and cluster stems). The results varied from 6 to 490 mg/kg D.W. [27]. Cluster stems were found as the richest sources of resveratrol; this is in accordance with the results of this study.

Gyongyi et al., (2017) reported that trans-resveratrol is undetectable in inflorescence and the highest content was found in cluster stems (344.5 mg/kg D.W.) after analyzing nine organs of *Vitis vinifera* cv. Merlot: canes, shoot tips, roots, buds, clusters at the veraison, inflorescences, mature berry skins, cluster stems, and seeds [20].

TPC values for the three types of waste did not vary widely, although vine leaves had the highest content (30.42 (±0.32) mg GAE/g D.W.).

The highest antioxidant activity recorded after the analysis of the extracts was found in the leaves (72.19%), although the highest amount of resveratrol was found in the vine shoots. The results are similar to those obtained by determining the total phenolic content. The lowest percentages were recorded in tendrils (14.69% DPPH inhibition) and vine shoots (52.72% DPPH inhibition).

4.2. Kinetics of Resveratrol Increase during Post-Pruning Storage

Accumulation on trans-resveratrol during the storage period is reported on in many studies [5,28,29]. Recent research showed that increases in trans-resveratrol content in shoot samples were due to biosynthesis pathway genes (PAL, C4H, 4CL, and STS), present in abundance, especially the STS gene (forming trans-resveratrol), which was induced over the first 4 weeks of storage.

Vine shoot storage time after pruning exposed an "important" trans-resveratrol accumulation, with a maximum 181-fold induction on day 70, and a maximum average for days 71–75, as can be seen in Table 2. To highlight the increase in trans-resveratrol content over the 90 days, the average results were calculated for 5 consecutive days; the results are displayed in Figure 1.

Day 1 of the analysis was when the vine waste was collected from the vineyard; the first sample was dried and then extracted with an ethanol–diethyl ether mixture. The minimum value of trans-resveratrol content in vine shoots was registered on day one of the analysis (14.93 mg/kg D.W.) and the maximum value was registered on day 70 of the analysis (2712.86 mg/kg D.W.).

Cebrián et al., (2017) reported an increase in the concentration of trans-resveratrol and other non-volatile phenolic compounds on samples of vine waste for two Spanish varieties (Airèn and Cencibel) after different times of storage (one month, three months, and six months). After one month, the concentration of trans-resveratrol in the Airèn variety was 77.10 ± 11.29 mg/kg D.W. After three months, the concentration almost doubled (151.60 ± 9.42 mg/kg D.W.), and a slight increase was recorded after six months (170.44 ± 3.82 mg/kg D.W). The sample from the Cencibel variety followed the same way regarding increasing the concentration of resveratrol, with 50.41 ± 2.74 mg/kg D.W. after one month. After three months, the concentration was four time higher (224.83 ± 29.64 mg/kg D.W.); after six months of storage, a slight increase was recorded (227.00 ± 6.03 mg/kg D.W.) [28]. For the Airèn variety, the increase in resveratrol content from one month to three months was 196% and 448% for the Cencibel variety. According to the results of this study, the trans-resveratrol content increases only until days 70–75, after which, it decreases slightly. However, the content of trans-resveratrol per kilogram of dry matter was much higher for the Fetească Neagră variety than for the Spanish varieties, which reached up to 923.10 mg/kg D.W. after one month. Trans-resveratrol increase can vary widely due to the synthesis of the monomer in plants, if influenced by various factors (variety of vine, climate, growing conditions, temperature, exposure to fungal and bacterial infections, UV irradiation season of pruning).

These considerable increase in trans-resveratrol content in vine shoot samples can be attributed to the structural genes (PAL, C4H, 4CL and STS), which are responsible for the biosynthesis of trans-resveratrol. They are induced during the storage period, and are active until the vine shoots are dry. It is believed that the drying process during the storage period might be "sensed" to be a stress signal by living plant tissues, resulting in biosynthesis of this monomer, and trans-resveratrol accumulates until the tissues are dry [14].

This is confirmed by the results that show that, at temperatures below 20 °C, trans-resveratrol accumulation is delayed. If the vine shoot samples are stored at −20 °C, trans-resveratrol is not induced [14].

Houille et al., (2015) reported that, over the first six weeks of storage at 20 °C after pruning, the concentration of trans-resveratrol increased highly (approximately 106-fold) in eight different grape vine varieties, as it follows, the Sauvignon Blanc—48-fold induction

(2908 mg/kg D.W.), Chardonnay—33-fold induction (3175 mg/kg D.W.), Côt—21-fold induction (3316 mg/kg D.W.), Grolleau—95-fold induction (3913 mg/kg D.W.), Chenin—106-fold induction (4631 mg/kg D.W.), Pinot Noir—19-fold induction (4725 mg/kg D.W.), Cabernet Franc—83-fold induction (4762 mg/kg D.W.), Gamay—51-fold induction (5100 mg/kg D.W.), and in Means—40-fold induction (4066 mg/kg D.W.) [14]. This study agrees with the present study concerning the increase of trans-resveratrol in the Fetească Neagră variety, which showed an 80-fold induction after 6 weeks (1205.11 mg/kg D.W.).

In the study by Ewald et al., (2017), the increase in trans-resveratrol concentration, depending on the storage time of the shoots, was studied. An increase between 400% and 1400% was observed after the first 6 months of storage. The values varied between 2096 mg/kg D.W. for the Regent variety and 7532 mg/kg D.W. for the Cabernet Sauvignon variety. The trans-resveratrol content dropped after 6 months in most of the varieties, except for the Regent variety, which continued rising [29]. The levels of increase in trans-resveratrol, obtained by Ewald et al., (2017), support our findings, with the mention that the increase of trans-resveratrol registered was between 52% and 18,063% compared to the amount of trans-resveratrol present in the sample from the first day (day of harvest).

A study conducted in Chile agrees with the results obtained in this work, regarding the increase of trans-resveratrol, depending on the retention time of the samples. Gorena et al., (2014) studied the trans-resveratrol increase in vine waste samples of the Pinot Noir variety harvested in 2012, for 8 months of storage, after cutting. A significant increase was registered after the second month. After the third month, the content of trans-resveratrol began to decrease; after that, it remained almost constant [5].

The grape variety also has a high influence on the amount of trans-resveratrol in vine shoots, according to Zhang et al., (2011), who analyzed 165 grape cane samples from large distribution centers and different major grape production regions. Trans-resveratrol content varied depending on the genotypes of vines; thus, *V. labrusca* and *V. vinifera* hybrids, as well as *V. labrusca,* had much lower content than *V. vinifera*. The content of this monomer can vary between the same genotype, depending on the purpose. Thus, in vine shoots from vine grapes for tables, the content of trans-resveratrol is much lower than from wine ones [13]. This could explain the large amounts of resveratrol obtained from extracts of the *Vitis vinifera* Fetească Neagră variety, which is a vine variety for wine.

4.3. Total Phenolic Content

Total phenolic content of the vine shoots was estimated using the Folin–Ciocalteu colorimetric method. The mean values calculated for 5 days in a row for TPC of vine shoots samples analyzed for 90 days are displayed in Table 2. As seen in Figure 2, the total phenolic content (TPC) values remained constant, between a 20 and 30 mg gallic acid equivalent (GAE)/g D.W., except for 5 days when higher values were recorded. The highest value was recorded on day 54 (82.42 mg GAE/g D.W.), and the lowest on day 57 (14.61 mg GAE/g D.W). Total polyphenol content remained constant, which could be attributed to polyphenols, other than trans-resveratrol, whose content may increase, decrease, or remain constant during the 90 days (piceatannol [14,17], apigenin, ellagic acid [30], viniferin, trans-vitisin, trans-piceid [31], gallic acid, ellagic acid, p-coumaric acid and others [32].

Cetin et al., (2011) reported that a TPC for grape canes ranged from 25.36 to 36.56 mg GAE/g D.W. in different vine varieties from Turkey: Alphonse Lavallée, Atasarısı, Cardinal, Hafızali, Horoz, Karası, Isabella, Italia, Sultani, Cekirdeksiz, Tekirdag, Cekirdeksiz, and Trakya Ilkeren [33]. Alexandru et al., (2014) reported that the values for total phenolic content for vine shoots ranged from 18.23 to 198 mg GAE/g D.W. for vineyards from Cuneo, Italy [34].

The TPC of the Portuguese vine shoot extract varieties Touriga Nacional and Tinta Roriz were analyzed by Moreira et al., (2018), using the three different extraction techniques—subcritical water extraction, microwave-assisted extraction, and conventional extraction.

The vine shoot variety Tinta Roriz had the highest total phenolic content (32.1 ± 0.9 GAE/g D.W.) and antioxidant activity [34].

Farhadi et al., (2016) reported higher values for the vine shoots (around 200 mg GAE/g D.W.) compared to the leaves (61 mg GAE/g D.W.) after analyzing the total phenolic content of different parts of the grape and vine (seed, skin, pulp, leaf, and vine shoots) for five different varieties native in West Azerbaijan (Ghara Shira, Hosseini, Agh Shani, Ghara Ghandome, and Ghara Shani) [35].

The lower amount of TPC in the extracts analyzed in this study may be attributed to the extracts being washed with sodium bicarbonate solution, and only the diethyl ether layer was analyzed, so the water-soluble polyphenols were not count. This theory is confirmed by the study by Angelov et al., (2016), where they analyzed the content of polyphenols in both fractions, "ethanol" fraction (with 295.1 mg·g^{-1} TPC content) and an "aqueous" fraction (with 103.0 TPC content), with a 143.7 mg·g^{-1} GAE of dry total extract [2].

4.4. Antioxidant Activity by the DPPH Method

Resveratrol is a phenolic compound that can act against free radicals, and it can act as a promoter of cellular antioxidant enzymes, such as peroxidase, glutathione, glutathione reductase, and glutathione-transferase, to induce neutralization of radical peroxides [36]. It is important to determine the antioxidant activities of vine shoots, due to future correlations with phenolic content, and to establish the important impacts of plant components to human health [22].

The antioxidant activities of the extracts were measured using the DPPH radical scavenging assay, as described by Liu et al., (2015) [23]. Several research studies were conducted to determine the antioxidant activity of vine shoots, via different methods (Barros et al., 2014; Ruiz-Moreno et al., 2015; Ju et al., 2016; Moreira et al., 2018; Karacabey and Mazza, 2010; Rajha et al., 2015); however, the most common was the DPPH method [32,37–42]. Although much data have been reported, antioxidant activity was reported in µM Trolox equivalent/mg extract or inhibition percentage (IC50), and a comparison of the results is not possible. Furthermore, even if the vine shoots are from the same variety, the analytical methods for extractions employed can widely vary, as reported in most studies (Ruiz-Moreno et al., 2015; Ju et al., 2016; Rajha et al., 2015; Farhadi et al., 2016) [35,37,40–42].

The storage times of the shoots after cutting revealed a slight increase in antioxidant activity starting on day 1 (31.90% DPPH inhibition), reaching a maximum of 76.52% DPPH inhibition on day 85, followed by a sharp decrease, as can be seen in Figure 3. To highlight the increase of DPPH inhibition over the 90 days, the average results were calculated for 5 consecutive days; the results are displayed in Figure 3. The increase in antioxidant activity follows the same trend as the increase in resveratrol, with the difference being that the maximum concentration of resveratrol is recorded 10 days before the maximum antioxidant activity, followed by a decrease in both cases. The different paths, regarding the increase of antioxidant activity toward trans-resveratrol, can be due to other polyphenols with antioxidant activities than this monomer, whose content can increase, decrease, or remain constant in vine shoot samples [14,32]. In any case, W=when the samples are dry, in both cases, a decrease is registered. That finding can be correlated to the findings of Houillé et al., who reported that trans-resveratrol and piceatannol accumulate until the tissues are dry [14]. This ascertainment is in agreement with the work by Guerrero et al., (2016), who reported that no correlation between antioxidant activity and total stilbene content was observed, due to the fact that non-phenolic antioxidants could be extracted from the vine shoot, and they could contribute to antioxidant activity [42].

4.5. Influence of Sample Sectioning on Trans-Resveratrol Increase, TPC, and Antioxidant Activity

The high increase of trans-resveratrol during the post-pruning period has been displayed in many studies in the field. Due to stilbene synthase gene transcription inducted by stilbenoid metabolism, this occurrence in vine shoots is explicit [14].

Notably, it was also found that the induction level was modulated by control storage temperatures. Based on STS induction properties, it was hypothesized that the growth of stilbenoid content could be raised by varying the external stress factors [43].

The effect of mechanical stress was tested by cutting the freshly-pruned vine shoots at different section lengths from pieces smaller than 1 cm, cut in sections to 10 cm and an uncut sample, over 42 days of post-harvest storage. Mechanical injury to freshly cut vine waste did not allow excessive accumulation of trans-resveratrol compared to the cut samples. It can be seen that only the sample cut into pieces came close in value to the uncut sample (Table ??). This result is confirmed by the results obtained by Billet et al., (2018), who also tested the effects of mechanical stress on trans-resveratrol accumulation by cutting freshly-pruned vine shoots into pieces from 0.5 to 10 cm. Only the sample cut into 0.5 cm allowed a higher accumulation of the monomer [17].

For the samples cut in different sizes, the highest TPC was found in the 4 cm sample section (35.55 mg GAE/g D.W.), and the lowest TPC content was found in the pieces of vine shoots (22.03 mg GAE/g D.W.).

The samples that were cut into pieces showed the highest antioxidant activity, of 79.24% DPPH inhibition. The uncut sample (52.72% DPPH inhibition) had almost the same antioxidant activity as the 1 cm sample (53.26% DPPH inhibition). The lowest antioxidant activity was for the 4 cm sample of 22.74% DPPH inhibition.

The 4 cm long sample had the higher mean value for TPC (3.555 mg GAE/g D.W.); the mean values for all samples for TPC did not vary widely.

5. Conclusions

The interest in resveratrol commercial supplements has increased due to its proven health benefits (e.g., anti-aging, anti-inflammatory, anti-carcinogenic, neuroprotective, and antioxidant agents), and that it could protect against, to a certain extent, heart and metabolic disorders.

The majority of resveratrol supplements available on the market are made of extracts obtained from the roots of Japanese knotweed *(Polygonum cuspidatum)*; some studies showed that emodin, a compound that could have a laxative effect, was found in partially purified extracts of resveratrol. Research studies show that resveratrol could be extracted from vine waste, which is abundant in this compound, without contamination of emodin.

Moreover, vine waste resulting from annual trimming is utilized as fertilizer by grinding or burning. Agricultural vine waste is an important source for phytochemicals, although economically it is unevaluated.

During the vineyard trimming process, a considerable amount of green mass (portions of shoots, leaves, tendrils) is removed. Vine shoots contain up to 10 times more trans-resveratrol than tendrils, and up to 40 times more than leaves. Although, besides vine shoots, the vineyard waste also contains trans-resveratrol, extraction and purification are not economically recommended.

To increase the extraction yield in the pruned vine shoots, these should be kept for 9–10 weeks at a controlled temperature, at around 20 °C. The increase is 18.063% on day 70 of storage compared to the first day of harvest.

Mechanical stress did not influence the increase of resveratrol content in vine shoots. A small difference was recorded between the sample that was cut in pieces smaller than 0.5 cm (1641.64 mg/kg D.W.) and the uncut samples (1667.49 mg/kg D.W.).

A correlation among the resveratrol content, TPC, and DPPH % inhibition could not be established, due to the fact that polyphenols and other compounds with antioxidant activity present in vine shoots, other than trans-resveratrol, can increase, decrease, or remain constant during the 90 days. The only similarity was that resveratrol content and DPPH % inhibition increased, with the difference being that the resveratrol content significantly increased until day 70, and then dropped from day 76; DPPH % inhibition slightly increased until day 85, and then decreased 50% over the next 5 days.

Author Contributions: Conceptualization, A.L.C. and G.G.; methodology, A.L.C.; validation, A.L.C. and G.G.; formal analysis, A.L.C.; investigation, A.L.C.; resources, A.L.C. and G.G.; data curation, A.L.C.; writing—original draft preparation, A.L.C.; writing—review and editing, A.L.C.; visualization, G.G.; supervision, G.G.; project administration, G.G.; funding acquisition, G.G. All authors have read and agreed to the published version of the manuscript.

Funding: This research received no external funding.

Institutional Review Board Statement: Not applicable.

Informed Consent Statement: Not applicable.

Data Availability Statement: All data related to this study are explicitly plotted in the figures in this article.

Conflicts of Interest: The authors declare no conflict of interest.

References

1. 2019 Statistical Report on World Vitiviniculture. Available online: http://oiv.int/public/medias/6782/oiv-2019-statistical-report-on-world-vitiviniculture.pdf (accessed on 15 March 2021).
2. Angelov, G.; Boyadzhiev, L.; Georgieva, S. Useful bioactive Substances from Wastes: Recovery of trans-resveratrol from grapevine stems. *Open Chem. Eng. J.* **2016**, *10*, 4–9. [CrossRef]
3. Lambert, C.; Richard, T.; Renouf, E.; Bisson, J.; Waffo-Teguo, P.; Bordenave, L.; Ollat, N.; Merillon, J.-M.; Cluzet, S. Comparative analyses of stilbenoids in canes of major *Vitis vinifera* L. *cultivars. J. Agric. Food Chem.* **2013**, *61*, 11392–11399. [CrossRef] [PubMed]
4. Schnee, S.; Queiroz, E.F.; Voinesco, F.; Marcourt, L.; Dubuis, P.-H.; Wolfender, J.-L.; Gindro, K. *Vitis vinifera* canes, a new source of antifungal compounds against Plasmopara viticola, Erysiphe necator, and Botrytis cinerea. *J. Agric. Food Chem.* **2013**, *61*, 5459–5467. [CrossRef] [PubMed]
5. Gorena, T.; Saez, V.; Mardones, C.; Vergara, C.; Winterhalter, P.; Von Baer, D. Influence of post-pruning storage on stilbenoid levels in *Vitis vinifera* L. Canes. *Food Chem.* **2014**, *155*, 256–263. [CrossRef] [PubMed]
6. Ehlting, J.; Hamberger, B.; Million-Rousseau, R.; Werck-Reichhart, D. Cytochromes P450 in phenolic metabolism. *Phytochem. Rev.* **2006**, *5*, 239–270. [CrossRef]
7. Ferrer, J.L.; Jez, J.M.; Bowman, M.E.; Dixon, R.A.; Noel, J.P. Structure of chalcone synthase and the molecular basis of plant polyketide biosynthesis. *Nat. Struct. Biol.* **1999**, *6*, 775–784. [PubMed]
8. Schöppner, A.; Kindl, H. Purification and properties of a stilbene synthase from induced cell suspension cultures of peanut. *J. Biol. Chem.* **1984**, *259*, 6806–6811. [CrossRef]
9. Sparvoli, F.; Martin, C.; Scienza, A.; Gavazzi, G.; Tonelli, C. Cloning and molecular analysis of structural genes involved in flavonoid and stilbene biosynthesis in grape (*Vitis vinifera* L.). *Plant Mol. Biol.* **1994**, *24*, 743–755. [CrossRef]
10. Schröder, G.; Brown, J.W.; Schröder, J. Molecular analysis of resveratrol synthase: cDNA, genomic clones and relationship with chalcone synthase. *Eur. J. Biochem.* **1988**, *172*, 161–169. [CrossRef]
11. Richter, H.; Pezet, R.; Viret, O.; Gindro, K. Characterization of 3 new partial stilbene synthase genes out of over 20 expressed in *Vitis vinifera* during the interaction with Plasmopara viticola. *Physiol. Mol. Plant Pathol.* **2005**, *67*, 248–260. [CrossRef]
12. Wang, W.; Tang, K.; Yang, H.R.; Wen, P.F.; Zhang, P.; Wang, H.L.; Huang, W.D. Distribution of resveratrol and stilbene synthase in young grape plants (*Vitis vinifera* L. Cv. Cabernet Sauvignon) and the effect of UV-C on its accumulation. *Plant Physiol. Biochem.* **2010**, *48*, 142–152. [CrossRef] [PubMed]
13. Zhang, A.; Fang, Y.; Li, X.; Meng, J.; Wang, H.; Li, H.; Zhang, Z.; Guo, Z. Occurrence and estimation of trans-resveratrol in one-year-old canes from seven major Chinese grape producing regions. *Molecules* **2011**, *16*, 2846–2861. [CrossRef] [PubMed]
14. Houillé, B.; Besseau, S.; Courdavault, V.; Oudin, A.; Glévarec, G.; Delanoue Guérin, L.; Simkin, A.J.; Papon, N.; Clastre, M.; Giglioli-Guivarc'h, N.; et al. Byosynthetic origin of trans-resveratrol acummulation in grape canes during pastharvest storage. *J. Agric. Food Chem.* **2015**, *63*, 1631–1638. [CrossRef] [PubMed]
15. Srinivas, G.; Babykutty, S.; Sothiadevan, P.; Srinivas, P. Molecular mechanism of emodin action: Transition from laxative ingredient to an antitumor agent. *Med. Res.* **2007**, *2715*, 591–608.
16. Berchová-Bimová, K.; Soltysiak, J.; Vach, M. Role of different taxa and cytotyses in heavy metals eabsorption in knotweeds (Fallopia). *Sci. Agric.* **2014**, *45*, 11–18.
17. Billet, K.; Houillé, B.; Besseau, S.; Mélin, C.; Oudin, A.; Papon, N.; Courdavault, V.; Clastre, M.; Giglioli-Guivarc'h, N.; Lanoue, A. Mechanical stress rapidly induces E-resveratrol and E-piceatannol biosynthesis in grape canes stored as a freshly-pruned byproduct. *Food chem.* **2018**, *240*, 1022–1027. [CrossRef] [PubMed]
18. Karacabey, E.; Bayindirli, L.; Artik, N.; Mazza, G. Modeling solid−liquid extraction kinetics of trans-resveratrol and trans-ε-viniferin from grape cane. *J. Food Process Eng.* **2013**, *36*, 103–112. [CrossRef]
19. Vergara, C.; von Baer, D.; Mardones, C.; Wilkens, A.; Wernekinck, K.; Damm, A.; Macke, S.; Gorena, T.; Winterhalter, P. Stilbene levels in grape cane of different cultivars in southern Chile: Determination by HPLC−DAD−MS/MS method. *J. Agric. Food Chem.* **2012**, *60*, 929–933. [CrossRef]

20. Gyöngyi, N.; Orsolya, H.; Attila, D.; Laszlo, K. Stilbenes in the different organs of *Vitis vinifera* cv. Merlot grafted on TK5BB rootstock. *OENO One* **2017**, *51*, 323–328.

21. Marshall, D.A.; Stringer, S.J.; Spiers, J.D. Stilbene, ellagic acid, flavonol, and phenolic content of muscadine grape (*Vitis rotundifolia* Michx) Cultivars. *Pharm Crops* **2012**, *3*, 69–77. [CrossRef]

22. Luque-Rodríguez, J.M.; Pérez-Juan, P.; Luque De Castro, M.D. Extraction of polyphenols from vine shoots of *Vitis vinifera* by superheated ethanol– water mixtures. *J. Agric. Food Chem.* **2006**, *54*, 8775–8781. [CrossRef] [PubMed]

23. Liu, F.; Antoniou, J.; Yue, L.Y.J.; Yokoyama, W.; Ma, J.; Zhong, F. Preparation of gelatin films incorporated with tea polyphenol nanoparticles for enhancing controlled-release antioxidant properties. *J. Agric. Food Chem.* **2015**, *63*, 3987–3995. [CrossRef] [PubMed]

24. Brand-Williams, W.; Cuvelier, M.E.; Berset, C. Use of a free radical method to evaluate antioxidant activity. *LWT—Food Sci. Technol.* **1995**, *28*, 25–30. [CrossRef]

25. Rayne, S.; Karacabey, E.; Mazza, G. Vine wastewaste as a source of trans-resveratrol and trans-viniferin: High-value phytochemicals with medicinal and anti-phyto-pathogenic applications. *Ind. Crops Prod.* **2008**, *27*, 335–340. [CrossRef]

26. Lachman, J.; Kotíková, Z.; Hejtmánková, A.; Pivec, V.; Pšeničnaja, O.; Šulc, M.; Střalková, R.; Dědina, M. Resveratrol and piceid isomers concentrations in grapevine shoots, leaves, and tendrils. *Hortic. Sci.* **2016**, *43*, 25–32. [CrossRef]

27. Melzoch, K.; Hanzlíková, I.; Filip, V.; Buckiová, D.; Šmidrkal, J. Resveratrol in parts of vine and wine originating from Bohemian and Moravian vineyard regions. *Agric. Conspec. Sci.* **2001**, *66*, 53–57.

28. Cebrián, C.; Sánchez-Gómez, R.; Salinas, M.R.; Alonso, G.L.; Zalacain, A. Effect of post-pruning vine-shoots storage on the evolution of high-value compounds. *Ind. Crops Prod.* **2017**, *109*, 730–736. [CrossRef]

29. Ewald, P.; Delker, U.; Winterhalter, P. Quantification of stilbenoids in grapevine canes and grape cluster stems with a focus on long-term storage effects on stilbenoid concentration in grapevine canes. *Food Res. Int.* **2017**, *100*, 326–331. [CrossRef]

30. Jesus, M.S.; Genisheva, Z.; Romaní, A.; Pereira, R.N.; Teixeira, J.A.; Domingues, L. Bioactive compounds recovery optimization from vine pruning residues using conventional heating and microwave-assisted extraction methods. *Ind. Crops Prod.* **2019**, *132*, 99–110. [CrossRef]

31. Wei, Y.; Li, P.; Ma, L.; Li, J. Separation and purification of four stilbenes from *Vitis vinifera* L. cv. cabernet sauvignon roots through high-speed counter-current chromatography. *S. Afr. J. Enol. Vitic.* **2014**, *35*, 226–233. [CrossRef]

32. Moreira, M.M.; Barroso, M.F.; Porto, J.V.; Ramalhosa, M.J.; Švarc-Gajić, J.; Estevinho, L.; Morais, S.; Delerue-Matos, C. Potential of Portuguese vine shoot wastes as natural resources of bioactive compounds. *Sci. Total Environ.* **2018**, *634*, 831–842. [CrossRef] [PubMed]

33. Çetin, E.S.; Altinöz, D.; Tarçan, E.; Baydar, N.G. Chemical composition of grape canes. *Ind. Crops Prod.* **2011**, *34*, 994–998. [CrossRef]

34. Alexandru, L.; Binello, A.; Mantegna, S.; Boffa, L.; Chemat, F.; Cravotto, G. Efficient green extraction of polyphenols from post-harvested agro-industry vegetal sources in Piedmont. *Comptes Rendus Chim.* **2014**, *17*, 212–217. [CrossRef]

35. Farhadi, K.; Esmaeilzadeh, F.; Hatami, M.; Forough, M.; Molaie, R. Determination of phenolic compounds content and antioxidant activity in skin, pulp, seed, cane and leaf of five native grape cultivars in West Azerbaijan province, Iran. *Food Chem.* **2016**, *199*, 847–855. [CrossRef] [PubMed]

36. Lastra, C.A.; Villegas, I. Resveratrol as an atioxidant and pro-oxidant agent: Mechanisms and clinical implications. *Biochem. Soc. Trans.* **2007**, *35*, 1156–1160. [PubMed]

37. Ruiz-Moreno, M.J.; Raposo, R.; Cayuela, J.M.; Zafrilla, P.; Pineiro, Z.; Moreno-Rojas, J.M.; Mulero, J.; Puertas, B.; Giron, F.; Guerrero, R.F.; et al. Valorization of grape stems. *Ind. Crops Prod.* **2015**, *63*, 152–157. [CrossRef]

38. Barros, A.; Gironés-Vilaplana, A.; Teixeira, A.; Collado-González, J.; Moreno, D.A.; Gil-Izquierdo, A.; Rosa, E.; Domínguez-Perles, R. Evaluation of grape (*Vitis vinifera* L.) stems from Portuguese varieties as a resource of (poly) phenolic compounds: A comparative study. *Food Res. Int.* **2014**, *65*, 375–384. [CrossRef]

39. Ju, Y.; Zhang, A.; Fang, Y.; Liu, M.; Zhao, X.; Wang, H.; Zhang, Z. Phenolic compounds and antioxidant activities of grape canes extracts from vineyards. *Span. J. Agric. Res.* **2016**, *14*, 18. [CrossRef]

40. Karacabey, E.; Mazza, G. Optimisation of antioxidant activity of vine wasteextracts using response surface methodology. *Food Chem.* **2010**, *119*, 343–348. [CrossRef]

41. Rajha, H.N.; Chacar, S.; Afif, C.; Vorobiev, E.; Louka, N.; Maroun, R.G. β-Cyclodextrin-assisted extraction of polyphenols from vine shoot cultivars. *J. Agric. Food Chem.* **2015**, *63*, 3387–3393. [CrossRef]

42. Guerrero, R.F.; Biais, B.; Richard, T.; Puertas, B.; Waffo-Teguo, P.; Merillon, J.M.; Cantos-Villar, E. Grapevine cane's waste is a source of bioactive stilbenes. *Ind. Crops Prod.* **2016**, *94*, 884–892. [CrossRef]

43. Vannozzi, A.; Dry, I.B.; Fasoli, M.; Zenoni, S.; Lucchin, M. Genome-wide analysis of the grapevine stilbene synthase multigenic family: Genomic organization and expression profiles upon biotic and abiotic stresses. *BMC Plant Biol.* **2012**, *12*, 1–22. [CrossRef] [PubMed]

![MDPI]

Article

Effect of Biostimulants on the Yield and Quality of Selected Herbs

Joanna Majkowska-Gadomska [1], Krzysztof Jadwisieńczak [2], Anna Francke [1] and Zdzisław Kaliniewicz [2,*

[1] Department of Agroecosystems and Horticulture, University of Warmia and Mazury in Olsztyn, 10-957 Olsztyn, Poland; majkowska-gadomska@uwm.edu.pl (J.M.-G.); afrancke@uwm.edu.pl (A.F.)
[2] Department of Heavy Duty Machines and Research Methodology, University of Warmia and Mazury in Olsztyn, 10-957 Olsztyn, Poland; krzysztof.jadwisienczak@uwm.edu.pl
* Correspondence: zdzislaw.kaliniewicz@uwm.edu.pl; Tel.: +48-89-523-39-34

Abstract: The aim of this study was to determine the effect of amino acid biopreparations on the yield of summer savory, marjoram, and lemon balm, and the concentrations of selected biochemical and mineral compounds in their herbage. The first experimental factor was plant species: summer savory (*Satureja hortensis* L.) var. *Saturn*, marjoram (*Origanum majorana* L.) var. *Miraż*, and lemon balm (*Melissa officinalis* L.). The second experimental factor was the effect exerted by two biopreparations, Calleaf Aminovital and Maximus Amino Protect, on herbage yield and quality. In the control treatment, plants were sprayed with water. The analyzed herb species differed considerably in yield and the concentrations of selected biochemical compounds and minerals. Lemon balm was characterized by the highest yield (1.73 kg m^{-2}). Marjoram var. *Miraż* was characterized by the highest concentrations of reducing sugars (0.89 g 100 g^{-1} FM) and L-ascorbic acid (39.7 mg 100 g^{-1} FM). Summer savory was most abundant in total N, K, and Ca. The tested biostimulants contributed to a decrease in nitrate concentrations in the studied plants. The interaction between the experimental factors significantly affected the content of nitrates and mineral compounds and total N, P, K, and Ca in the herbage of the analyzed plant species.

Keywords: plant growth regulators; plant properties; biochemical compounds; mineral compounds; crop quality

Citation: Majkowska-Gadomska, J.; Jadwisieńczak, K.; Francke, A.; Kaliniewicz, Z. Effect of Biostimulants on the Yield and Quality of Selected Herbs. *Appl. Sci.* **2022**, *12*, 1500. https://doi.org/10.3390/app12031500

Academic Editors: Georgiana Gabriela Codină and Mircea Oroian

Received: 29 December 2021
Accepted: 28 January 2022
Published: 30 January 2022

Publisher's Note: MDPI stays neutral with regard to jurisdictional claims in published maps and institutional affiliations.

1. Introduction

The improvement of human health is a priority for many individuals dealing with food, including herb, vegetable, and fruit growers. Mineral elements play an important role in healthy growth and development, help maintain fitness, and prevent infections [1]. Among crops, herbs are valuable sources of readily absorbable mineral substances [2,3]. Herbs have gained popularity among consumers due to their medicinal properties and unique aroma [4]. As a source of biologically active substances, bioavailable micronutrients, and macronutrients, herbs can also be used in the prevention and treatment of many fungal diseases. The high bioavailability and optimal ratios of nutrients have contributed to the growing demand for plant raw materials [4–8]. The Polish market offers a wide range of herbs and spices. However, despite their medicinal properties, they are used mostly for culinary purposes, to enhance the taste, aroma, color, and appearance of dishes [9]. Summer savory (*Satureja hortensis* L.), marjoram (*Origanum majorana* L.), and lemon balm (*Melissa officinalis* L.) are well-known and highly valued species of the family *Lamiaceae*. Their herbage is used in both fresh and dried forms. In Poland, these herbs are used mostly to flavor foods and beverages [9]. The leaves of summer savory and marjoram have antiseptic, anti-inflammatory, analgesic, antioxidant, and antimicrobial properties. Lemon balm also exerts calming and antidepressant effects [4,8,10,11].

The chemical composition and nutrient content of the herbage of medicinal plants are determined by genetic and environmental factors [6]. According to Gugała et al. [12], integrated plant production technologies are insufficient to utilize the full biological potential of herb species due to the poor absorption of nutrients by leaves and difficult transport through plant tissues during periods of climatic or biotic stress. The increased consumption of organic herbs and spices in fresh form has contributed to the development of new cultivation methods. These new methods aim to optimize the growth and development of plants [11,13–16]. Amino acid biostimulants can counteract adverse environmental effects. Amino acids are the building blocks of proteins, and their supply decreases energy expenditure for nitrogen (N) synthesis [17]. According to Trawczyński [18], Maciejewski et al. [9], and Sawicka et al. [19], biostimulants support biochemical, morphological, and physiological processes in plants. Biostimulants promote the vegetative growth of herbaceous plants. Amino acid biostimulants available on the market enhance enzyme activity in plants and improve their stress tolerance.

Therefore, the research hypothesis postulates that the tested biostimulants may have a positive effect on the morphological traits and yield of herbs. Biostimulants are expected to increase the content of selected compounds as well as macronutrients and micronutrients in herbal raw material. In view of the above, the objective of this study was to determine the effect of amino acid biopreparations on the morphological traits and yield of summer savory var. *Saturn*, marjoram var. *Miraż*, and lemon balm, and the concentrations of selected nutrients and chemical elements in their herbage.

2. Materials and Methods

2.1. Study Site and Experimental Factors

The experiment was conducted in a heated greenhouse equipped with movable flood tables, at the University of Warmia and Mazury in Olsztyn (20°29′ E, 53°45′ N; 125 m a.s.l.), between 28–30 April and 30 June 2017–2018. The experiment had a randomized block design with three replications. The first experimental factor was plant species: summer savory (*Satureja hortensis* L.) var. *Saturn*, marjoram (*Origanum majorana* L.) var. *Miraż*, and lemon balm (*Melissa officinalis* L.) (Seed Company W. Legutko, Nad Stawem, Poland). The herbs selected for the study are most widely cultivated in Poland. The second experimental factor was the effect exerted by two biopreparations, Calleaf Aminovital (Calfert®, Warsaw, Poland), and Maximus Amino Protect (Ekoplon, Grabki Duże, Poland). The products were used in diluted form.

Calleaf Aminovital (according to the product label) is an organic foliar fertilizer in liquid form, with an amino acid-balanced composition, which contains large amounts of polypeptides and free amino acids. The product is obtained in the process of the hydrolysis of natural proteins.

Maximus Amino Protect (according to the product label) is a combination of free amino acids, peptides, and potassium phosphate, which has strong biostimulatory and antistress effects. Potassium phosphite stimulates the immune system of plants and prevents fungal diseases. Peptides improve the biological effectiveness of the fertilizer—they support the active uptake and penetration of micronutrients and plant protection products. It is practically advisable to use this fertilizer as an addition to treatments against fungi attacking plants (fungicide treatments), as it significantly improves the effectiveness of these treatments. In the control treatment, plants were sprayed with water.

In 2017–2018, in the last days of April, 10 seeds of each herb species were sown in 0.7 dm^3 pots. Each treatment consisted of 64 pots and had an area of 1 m^2. The experiment was performed in triplicate. The horticultural substrate used in the experiment was characterized by salt concentration of 1.5 g dm^{-3}, pH 5.9, and the following chemical composition: N-NO$_3$—112 mg dm^{-3}, P—80 mg dm^{-3}, K—143 mg dm^{-3}, Ca—1240 mg dm^{-3}, and Mg—383 mg dm^{-3}. Crop protection chemicals were not applied. Yellow sticky traps were placed over the pots to control the number of pests (*Trialeurodes vaporariorum* and *Sciara militaris*). *Encarsia formosa* (three individuals per m^2) were used for the biological control of *Trialeurodes vaporariorum*. Leaves were harvested at maturity.

2.2. Greenhouse Microclimate Conditions

Greenhouse microclimate conditions were modified to meet the needs of plants in different growth stages [20]. The microclimate was controlled by a climate computer, and temperature parameters were manually adjusted. The temperature was maintained at 24 °C during seed germination. To increase air humidity, the pots were covered with non-woven fabric. The non-woven fabric was removed after the seeds had germinated. At the same time, the temperature was lowered to 20 °C during the day and 18 °C at night. During the growing season, cultivation measures were limited to plant watering and pruning at 5 weeks after sowing to promote growth.

The biostimulants were applied three times, at 14-day intervals, beginning at 2 weeks after seedling emergence, at a dose of 0.3%. The doses given in amounts per ha were diluted in water at 500 L ha^{-1}. In order to prepare a smaller amount of working solution, the recommended dose was adjusted proportionally. Plants were supplied with water from the water mains.

2.3. Harvest and Yield of Herbs

A once-over harvest of herbage was carried out. The plants were cut manually, 1.5 cm above the ground, in the vegetative growth stage. Plant height and plant weight per pot were determined, and average samples of herbage were collected from marketable yield in each treatment. Plant height was measured using a 1 m ruler. Fresh weight was determined using the Radwag PST 750 R2 digital balance (Radwag, Radom, Poland). The weight and height of fresh herbs were determined separately for each pot. The results were then summed up and averaged. Marketable yield was identical to total yield.

2.4. Chemical Analysis of Herbs

Immediately after cutting and weighing the herbs, all plant material was dried. The herbs were dried for 12 h at 45 °C in a Binder Avantgarde Line laboratory dryer with air circulation. Herbage samples were analyzed to determine the content of selected biochemical compounds. The chemical composition of herbage was analyzed by determining the content of air-dry matter and dry matter (DM) by drying the samples at 105 °C to a constant weight [21] in a Pol-Eko Aparatura SLW 535 SD laboratory dryer. To calculate the percentage of DM in air-dried samples, the weight of the sample after drying was divided by the weight of the sample before drying and multiplied by 100. All herb samples were weighed in containers (in duplicate). The containers were placed in an oven at 60 °C for 72 h. The containers were re-weighed and the weight of the samples was determined. Weight loss is an indicator of the amount of water and DM.

Total sugars and reducing sugars were determined as described by Luff-Schoorl [22], L-ascorbic acid was determined by the method proposed by Tillmans and modified by Pijanowski [23], and total acidity [24] and nitrates (V) were determined colorimetrically with the TECHCOMP UV2310II spectrometer, using salicylic acid [25].

The concentrations of macronutrients were determined in dry and wet mineralized plant materials in three replications. Plants were dried for 24 h at 65 °C in a Binder ED400 dryer (Binder GmbH, Tuttlingen, Germany), and were ground in a Grindomix GM300 knife mill (Retsch GmbH, Haan, Germany). To determine macronutrient content, herbage samples were wet mineralized in H_2SO_4 with the addition of H_2O_2 as the oxidizing agent, using the SpeedDigester K-439 unit (Büchi Labortechnik AG, Flawil, Switzerland). To determine micronutrient content, herbage samples were wet mineralized in a mixture of HNO_3 + $HClO_4$ + HCl using a CEM Mars 5 Digestion Oven (CEM Corporation, Matthews, NC, USA).

Herbage samples were analyzed to determine the content of total nitrogen (N_{total}) by the Kjeldahl method, phosphorus (P) by the colorimetric method (UV-1201V spectrophotometer, Shimadzu Corporation Kyoto, Tokyo, Japan), potassium (K) and calcium (Ca) by atomic emission spectrometry (AES) (Flame Photometers, BWB Technologies Ltd.,

Newbury, UK), and magnesium (Mg) by atomic absorption spectrometry (AAS) (AAS1N, Carl Zeiss Jena, Jena, Germany).

2.5. Statistical Analysis

The results were presented as means for the years of study since only minor differences were noted. A two-way analysis of variance (ANOVA) was performed to determine significant differences between each treatment and the control treatment for both experiments. The differences between treatment means were pooled by Tukey's test at a 0.05 probability level. The analysis was performed using SAS (Statistical Analysis System) software, ver. 12 (TIBCO Software Inc. Statistica, Paolo Alto, CA, USA).

3. Results and Discussion

Differences in the growth rate and yields of the analyzed herb species are presented in Table 1 and Figure 1. Lemon balm was characterized by the significantly highest fresh herbage weight per pot (27.8 g). At the time of harvest, summer savory plants were the tallest, and marjoram plants were the shortest. The use of the Maximus Amino Protect biostimulant in the cultivation of the tested plant species has a positive effect on the height and weight of plants, increasing them by 3.4% and 1.4%, respectively, on average. When applied in small concentrations, these substances enhance nutrition efficiency, abiotic stress tolerance, and/or crop quality traits, regardless of their nutrient content. Exogenously applied biostimulants have a similar mode of action to that of plant hormones such as auxins, gibberellins, and cytokinins [26]. Similar results can be obtained by cultivating garden savory [10] and bell peppers [27] at different rates of nitrogen–potassium fertilizers. According to El-Gohary et al. [28], NPK and amino acids have a pronounced effect on all tested parameters (fresh and dry herbage of *Satureja hortensis* L., essential oil content and its constituents).

The use of an appropriate biostimulant can improve the vitality of roots and shoots, thus increasing crop yields. The total fresh herbage yield was 1.07 kg m^{-2} (*Satureja hortensis* L.), 1.35 kg m^{-2} (*Origanum majorana* L.), and 1.73 kg m^{-2} (*Melissa officinalis* L.) on average. Crop yield is significantly affected by the cultivated species. The obtained results corroborate the findings of other authors, indicating that plants characterized by different growth rates of aboveground and underground organs may differ in the yield of edible parts [10,15].

There are differences in the chemical composition of the analyzed herb species (Table 2, Figure 2). According to other authors, DM content is highest in marjoram (20.4%), savory, lemon balm, hyssop, and chervil (16.4% on average), and lowest in basil and tarragon (12.9% on average) [29]. Fresh herb species tested in this experiment had a relatively low DM content ranging from 12.2% in *Satureja hortensis* L. to 16.3% in *Origanum majorana* L. Biostimulants had no significant effect on the DM content of the analyzed herbs. Similar DM content was reported by Nurzyńska-Wierdak and Zawiślak [30] with 9.4% on average in sweet basil and lemon balm, Kazimierczak et al. [13,31] with 22.1 to 26.0% in lemon balm, and Nurzyńska-Wierdak et al. [15] with 13.5 to 17.4% in lemon balm and 11.9 to 17.1% in marjoram, depending on the sowing date and pot size. In a study by Telesiński et al. [32], summer savory and marjoram had higher DM contents (21.4% and 30.3%, respectively).

Table 1. *p*-Value of ANOVA for plant height, fresh herbage weight, and yield in the analyzed herb species.

Factor	Plant Height	Fresh Herbage Weight per Pot	Yield
Treatment (A)	0.922	0.991	0.991
Species (B)	<0.001	<0.001	<0.001
Interaction (A × B)	0.834	0.351	0.351

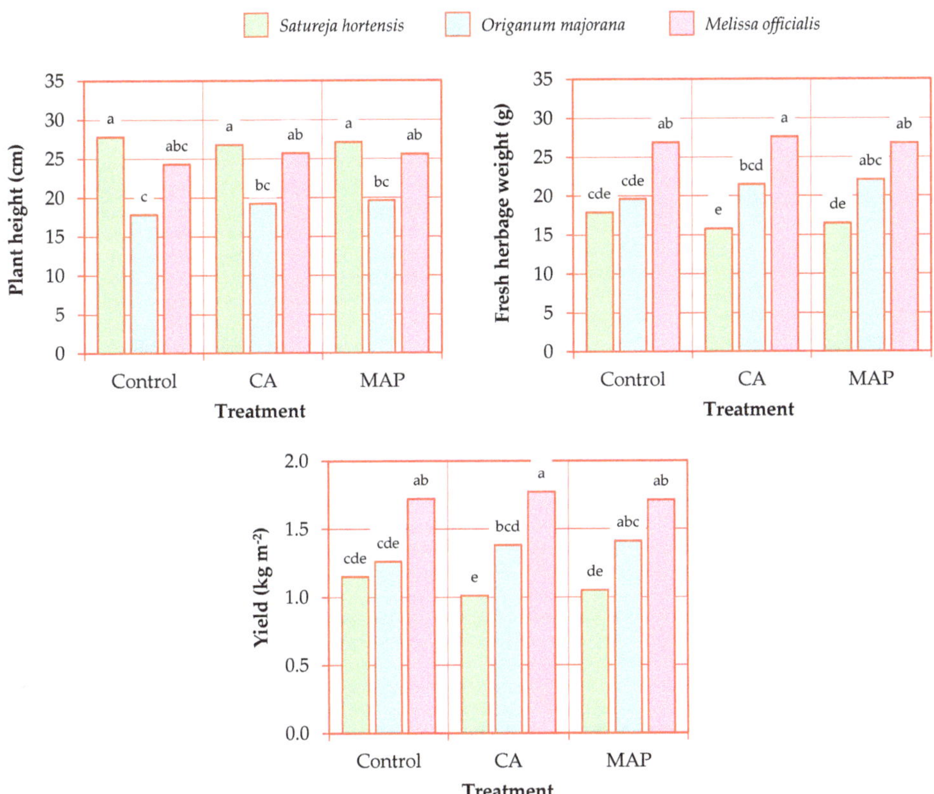

Figure 1. Effect of biostimulants on plant height, fresh herbage weight, and yield in the analyzed herb species: a–e—various letters denote significant differences at $p < 0.05$ (Tukey's test); CA—application of Calleaf Aminovital; MAP—application of Maximus Amino Protect.

Table 2. *p*-Value of ANOVA for the nutrient composition of herbage in the analyzed herb species.

Factor	c_{dm}	c_{ts}	c_{rs}	c_O	c_L	c_N
Treatment (A)	0.061	0.381	0.791	0.379	0.803	<0.001
Species (B)	0.096	0.887	<0.001	0.324	<0.001	0.016
Interaction (A × B)	0.520	0.340	0.145	0.325	0.318	<0.001

c_{dm}—Dry matter content, c_{ts}—total sugar content, c_{rs}—reducing sugar content, c_O—organic acid content, c_L—L-ascorbic acid content, c_N—nitrate (V) content.

There were no statistically significant differences in the content of total sugars and reducing sugars between the tested herb species; the content of total sugars ranged from 0.52 to 1.61 g 100 g^{-1} FM, and the content of reducing sugars ranged from 0.31 to 1.01 g 100 g^{-1} FM. The concentrations of total sugars and reducing sugars were the highest in marjoram herbage and the lowest in summer savory herbage. Nurzyńska-Wierdak et al. [15] reported similar concentrations of total sugars and reducing sugars, i.e., 0.56–1.49 and 0.25–0.32 g 100 g^{-1} FM, respectively, in lemon balm, and 0.93–1.00 and 0.17–0.42 g 100 g^{-1} FM, respectively, in marjoram.

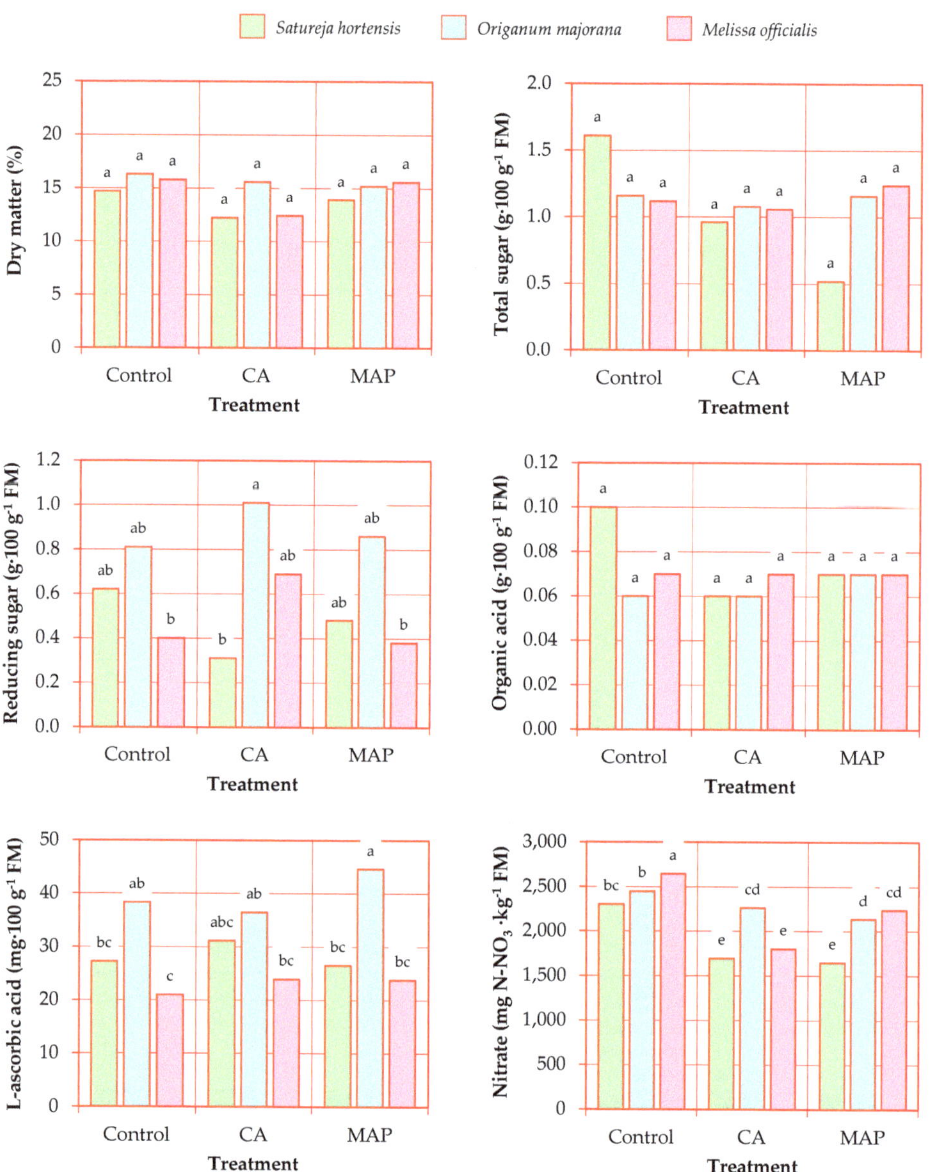

Figure 2. Effect of biostimulants on the nutrient composition of herbage in the analyzed herb species: a–e—various letters denote significant differences at $p < 0.05$ (Tukey's test); CA—application of Calleaf Aminovital; MAP—application of Maximus Amino Protect.

According to Frąszczak et al. [33], the concentration of L-ascorbic acid in herbs is significantly affected by the species and cultivation method. In the present study, marjoram was characterized by the highest concentration of L-ascorbic acid (39.7 mg 100 g^{-1} FM on average). The L-ascorbic acid content of the herbage was not significantly modified by the biostimulants. The concentration of L-ascorbic acid increased in marjoram treated with Maximus Amino Protect (44.6 mg 100 g^{-1} FM), whereas it was significantly lower in lemon balm in all treatments and in summer savory in the control treatment and in the treatment with Maximus Amino Protect. Similar L-ascorbic acid concentrations were determined by Kazimierczak et al. [31] in lemon balm (32.1–76.3 mg 100 g^{-1} FM depending on the cultivation method), and Nurzyńska-Wierdak et al. [15] (9.8–79.4 mg 100 g^{-1} FM in lemon balm and 23.6–74.5 mg 100 g^{-1} FM in marjoram). In a study by Frąszczak et al. [33], the L-ascorbic acid content of lemon balm herbage ranged from 4.2 to 5.3 mg 100 g^{-1} FM.

The concentrations of organic acids in the analyzed herb species were not significantly affected by the experimental factors or their interaction. Summer savory herbage harvested in the control treatment had the highest organic acid content (0.10 g 100 g^{-1} FM). The concentration of organic acids was the lowest (0.06 g 100 g^{-1} FM) in summer savory and marjoram treated with Calleaf Aminovital.

According to Commission Regulation (EU) No. 1881/2006 of 19 December 2006 [34], which set the maximum levels for certain contaminants in foodstuffs, the maximum nitrate levels in lettuce plants grown in a greenhouse and in the field are 3500–4500 N-NO$_{-3}$ kg^{-1} FM and 2500–4000 N-NO$_{-3}$ kg^{-1} FM, respectively. Such standards have not been established for herbs and spices. In the current study, the nitrate content of herbage in the analyzed herb species was relatively low. Proper plant nutrition promotes good plant development and growth, as well as an adequate circulation of matter and the reduction of harmful residues. Nitrate levels were significantly affected by herb species, ranging from 1880 N-NO$_{-3}$ kg^{-1} FM in summer savory herbage to 2282 N-NO$_{-3}$ kg^{-1} FM in marjoram herbage on average. The use of Calleaf Aminovital and Maximus Amino Protect in the cultivation of spice plants reduced the accumulation of nitrates. The average differences were 22.8% and 18.7%, respectively, relative to control treatment plants. An analysis of the interaction between the experimental factors revealed that the significantly highest amounts of nitrates were accumulated in lemon balm in the control treatment (2646 N-NO$_{-3}$ kg^{-1} FM) and the lowest amounts were in summer savory treated with Maximus Amino Protect (1646 N-NO$_{-3}$ kg^{-1} FM). Telesiński et al. [32] reported similar nitrate levels in summer savory (1700 N-NO$_{-3}$ kg^{-1} FM).

Macronutrients play important functions in the body, including the regulation and maintenance of the acid–base balance. The daily requirement for these ingredients is over 100 mg. The mineral status of plants is important not only for the nutritional value of food, but also for the growth, development, and yielding of crops [1]. Plants are exposed to the influence of adverse environmental factors, which are known as abiotic stresses, excessive soil salinity, and the absence of mineral salts [3]. The concentrations of selected macronutrients in the analyzed herb species are presented in Figure 3. Herb species had a significant effect (Table 3, Figure 3) on total N content (14.4 g kg^{-1} DM on average). Summer savory herbage was characterized by the significantly highest total N content (18.5 g kg^{-1} DM on average). Maximus Amino Protect had a positive influence on total N levels in the examined herb species, which reached 16.1 g kg^{-1} DM on average. However, the observed differences are not statistically significant. The total N content of summer savory treated with Maximus Amino Protect was significantly higher (24.8 g kg^{-1} DM) than in the remaining treatments. Lower total N concentrations were noted in lemon balm herbage in all treatments, and in summer savory treated with Calleaf Aminovital. Different total N levels were reported by Seidler-Łożykowska et al. [35]—2.5–3.8% in summer savory and 2.0–2.6% in marjoram, depending on the cultivation method. Dzida and Jarosz [36] found that the total N content of summer savory herbage ranged from 2.9 to 4.5% DM, depending on fertilization. In the experiment conducted by Skubij et al. [3], nitrogen rate and plant growth stage had a significant effect of the total nitrogen content of summer savory var. *Saturn*.

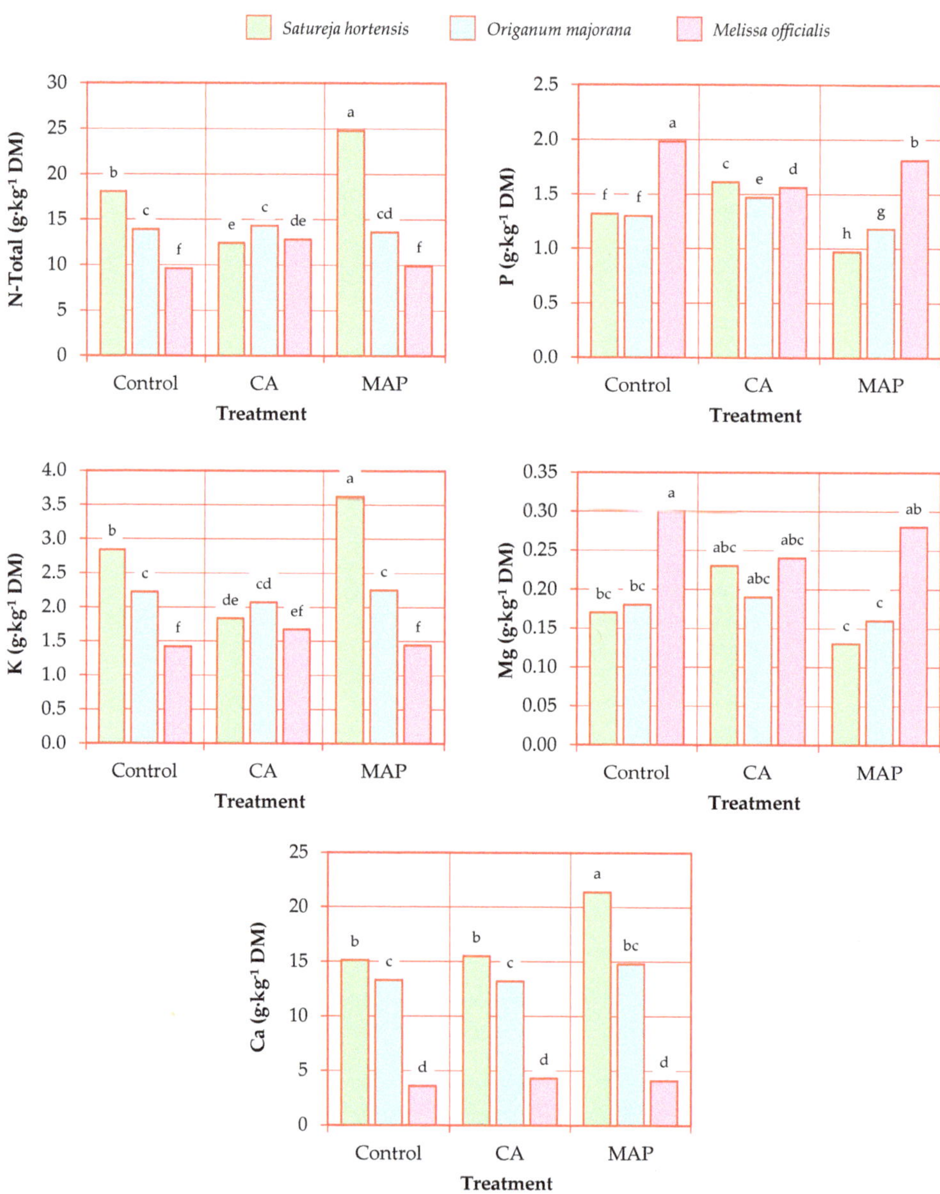

Figure 3. Effect of biostimulants on the mineral composition of herbage in the analyzed herb species: a–h—various letters denote significant differences at $p < 0.05$ (Tukey's test); CA—application of Calleaf Aminovital; MAP—application of Maximus Amino Protect.

Table 3. *p*-Value for ANOVA on the mineral composition of herbage in the analyzed herb species.

Factor	N-Total	P	K	Mg	Ca
Treatment (A)	0.359	0.211	0.209	0.586	0.610
Species (B)	<0.001	<0.001	<0.001	<0.001	<0.001
Interaction (A × B)	<0.001	<0.001	<0.001	0.057	<0.001

In a study by Seidler-Łożykowska et al. [35], the P content of herbage ranged from 0.97 to 1.61 g kg^{-1} DM in *Satureja hortensis* L. and from 1.30 to 1.47 g kg^{-1} DM in *Origanum majorana* L. In the present study, the P content of herbage was the lowest in summer savory treated with Maximus Amino Protect (0.97 g kg^{-1} DM) and the highest in lemon balm in the control treatment (1.98 g kg^{-1} DM).

The average K content of herbage ranged from 1.51 g kg^{-1} DM in lemon balm to 2.76 g kg^{-1} DM in summer savory. An analysis of the interaction between the experimental factors revealed that summer savory treated with Maximus Amino Protect was characterized by the significantly highest K content, whereas K concentrations were lower in lemon balm herbage in all treatments. Seidler-Łożykowska et al. [35] demonstrated that the levels of this macronutrient varied across species and reached 2.4–4.1% in summer savory and 1.6–2.7% in marjoram.

According to Kudelka and Kosowska [37], the content of Mg and Ca in lemon balm, thyme, and cayenne pepper is lower than that in basil. In the present study, differences were found in Mg content between the analyzed herb species, and it was higher in lemon balm than in summer savory and marjoram. Similar results were reported by Dzida and Jarosz [10,36] for marjoram (0.15–0.31% DM) and summer savory herbage (0.32% DM to 0.46% DM). In the cited study, Mg concentrations were higher (1.72 g kg^{-1} DM on average) in all analyzed herb species, and the Mg content of lemon balm ranged from 0.60 to 3.22 g kg^{-1} DM in [37]. In a study by Seidler-Łożykowska et al. [35], Mg concentrations reached 0.41–0.71% in summer savory and 0.27–0.39% in marjoram, and were comparable with those noted in the present experiment.

Raczuk et al. [38] found that the average Ca content of herbage in the analyzed herb species was 19.6 g kg^{-1} DM, ranging from 10.2 to 13.0 g kg^{-1} DM in lemon balm. In the current study, Ca concentrations varied widely across species, from 4.0 g kg^{-1} DM in lemon balm to 17.3 g kg^{-1} DM in summer savory. Maximus Amino Protect contributed to an increase in the Ca content of herbage. Summer savory treated with Maximus Amino Protect was characterized by the highest Ca content (21.4 g kg^{-1} DM). The concentrations of Ca were significantly lower in lemon balm in all treatments (3.6–4.3 g kg^{-1} DM). In the work of Dzida and Jarosz [10,36], Ca content ranged from 0.7 to 1.9 g kg^{-1} DM in summer savory herbage, and from 0.7 to 1.3 g kg^{-1} DM in marjoram herbage, whereas Seidler-Łożykowska et al. [35] determined Ca concentrations at 1.0–2.4 g kg^{-1} DM in summer savory and 1.5–2.9 g·kg^{-1} DM in marjoram. New products stimulating plant growth, based on amino acid preparations, applied at different doses, also contributed to an increase in the nutrient content of winter wheat grain, in particular copper (31–50%), as well as sodium, calcium, and molybdenum [39]. Biostimulants based on amino acids, tested in the present study, can be recommended for improving efficiency in agricultural production.

4. Conclusions

An analysis of the influence of the studied biostimulants on selected plant species revealed their beneficial interaction. The biostimulants had a positive effect on selected biometric and quality characteristics of the studied plants. They were particularly effective in reducing nitrate levels in herbal raw material. Therefore, they can be recommended for the commercial cultivation of herbs.

Marjoram var. *Miraż* was characterized by higher concentrations of DM, reducing sugars, and L-ascorbic acid than the other herb species.

The greatest amounts of minerals were accumulated by summer savory herbage (total N, K, and Ca) and lemon balm herbage (P and Mg).

Author Contributions: Conceptualization, J.M.-G.; methodology, J.M.-G. and A.F.; software, Z.K.; validation, A.F. and Z.K.; formal analysis, J.M.-G. and Z.K.; investigation, J.M.-G. and A.F.; resources, J.M.-G. and K.J.; data curation, J.M.-G.; writing—original draft preparation, J.M.-G., A.F. and K.J.; writing—review and editing, J.M.-G. and Z.K.; visualization, Z.K.; supervision, J.M.-G.; project administration, J.M.-G.; funding acquisition, J.M.-G. and K.J. All authors have read and agreed to the published version of the manuscript.

Funding: The results presented in this paper were obtained as part of a comprehensive study financed by the University of Warmia and Mazury in Olsztyn, Faculty of Agriculture and Forestry, Department of Agroecosystems and Horticulture 30.610.016-110 and Faculty of Technical Sciences, Department of Heavy Duty Machines and Research Methodology 16.610.001-110. This project was financially supported by the Minister of Education and Science in the range of the program entitled "Regional Initiative of Excellence" for the years 2019–2022, Project No. 010/RID/2018/19, amount of funding: 12,000,000 PLN.

Institutional Review Board Statement: Not applicable.

Informed Consent Statement: Not applicable.

Data Availability Statement: The data presented in this study are available on request from the corresponding author.

Conflicts of Interest: The authors declare no conflict of interest.

References

1. Grusak, M.A. Enhancing mineral content in plant food products. *J. Am. Coll. Nutr.* **2002**, *21*, 178S–183S. [CrossRef] [PubMed]
2. Özcan, M.M. Mineral contents of some plants used as condiments in Turkey. *Food Chem.* **2004**, *84*, 437–440. [CrossRef]
3. Skubij, N.; Dzida, K.; Jarosz, Z.; Pitura, K.; Jaroszuk-Sierocińska, M. Nutritional value of savory herb (*Satureja hortensis* L.) and plant response to variable mineral nutrition conditions in various phases of development. *Plants* **2020**, *9*, 706. [CrossRef] [PubMed]
4. Komaitis, M.E.; Ifanti-Papatragianni, N.; Melissari-Panagiotou, E. Composition of the essential oil of marjoram (*Origanum majorana* L.). *Food Chem.* **1992**, *45*, 117–118. [CrossRef]
5. Newerli-Guz, J. Factors shaping the quality of herbs and spices from organic and conventional crops. *Work. Mater. Fac. Manag. Univ. Gdańsk* **2010**, *2*, 451–459.
6. Rowayshed, G.H.; Abd-Elhameed, A.A.; Abd-Elghany, M.E.A.; Shahat, A.A.; Younes, O.A.A. Effective chemical compounds and antibacterial activities of marjoram leaves, teucrium leaves and fennel fruits essential oils. *Middle East J. Appl. Sci.* **2014**, *4*, 637–647.
7. Bina, F.; Rahimi, R. Sweet marjoram: A review of ethnopharmacology, phytochemistry and biological activities. *J. Evid. Based Complementary Altern. Med.* **2017**, *22*, 175–185. [CrossRef] [PubMed]
8. Połumackanycz, M.; Wesołowski, M.; Viapiana, A. Health benefits of lemon balm (*Melissa officinalis* L.). *Farmakognozja* **2019**, *75*, 659–663.
9. Maciejewski, T.; Szukała, J.; Jarosz, A. Influence of biostymulator Asahi SL i Atonik SL on qualitative tubers of potatoes. *J. Res. Appl. Agric. Eng.* **2007**, *52*, 109–112.
10. Dzida, K.; Jarosz, Z. Yielding and chemical composition of *Origanum majorana* L. depending of different nitrogen-potassium fertilization. *Acta Agroph.* **2006**, *7*, 561–566.
11. Majkowska-Gadomska, J.; Wierzbicka, B. Effect of the biostimulator Asahi SL on the mineral content of eggplant (*Solanum melongenum* L.) grown in an unheated plastic tunnel. *J. Elem.* **2013**, *18*, 269–276. [CrossRef]
12. Gugała, M.; Zarzecka, K.; Sikorska, A.; Mystkowska, I.; Dołęga, H. Effect of herbicydes and growth biostimulants on weed reduction and yield of edible potato. *Fragm. Agron.* **2017**, *34*, 59–66.
13. Kazimierczak, R.; Hallmann, E.; Ardasińska, B.; Łoś, B.; Rembiałkowska, E. The impact of organic and conventional crop production systems on phenolic compounds content in medicinal plants. *J. Res. Appl. Agric. Eng.* **2011**, *57*, 198–203.
14. Paradiković, N.; Vinković, T.; Vrček, I.V.; Žuntar, I.; Bojić, M.; Medić-Šarić, M. Effect of natural biostimulants on yield and nutritional quality: An example of sweet yellow pepper (*Capsicum annuum* L.) plants. *J. Sci. Food Agric.* **2011**, *91*, 2146–2152. [CrossRef]
15. Nurzyńska-Wierdak, R.; Rożek, E.; Bolanowska, K. Growth and yield of chervil, rocket and parsley plants depending on the cultivation method. *Ann. UMCS Sec. EEE Hortic.* **2012**, *22*, 1–12.
16. Nurzyńska-Wierdak, R. Lemon balm (*Melissa officinalis* L.)—Chemical composition and biological activity. *Ann. UMCS Sec. EEE Hortic.* **2013**, *23*, 25–35.
17. Mystkowska, I.T. Biostimulators as a factor affecting the yield of edible potato. *Acta Agroph.* **2018**, *25*, 307–315. [CrossRef]
18. Trawczyński, C. Effect of amino biostimulators on-Tecamin-on potato yield and quality. *Ziemn. Pol.* **2014**, *3*, 29–34.
19. Sawicka, B.; Barbaś, P.; Dąbek-Gad, M. The problem of weed infestation in conditions of applying the growth bioregulators and foliar fertilization in potato cultivation. *Nauka Przyr. Tech.* **2011**, *5*, 1–12. [CrossRef]
20. Osińska, E.; Rosłon, W. *Zioła Uprawa i Zastosowanie [Herbs, Cultivation and Application]*; Hortpress: Warsaw, Poland, 2016; pp. 1–164. (In Polish)
21. EN 12145:1996. *Fruit and Vegetable Juices—Determination of Total Dry Matter—Gravimetric Method with Loss of Mass on Dryin*; Comite Europeen de Normalisation: Brussels, Belgium, 1996.
22. PN-A-75101-07:1990. *Fruit and Vegetable Products—Sample Preparation and Methods of Physicochemical Analyses—Determination of Sugar Content and Non-Sugar Extract Content*; Polish Committee for Standardization: Warsaw, Poland, 1990.

23. PN-A-04019:1998. *Food Products—Determination of Vitamin C Content*; Polish Committee for Standardization: Warsaw, Poland, 1998.
24. PN-A-75101.04:1990. *Fruit and Vegetable Products—Preparation of Samples and Physical and Chemical Test Methods—Determination of Total Acidity*; Polish Committee for Standardization: Warsaw, Poland, 1990.
25. Krauze, A.; Domska, D. *Ćwiczenia Specjalistyczne z Chemii Rolnej [Specialist Exercises in Agricultural Chemistry]*; ART: Olsztyn, Poland, 1991; pp. 64–65. (In Polish)
26. Yaronskaya, E.; Vershilovskaya, I.; Poers, Y.; Alavady, A.E.; Averina, N.; Grimm, B. Cytokinin effects on tetrapyrrole biosynthesis and photosynthetic activity in barley seedlings. *Planta* **2006**, *224*, 700–709. [CrossRef]
27. Bello, A.S.; Saadaoui, I.; Ahmed, T.; Hamdi, H.; Cherif, M.; Dalgamouni, T.; Al Ghazal, G.; Ben-Hamadou, R. Enhancement in bell pepper (*Capsicum annuum* L.) plants with application of *Roholtiella* sp. (Nostocales) under soilless cultivation. *Agronomy* **2021**, *11*, 1624. [CrossRef]
28. El-Gohary, A.E.; El Gendy, A.G.; Hendawy, S.F.; El-Sherbeny, S.E.; Hussein, M.S.; Geneva, M. Herbage yield, essential oil content and composition of summer savory (*Satureja hortensis* L.) as affected by sowing date and foliar nutrition. *Genet. Plant Physiol.* **2015**, *5*, 170–178.
29. Jadczak, D.; Grzeszczuk, M. Spice herbs—biological value of selected species. *Panacea* **2008**, *2*, 15–17. (In Polish)
30. Nurzyńska-Wierdak, R.; Zawiślak, G. Bioactive compounds and antioxidant activity of sweet basil (*Ocimum basilicum* L.) and lemon balm (*Melissa officinalis* L.). *Ann. UMCS Sec. EEE Hortic.* **2016**, *26*, 43–51.
31. Kazimierczak, R.; Hallmann, E.; Kazimierczyk, M.; Rembiałkowska, E. Antioxidants content in chosen spice plants from organic and conventional cultivation. *J. Res. Appl. Agric. Eng.* **2010**, *53*, 164–170.
32. Telesiński, A.; Grzeszczuk, M.; Jadczak, D.; Wysocka, G.; Onyszko, M. Assessment of changes in content of nitrates (V) in selected spice herbs depending on their preservation method and storage time. *Zywn. Nauk. Technol. Jakosc* **2013**, *5*, 168–176. [CrossRef]
33. Frąszczak, B.; Gąsecka, M.; Golcz, A.; Zawirska-Wojtasiak, R. The chemical composition of lemon balm and basil plants grown under different lights condition. *Acta Sci. Pol. Hortorum Cultus* **2015**, *14*, 93–104.
34. Official Journal of the European Union. Commission Regulation (EC) No 1881/2006 of 19 December 2006 Setting Maximum Levels for Certain 424 Contaminants in Foodstuffs. 2006. Available online: https://eur-lex.europa.eu/legal-content/EN/ALL/?uri=CELEX%3A32006R1881 (accessed on 28 December 2021).
35. Seidler-Łożykowska, K.; Kozik, E.; Golcz, A.; Mieloszyk, E. Macroelements and essential oil content in the raw material of the selected medicinal plant species from organic cultivation. *J. Res. Appl. Agric. Eng.* **2016**, *51*, 161–163.
36. Dzida, K.; Jarosz, Z. Influence of nitrogen-potassium fertilization on the yield and on the nutrient content in *Satureja hortensis* L. *Acta Agroph.* **2006**, *7*, 879–884.
37. Kudełka, W.; Kosowska, A. Components of spices and herbs determining their functional properties and their role in human nutrition and prevention of diseases. *Zesz. Nauk. Uniw. Ekonom. Krak.* **2008**, *781*, 83–111.
38. Raczuk, J.; Biardzka, E.; Daruk, J. The content of Ca, Mg, Fe and Cu in selected species of herbs and herbs' infusions. *Rocz. Panstw. Zakl. Hig.* **2008**, *59*, 33–40. [PubMed]
39. Popko, M.; Michalak, I.; Wilk, R.; Gramza, M.; Chojnacka, K.; Górecki, H. Effect of the new plant growth biostimulants based on amino acids on yield and grain quality of winter wheat. *Molecules* **2018**, *23*, 470. [CrossRef] [PubMed]

applied
sciences

Article

Study on Contamination with Some Mycotoxins in Maize and Maize-Derived Foods

Alina Mihalcea and Sonia Amariei *

Faculty of Food Engineering, Stefan cel Mare University of Suceava, 720229 Suceava, Romania; alina.filip@fia.usv.com
* Correspondence: sonia@usm.ro; Tel.: +40-740-311-291

Abstract: Crops can be contaminated by fungi which produce mycotoxins. Many fungal strains are responsible for producing varied mycotoxins. The research carried out so far has described over 400 different mycotoxins. They have chemical and physical properties that significantly differ, and they are produced by several different existing fungi. The intake of mycotoxins through food can be achieved directly, by feeding on contaminated food, or indirectly from foods of animal origin. The mycotoxin contamination of food and food products for certain animals is a phenomenon studied worldwide, in countries in Europe but also in Asia, Africa and America. The purpose of this study is to develop an evaluation of the mycotoxins prevalent in corn and corn-derived products produced in Romania. A total of 38 maize samples and 19 corn-derivative samples were investigated for the presence of mycotoxins specific to these products, such as deoxynivalenol, zearalenone and fumonisins. Fumonisins had the highest presence and zearalenone had the lowest. The limits determined for the three mycotoxins were always in accordance with legal regulations.

Keywords: corn; deoxynivalenol; food; fumonisins; mycotoxins; zearalenone

Citation: Mihalcea, A.; Amariei, S. Study on Contamination with Some Mycotoxins in Maize and Maize-Derived Foods. *Appl. Sci.* **2022**, *12*, 2579. https://doi.org/10.3390/app12052579

Academic Editors: Alessandra Biancolillo and Chiara Cavaliere

Received: 17 January 2022
Accepted: 28 February 2022
Published: 2 March 2022

Publisher's Note: MDPI stays neutral with regard to jurisdictional claims in published maps and institutional affiliations.

1. Introduction

Corn (*Zea mays* L.) is native to Central America, cultivated worldwide as a food, industrial and fodder plant and was first mentioned in Romania in the late seventeenth century. Maize is a plant of the *Poaceae* family. The annual harvest of maize in 2020 was 1192 million tons, as maize is one of the most commonly used cereals in food [1]. Mycotoxins infect corn in the field or during the storage stage [2]. Mycotoxin contamination is the main problem with maize production. Mycotoxins are formed by filamentous fungi of the genus *Ascomycota*, which affects human and animal health and causes chronic and acute diseases [3–6]. Corn is most frequently contaminated by *Aspergillus* sp. and *Fusarium* sp. [7].

The most important mycotoxins that contaminate maize throughout its development cycle and storage period are fumonisins (FUM), deoxynivalenol (DON) and zearalenone (ZEA) [8]. Fumonisins are formed by *Fusarium verticillioides*, *F. proliferatum* and *F. fujikuroi*; zearalenone are formed by *Fusarium graminearum*, *F. culmorum*, *F. cerealis* and *F. equiseti*; deoxynivalenol are produced by *Fusarium graminearum* and *F. culmorum* [9]. The development of fungi and, by extension, mycotoxins, is favored by high temperatures, high humidity and prolonged storage [10]. The possibility of mycotoxin production rises if inappropriate agricultural practices are applied.

Mycotoxins cannot be detected in the visible field (VIS) but can be recognized in ultraviolet (UV) light; moreover, they do not have a distinctive smell and do not change the organoleptic properties of foodstuff [11]. The consumption of corn contaminated with mycotoxins by humans or animals can generate multiple serious toxic effects [12]. Exposure to mycotoxins leads to mycotoxicosis (ergotism, food poisoning and aflatoxins), a disease affecting human organs which, under certain conditions, may cause death [11,13].

Mycotoxicosis may be acute or chronic [2]. The main ways of exposing the human body to mycotoxins are by ingestion, inhalation and contact.

The European Union has established by legislation limits on the content of mycotoxins in all categories of foodstuffs, including corn and corn-derived aliments, to protect the population's health. Regulation (EC) No. 1881/2006 [14] sets the maximum levels for aflatoxins, fumonisins, zearalenone and deoxynivalenol in food. This regulation was amended by (EC) Regulation No. 1126/2007 [15] and (EC) Regulation No. 165/2010 [16]. Prevention, detoxification and decontamination measures are needed to combat mycotoxin contamination of food [17]. The objective of this study is to assess the contamination with three mycotoxins, namely deoxynivalenol, zearalenone and fumonisins, of maize and maize-derived products from local producers in the eastern part of Romania.

2. Materials and Methods

2.1. Biological Material

For the study of the deoxynivalenol, zearalenone and fumonisin contamination of corn and corn-derived foods, 57 samples were taken from the Romanian market, from the local producers of corn or corn products (breakfast cereals, cornflakes, canned corn, puffs and puff pastry). The grain maize samples were taken from various producers in eastern Romania. Samples were taken after the harvesting and drying of corn kernels and before storing them in silos. A total of 38 samples of grain maize were analyzed: 16 samples for the detection of deoxynivalenol, 12 samples for the detection of zearalenone and 10 samples for the investigation of fumonisins. The samples of food products derived from corn were taken from local producers in the eastern part of Romania and all products were autochthonous. A total of 19 samples were analyzed: 5 samples for the detection of deoxynivalenol (2 samples of corn flakes and 3 samples of maize breakfast cereals), 10 samples for the investigation of zearalenone (3 samples of preserved maize, 3 samples of corn flakes, 3 samples of maize breakfast cereals and 1 sample puffed) and 4 samples for the detection of fumonisins (2 canned maize samples and 2 corn flakes samples).

2.2. Sampling Method

Mycotoxins are distributed unevenly in a product, so every measure was taken at sampling so to ensure that the sample was representative. In the case of the grain maize samples, a number of 100 incremental samples were taken, making up the 10 kg aggregate sample. For products derived from maize, 3 incremental samples were taken each, making up the aggregate sample, weighing 1 kg. The analysis of the samples was made in one replication. When selecting a sample for the determination of a particular mycotoxin, the evaluations performed by each economic operator in the previous years regarding the risk of mycotoxins were taken into account, while also taking into account the mycotoxin most frequently identified in the previous years.

2.3. Laboratory Method

ELISA (enzyme-linked immunosorbent assay) kits provided by ProGnosis Biotech S.A were utilized to analyze the content of deoxynivalenol, zearalenone and fumonisins in the samples. The kits conformed to the specifications of EN ISO 14675: 2003. ProGnosis Biotech S.A is ISO 9001: 2015 was certified by TÜV Hellas (TÜV NORD). Bio-Shield deoxynivalenol, B2648/B2696, Bio-Shield zearalenone, B2748/B2796 and Bio-Shield fumonisin, B2848/B2896, are ELISA tests used to establish the content of deoxynivalenol, zearalenone and fumonisins in food and animal feed [18]. Other materials utilized include a grinding device, balance, graduated cylinder, distilled water, filter paper, filter funnel, laboratory tubes, micropipettes, mixer and spectrophotometer. The aggregate sample was homogenized and ground in the laboratory. The toxins were extracted from the sample ground with distilled water; the sample was then ground to obtain fine particles weighing 20 g and then added to 100 mL of distilled water. The mixture obtained was homogenized with the mixer, filtered and diluted 5 times with distilled water. The standards of the mycotoxin or the samples, and the detection solution, were added to padded wells. The enzyme conjugate and antimycotoxin antibodies were added to each well and incubated for 5 min at room temperature. Discard the liquid from the wells, add the wash buffer, then incubate the chromogenic substrate for 3 min in the dark at room temperature. The conjugate connects to the antibody, and the accession places padding not already occupied by mycotoxin in standards or samples. The addition of a chromogen substrate leads to the appearance of a blue complex. Sulfuric acid must be added, which stops the development of the color, which becomes yellow. The photometric reading was made at 450 nm [19–21]. The competitive format of the ELISA was the most used [22].

2.4. Statistical Analysis

All results are presented as mean standard deviation. The values obtained were processed by using the SPSS 25.0 (trial version) software (IBM, New York, NY, USA). The difference between the samples was established by an analysis of variance (ANOVA) using the Turkey's test at a 5% significance level.

3. Results

3.1. Deoxynivalenol

Deoxynivalenol (Figure 1) is also named vomitoxin and it is produced by the *Fusarium* species, which is adapted to various pedoclimatic conditions, and found in Romania in the plateaus of Moldova and Dobrogea or in the wetlands of the Danube Delta. *Fusarium graminearum* can infect maize and wheat, but it was also isolated from rice, coffee, peas and tomatoes [23,24]. *Fusarium culmorum* can infect plants of the *Gramineae* family (wheat, rye, rice, oats and corn) but also plants of the *Solanaceae*, *Leguminosae* and *Cucurbitaceae* family. Deoxynivalenol is one of the least toxic of the trichothecenes, and its appearance is common, but in combination with other mycotoxins (fumonisins, T2 toxin and beauvericin) it can have much more serious effects [25].

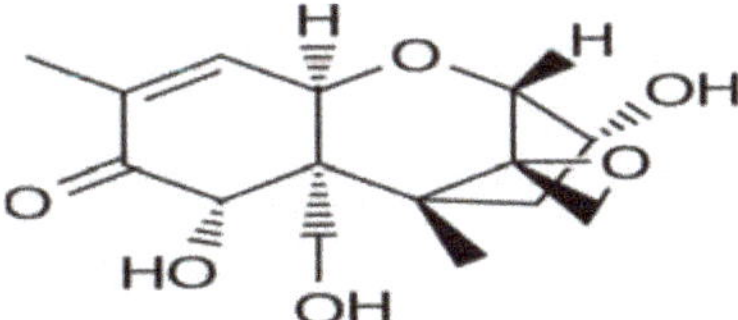

Figure 1. Deoxynivalenol chemical structure.

Deoxynivalenol is a very stable trichothecene at high temperatures of 120 °C, and less stable at 180 °C. It may be dissolved in water, ethanol, methanol, chloroform, acetonitrile and ethyl acetate [26].

Analyzing the results presented in Table 1 and Figures 2 and 3, we can say that in 25% of the analyzed samples of corn grains, deoxynivalenol was identified in 20% of the samples of corn products. The determined values do not exceed the limits established by Regulation (EC) No. 1881/2006 with the subsequent modifications (for maize grains, the maximum of 1750 μg/kg, and for maize-derived foods, the maximum of 500 μg/kg). The deoxynivalenol level is 200 μg/kg in cereal-based aliments for children, 750 μg/kg in pasta and 500 μg/kg in bread [24]. The EFSA recently established a PMTDI of 1 μg/kg bw/day for deoxynivalenol and its metabolites [27,28].

Table 1. Study on the incidence of deoxynivalenol in maize and maize-derived foods in Romania.

Sample Number	The Analyzed Product	Country of Origin	Limit of Detection µg/kg	Limit of Quantification µg/kg	Deoxynivalenol µg/kg	Limit µg/kg
1.	Maize01	Romania	40.720	67.198	<67.198 [e]	1750
2.	Maize02	Romania	40.720	67.198	70.400 [d]	1750
3.	Maize03	Romania	40.720	67.198	<40.720 [g]	1750
4.	Maize04	Romania	40.720	67.198	77.100 [c]	1750
5.	Maize05	Romania	40.720	67.198	<40.720 [g]	1750
6.	Maize06	Romania	40.720	67.198	<40.720 [g]	1750
7.	Maize07	Romania	40.720	67.198	<67.198 [e]	1750
8.	Maize08	Romania	40.720	67.198	<67.198 [e]	1750
9.	Maize09	Romania	40.720	67.198	<40.720 [g]	1750
10.	Maize10	Romania	40.720	67.198	92.900 [b]	1750
11.	Maize11	Romania	40.720	67.198	<40.720 [g]	1750
12.	Maize12	Romania	40.720	67.198	152.200 [a]	1750
13.	Maize13	Romania	40.720	67.198	<40.720 [g]	1750
14.	Maize14	Romania	40.720	67.198	<40.720 [g]	1750
15.	Maize15	Romania	40.720	67.198	<40.720 [g]	1750
16.	Maize16	Romania	40.720	67.198	<67.198 [e]	1750
17.	Cornflakes01	Romania	18.610	37.230	<18.610 [h]	500
18.	Cornflakes02	Romania	18.610	37.230	<18.610 [h]	500
19.	Breakfast cereals01	Romania	18.610	37.230	<18.610 [h]	500
20.	Breakfast cereals02	Romania	18.610	37.230	63.990 [f]	500
21.	Breakfast cereals03	Romania	18.610	37.230	<18.610 [h]	500

The differences between samples were established by analysis of variance (ANOVA) using Turkey's test at a 5% significance level. Different superscript letters after the values indicated a statistically significant difference at $p < 0.05\%$.

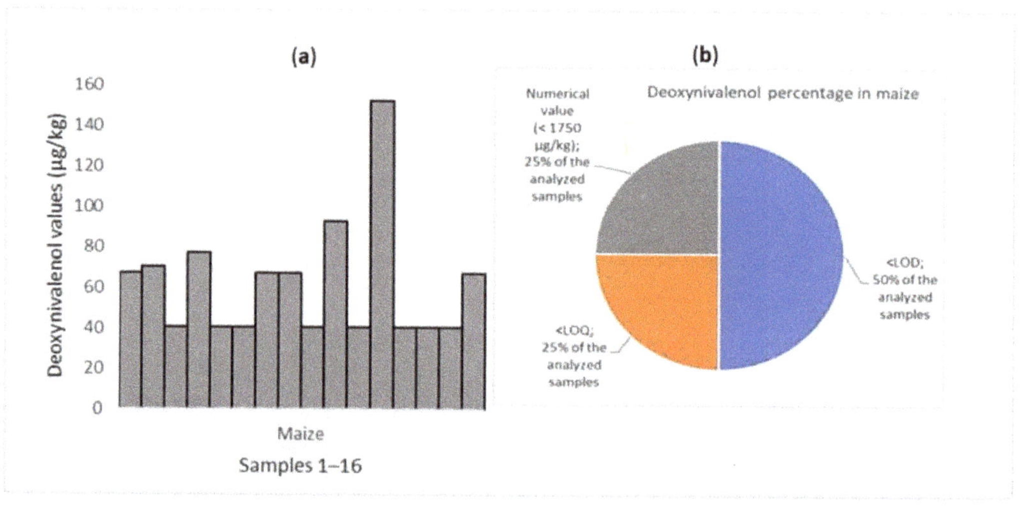

Figure 2. (**a**) Deoxynivalenol in maize samples; (**b**) Deoxynivalenol percentage in maize samples.

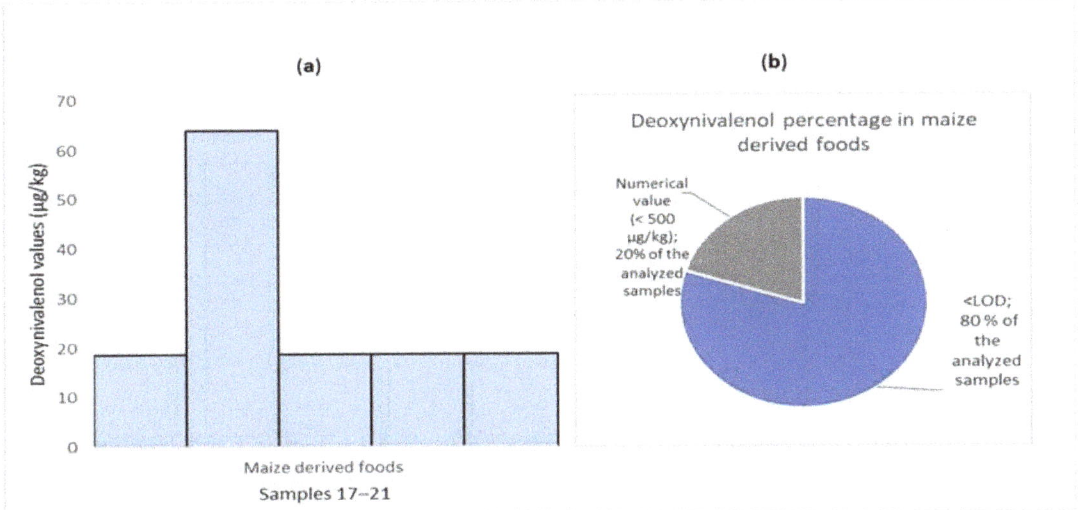

Figure 3. (**a**) Deoxynivalenol in maize-derived food samples; (**b**) Deoxynivalenol percentage in maize-derived food samples.

The difference between the samples analyzed was significantly different (95% confidence level). In the maize category, the results obtained for samples 1, 7, 8 and 16 are not significant. The same was observed for samples 3, 5, 6, 9, 11, 13, 14 and 15. On the other hand the results obtained for the remaining four samples (samples 2, 4, 10, 12) are significantly different ($p < 0.5\%$). The two cornflakes samples presented similar results. From the three breakfast cereals studied, only one (sample 2) was found to be significantly different (at 95% confidence level).

Deoxynivalenol is the most commonly identified mycotoxin in cereals, with a mean percentage between 50% and 76% in Asia and Africa. Higher concentrations have been identified in Europe and Asia [29]. In North America, the reported deoxynivalenol levels are similar to those in the rest of the world. Increased levels of deoxynivalenol are recorded in years of severe infections caused by *Fusarium* sp. [30].

Exposure to deoxynivalenol is usually made by aliments. The intake of contaminated foodstuffs may cause nausea, vomiting, inappetence and diarrhea. When foodstuffs consumed daily, for a longer period of time, are contaminated by deoxynivalenol, manifestations such as weight decline, anorexia and histopathological lesions in the liver occur [23]. In India, the intake of contaminated food has led to acute gastric symptoms, vomiting and diarrhea [4]. There is no evidence for the carcinogenic potential of deoxynivalenol, which is why the IARC classifies this mycotoxin as noncarcinogenic to humans (Group 3) [31].

For the deoxynivalenol parameter in samples in Romania, the determined values, compared to other studies in other parts of the globe, as presented in Table 2, are low. Moreover, the incidence in food samples is reduced.

Table 2. Studies on the incidence of deoxynivalenol in maize and maize-derived foods.

Country	The Analyzed Product	No. Samples	Deoxynivalenol Incidence%	Deoxynivalenol Mean µg/kg	Deoxynivalenol Range µg/kg	Reference
Romania	Corn	16	25.00	61.70	n/a–152.20	
	Corn flakes, Breakfast cereals	5	20.00	27.69	n/a–63.99	

Table 2. *Cont.*

Country	The Analyzed Product	No. Samples	Deoxynivalenol Incidence%	Deoxynivalenol Mean µg/kg	Deoxynivalenol Range µg/kg	Reference
Serbia	Corn flakes	15	40.00	255.00	n/a–878.00	[32]
	Corn flour	56	42.90	101.00	n/a–931.00	[32]
Hungary	Corn	29	86.00	1872.00	225.00–2963.00	[33]
	Corn	106	92.30	449.00	263.10–2777.40	[34]
Belgium	Corn	106	100.00	557.50	337.40–5322.40	[34]
	Corn	106	64.70	186.50	121.30–2110.50	[34]
Germany	Corn	120	100.00	216.50	n/a–1097.20	[35]
Switzerland	Corn	19	100.00	135.60	78.00–299.00	[36]
China	Corn	44	65.90	831.00	5.80–9843.30	[37]
Brazil	Pastry	36	100.00	591.00	60.00–1720.00	[38]

3.2. Zearalenone

Fusarium fungi produce zearalenone (Figure 4), and the main species producing zearalenone are *F. roseum*, *F. tricinctum*, *F. oxysporum*, *F. graminearum*, *F. moniliforme*, *F. culmorum*, *F. avenaceum*, *F. crookwellense*, *F. nivale*, *F. semitectum*, *F. solani* and *F. echiseti*, which are spread throughout the world, especially in temperate climates [1,39]. Zearalenone infection is reduced in cereals at harvest and raises in storage conditions if humidity reaches 30–40% [1]. Zearalenone is stable at high temperatures (80 °C–120 °C) and in neutral media. It may be dissolved in methyl chloride, dimethylformamide alcohols and ethers [40].

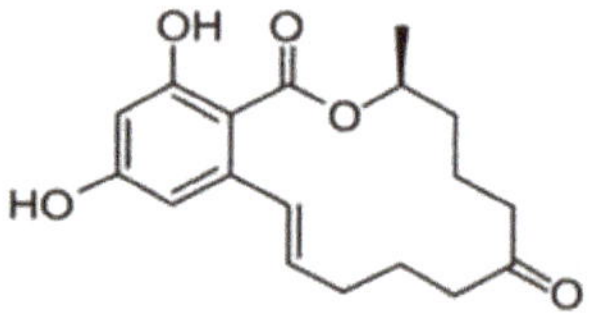

Figure 4. Zearalenone chemical structure.

The maximum levels of zearalenone are established by Regulation (EC) 1881/2006 for several food categories: 350 µf/kg maize, other cereals 100 µg/kg, bread, pastry, breakfast cereals 50 µg/kg and food for infants 20 µg/kg [41–43].

From all the maize samples, only 12 are significantly different ($p < 0.05\%$). No significant difference was found at a 95% confidence level between the cornflakes, breakfast cereals and puffs samples. Moreover, the results obtained showed no significant difference ($p < 0.05\%$) for the maize and canned corn samples.

The study on the presence of zearalenone was performed on 22 samples of corn grains and corn products (canned corn, corn flakes, breakfast cereals and puff pastry) produced in Romania, as presented in Table 3. Of the 12 corn samples analyzed, 8% were contaminated with zearalenone, and none of the samples of the maize-based foods were contaminated with zearalenone. The determined values did not exceed the maximum limit established by Regulation (EU) No. 1881/2006 (for maize grains, the maximum of 350 µg/kg and for food based on maize, the maximum of 50–100 µg/kg), as presented in Figures 5 and 6.

Table 3. Study on incidence of zearalenone in maize and maize-derived foods in Romania.

Sample Number	The Analyzed Product	Country of Origin	Limit of Detection µg/kg	Limit of Quantification µg/kg	Zearalenone µg/kg	Limit µg/kg
1.	Maize01	Romania	2.060	2.456	<2.060 [b]	350
2.	Maize02	Romania	2.060	2.456	<2.060 [b]	350

Table 3. *Cont.*

Sample Number	The Analyzed Product	Country of Origin	Limit of Detection µg/kg	Limit of Quantification µg/kg	Zearalenone µg/kg	Limit µg/kg
3.	Maize03	Romania	2.060	2.456	<2.060 [b]	350
4.	Maize04	Romania	2.060	2.456	<2.060 [b]	350
5.	Maize05	Romania	2.060	2.456	<2.060 [b]	350
6.	Maize06	Romania	2.060	2.456	<2.060 [b]	350
7.	Maize07	Romania	2.060	2.456	<2.060 [b]	350
8.	Maize08	Romania	2.060	2.456	<2.060 [b]	350
9.	Maize09	Romania	2.060	2.456	<2.060 [b]	350
10.	Maize10	Romania	2.060	2.456	<2.060 [b]	350
11.	Maize11	Romania	2.060	2.456	<2.060 [b]	350
12.	Maize12	Romania	2.060	2.456	18.565 ± 1.411 [a]	350
13.	Canned corn01	Romania	2.060	2.456	<2.060 [b]	100
14.	Canned corn02	Romania	2.060	2.456	<2.060 [b]	100
15.	Canned corn03	Romania	2.060	2.456	<2.060 [b]	100
16.	Cornflakes01	Romania	1.860	3.730	<1.860 [c]	100
17.	Cornflakes02	Romania	1.860	3.730	<1.860 [c]	100
18.	Cornflakes03	Romania	1.860	3.730	<1.860 [c]	100
19.	Breakfast cereals01	Romania	1.860	3.730	<1.860 [c]	50
20.	Breakfast cereals02	Romania	1.860	3.730	<1.860 [c]	50
21.	Breakfast cereals03	Romania	1.860	3.730	<1.860 [c]	50
22.	Puffs01	Romania	1.860	3.730	<1.860 [c]	100

The differences between samples were established by analysis of variance (ANOVA) using Turkey's test at a 5% significance level. Different superscript letters after the values indicated statistically significant differences at $p < 0.05\%$.

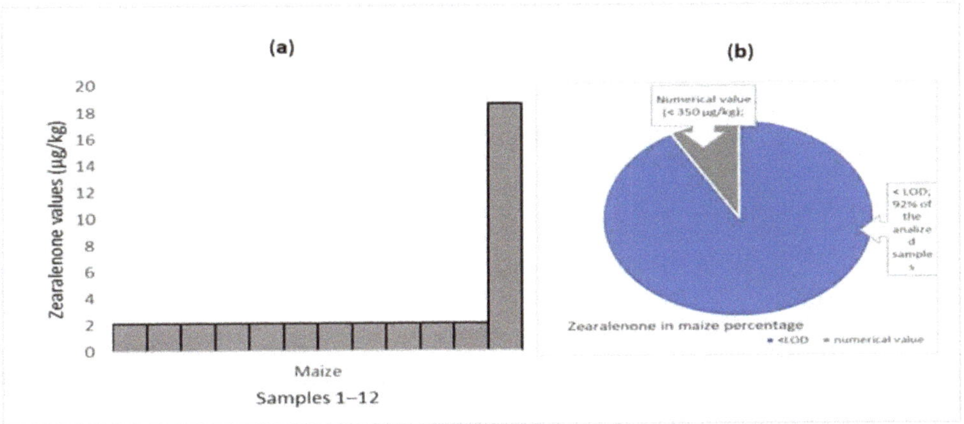

Figure 5. (**a**) Zearalenone in maize samples; (**b**) Zearalenone percentage in maize.

Risk assessments for the intake of foods contaminated with zearalenone have been performed based on existing exposure data from countries in Europe and Asia. Studies have identified several situations where the daily intake of zearalenone exceeds the TDI as stated by the European Union [43]. According to the IARC, zearalenone is a Group 3 carcinogen [34]. Zearalenone is the lactone of resorcin acid whose chemical structure resembles that of steroid hormones. Zearalenone has nonsteroidal estrogenic action and affects the conception, ovulation and development of the fetus at concentrations over 1 mg/kg [44]. Zearalenone can cause hyperestrogenism and it primarily affects reproductive functions.

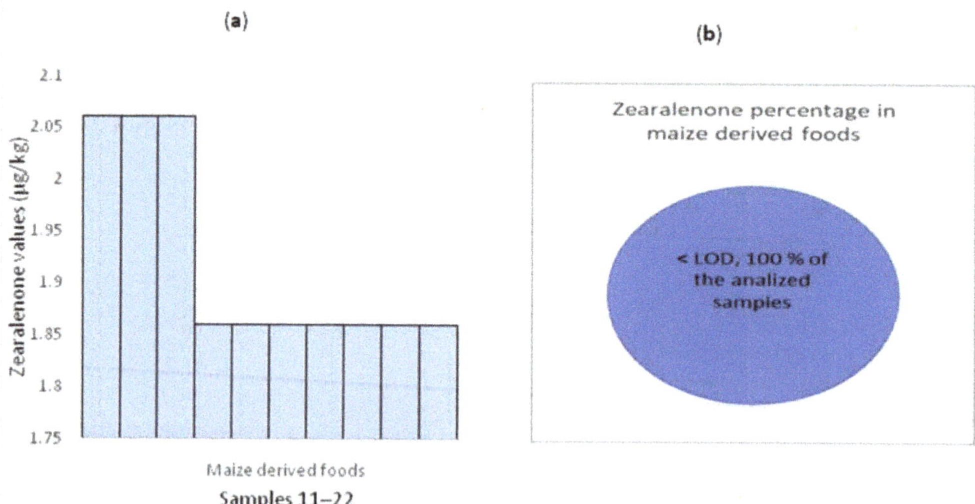

Figure 6. (**a**) Zearalenone in maize-derived food samples; (**b**) Zearalenone percentage in maize-derived food samples.

Table 4 shows the specific studies on the presence of zearalenone in maize and maize-derived foods worldwide from 2014 to 2020. The determined values, compared to other studies in other areas of the globe, as presented in Table 4, are low. Moreover, the incidence in food samples is reduced.

Table 4. Studies on the incidence of zearalenone in maize and maize-derived foods.

Country	The Analyzed Product	No. Samples	Zearalenone Incidence%	Zearalenone Mean μg/kg	Zearalenone Range μg/kg	Reference
Romania	Corn	12	8.00	3.44	nd–18.565	
	Corn flakes, breakfast cereals, canned corn, puffs	10	0.00	nd	nd	
Serbia	Maize	90	0.00	0.00	0.00	[45]
Hungary	Maize	29	41.00	267.00	nd–565.00	[33]
	Maize	106	64.80	100.50	71.20–1085.60	[34]
Belgium	Maize	106	40.70	158.50	0.00–1085.60	[34]
	Maize	106	42.40	175.50	0.00–2791.60	[34]
Germany	Maize	120	96.00	819.00	nd–3910.00	[35]
Brazil	Maize	40	95.00	22.20	1.80–99.00	[46]
Egypt	Maize	30	33.34	2.39	1.13–3.70	[47]
	Maize	50	94.00	109.10	0.20–3613.00	[35]
China	Maize	44	13.60	50.80	40.70–1056.80	[48]

3.3. Fumonisins

Fumonisins (Figures 7 and 8) are also called *Fusarium* toxins. The main producing species for fumonisins are *Fusarium verticillioides, F. proliferatum, F. sacchari, F. fujikuroi* and

F. subglutinans. [49]. The twenty-eight fumonisin analogs identified so far are split into four groups: fumonisins A, B, C and P. Group Fumonisins B includes FB1, FB2 and FB3. Of these, FB1 is the most frequent and has the highest concentrations [50]. Fumonisins have been identified in several food categories including cereals [51], foods derived from cereals [52,53], asparagus, garlic [54], grapes, beer, milk and raisins [55].

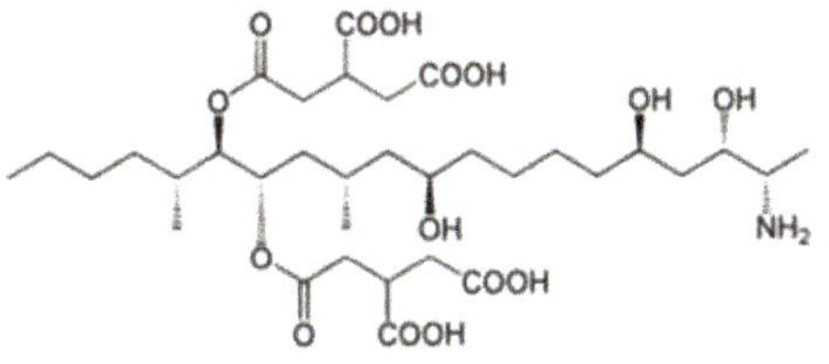

Figure 7. Fumonisin B1 chemical structure.

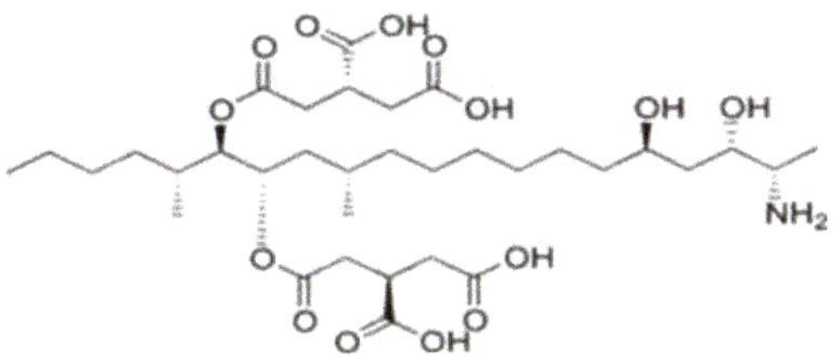

Figure 8. Fumonisin B2 chemical structure.

The study on the presence of fumonisins was performed on 14 samples of maize and maize-derived foods (canned corn and corn flakes) produced in Romania, as presented in Table 5.

Table 5. Study on incidence of fumonisins in corn and corn-derived foods in Romania.

Sample Number	The Analyzed Product	Country of Origin	Limit of Detection µg/kg	Limit of Quantification µg/kg	Fumonisins µg/kg	Limit µg/kg
1.	Maize01	Romania	21.722	30.644	163.551 ± 49.065 [b]	4000
2.	Maize02	Romania	21.722	30.644	1009.360 ± 302.804 [a]	4000
3.	Maize03	Romania	21.722	30.644	38.318 ± 11.495 [f]	4000
4.	Maize04	Romania	21.722	30.644	<21.722 [h]	4000
5.	Maize05	Romania	21.722	30.644	<21.722 [h]	4000
6.	Maize06	Romania	21.722	30.644	<21.722 [h]	4000
7.	Maize07	Romania	21.722	30.644	45.614 ± 13.684 [e]	4000
8.	Maize08	Romania	21.722	30.644	104.368 ± 31.315 [c]	4000
9.	Maize09	Romania	21.722	30.644	79.824 ± 23.947 [d]	4000
10.	Maize10	Romania	21.722	30.644	39.260 ± 12.640 [f]	4000
11.	Canned corn01	Romania	25.900	51.810	<25.900 [g]	1000
12.	Canned corn02	Romania	25.900	51.810	<25.900 [g]	1000
13.	Cornflakes01	Romania	25.900	51.810	73.000 ± 5.590 [d]	800
14.	Cornflakes02	Romania	25.900	51.810	<25.900 [g]	800

The differences between samples were established by analysis of variance (ANOVA) using Turkey's test at a 5% significance level. Different superscript letters after the values indicate statistically significant differences at $p < 0.05\%$.

From all the maize samples, samples 4, 5 and 6 are not significantly different ($p < 0.05\%$). All the other samples are significantly different from each other. No significant (at 95% confidence level) difference was found between the canned corn samples. The result obtained for the two cornflakes samples showed a significant difference ($p < 0.05\%$) between them.

The mycotoxins identified in this research, in the largest percentage of the analyzed samples (70%), were fumonisins, as presented in Figures 9 and 10. The determined values are within the allowed legal limits. In 70% of the samples of unprocessed corn, fumonisins were detected; in those of products processed from corn, only 25% indicated their presence.

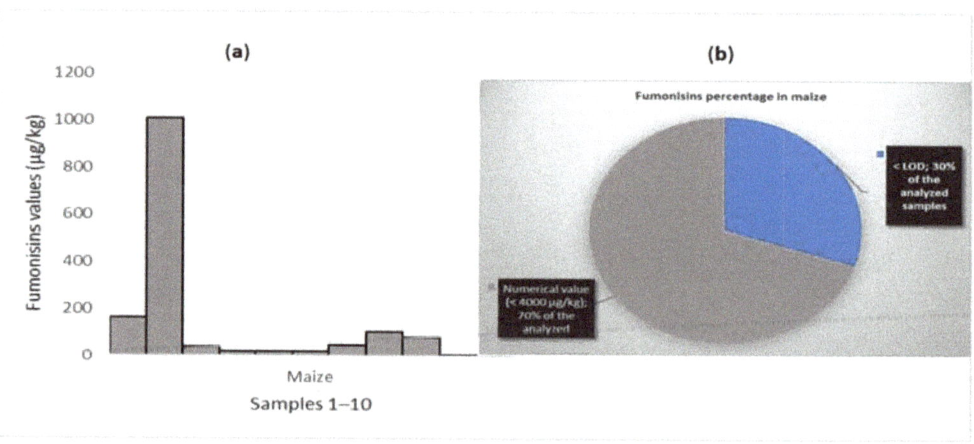

Figure 9. (**a**) Fumonisins in maize samples; (**b**) Fumonisins percentage in maize.

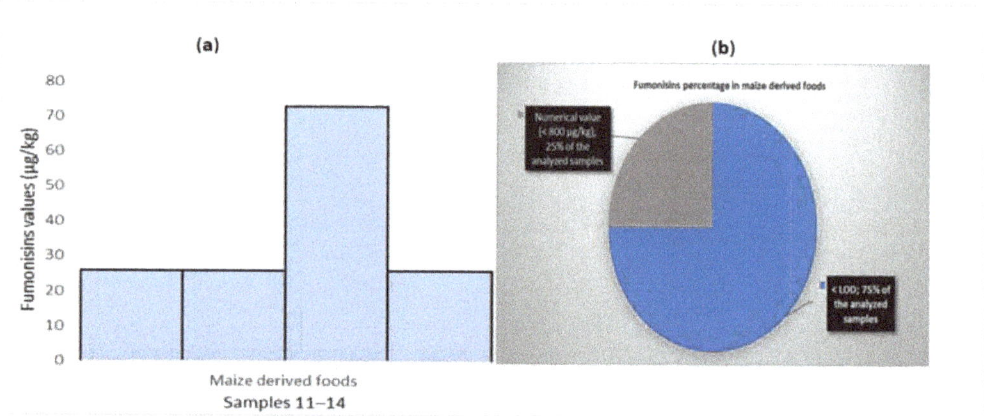

Figure 10. (**a**) Fumonisins in maize-derived foods; (**b**) Fumonisins percentage in maize-derived foods.

JECFA has provisionally settled a PMTDI of 2 µg/kg bw/day for fumonisins alone or in combination [56]. The maximum levels of fumonisins are set by Regulation (EC) 1881/2006 for several food categories: 4000 µg/kg for maize, 800 µg/kg for breakfast cereals and 200 µg/kg for food for children [57].

The fumonisin values from the samples investigated, compared to samples from other areas of the globe, as shown in Table 6, are higher. The impact on the food samples in Romania is also higher.

Table 6. Studies on the occurrence of fumonisins in maize and maize-derived foods.

Country	The Analyzed Product	No. Samples	Fumonisins Incidence%	Fumonisins Mean µg/kg	Fumonisins Range µg/kg	Reference
Romania	Corn, canned corn, corn flakes	10 4	70.00 25.00	154.55 37.68	nd–1009.36 nd–73.00	
Serbia	Corn	90	100.00	1730.00	520.00–5800.00	[45]

Table 6. *Cont.*

Country	The Analyzed Product	No. Samples	Fumonisins Incidence%	Fumonisins Mean µg/kg	Fumonisins Range µg/kg	Reference
	Corn	106	2.50	1.50	0.00–70.20	
	Corn	106	19.80	61.10	0.00–1362.90	
	Corn	106	61.20	247.40	54.00–4414.90	
	Corn	106	nd	nd	nd	
	Corn	106	4.90	9.00	0.00–412.60	
	Corn	106	24.70	61.60	0.00–1427.40	
Belgium	Corn	106	nd	nd	nd	[34]
	Corn	106	7.40	3.40	0.00–90.50	
	Corn	106	18.80	18.00	0.00–451.20	
	Corn	106	2.50	1.30	0.00–70.20	
	Corn	106	19.80	73.60	0.00–1782.80	
	Corn	106	61.20	327.00	58.70–6293.50	
Italy	Corn	697	100.00	10900.00	25.00–77000.00	[58]
Brazil	Corn	40	100.00	2338.50	230.00–6450.00	[46]
Spain	Corn	92	100.00	2610.00	337.00–10613.00	[59]
Spain	Corn flakes	47	21.00	42.00	nd–67.00	
Canada	Corn	10	100.00	4.64	0.73–10.21	[60]
	Corn flakes	14	100.00	104.10	1.00–171.00	
China	Corn flakes	14	93.00	14.20	<0.27–25.60	[61]
	Corn flakes	14	93.00	17.30	<0.27–31.50	

4. Discussion

Strategies to prevent mycotoxin contamination prior to harvesting are particularly important. Good agricultural practices and good production practices are used by producers around the world. Crop rotation, the use of approved substances such as herbicides, fungicides and insecticides (insects are vectors for fungal dissemination) and storage in appropriate conditions for temperature and humidity contributes substantially to reducing the risk of contamination of cereals in the first stages of the food chain [62,63]. The implementation of HACCP (hazard analysis and critical control points), through clear instructions for monitoring the presence and level of mycotoxins in food, allows safe foods to be placed on the market.

One of the greatest dangers associated with food safety is mycotoxins. They have negative effects on all biological systems, from affecting clinical conditions, cellular systems and the metabolism, to decreased animal production. Mycotoxins cause major economic losses through direct action, inducing disease indirectly through damage to the immune system, or qualitatively affecting contaminated products.

The world's need for goods commonly used in the manufacture of food for humans and animals, such as corn, has grown constantly in the last years due to higher demand from manufacturers and consumers. The RASFF annual report (Rapid Alert System for Food and Feed) for 2020 presents 400 notifications regarding mycotoxin (down by 23%) contamination of food products in the European Union [64]. Aflatoxins are the most detected mycotoxins in food in the EU (367 notifications), also detected in dried figs from Turkey (58 notifications) and followed by groundnuts from the United States (29 notifications) [64]. The formation of mycotoxins is a composite process influenced by a multitude of factors, mainly environmental conditions under the influence of climate change.

Comparing the results of this study with previous research, for deoxynivalenol, the incidence in the analyzed samples is 25% for maize, while in Hungary the incidence of deoxynivalenol is 86% [33]; in Belgium, the incidence in the studies carried out varies from 64.7% to 92.3% or 100% [34]. Studies carried out in Germany and Switzerland show a deoxynivalenol incidence of 100%, while in China the incidence is 65.9% [35–37]. For the analyzed products derived from maize, the incidence of deoxynivalenol was 20%, while other studies show an incidence of 40% for maize flakes and 42.9% for maize flour in Serbia, and 100% for pastry products in Brazil [32,38]. The maximum value identified for deoxynivalenol in this maize study is 152.2 µg/kg, well below the maximum values identified in other studies, while it is 9843.3 µg/kg in China and 5322.4 µg/kg in Belgium [34,37]. In products derived from maize, deoxynivalenol had a maximum of 63.99 µg/kg; other studies indicate a maximum of 931 µg/kg in Serbia and 1720 µg/kg in Brazil [32,38].

In the tested samples of maize, the incidence for zearalenone was 8%. In previous studies on maize, zearalenone had an incidence of 0% in Serbia [45], 13.6% and 94% in China [48], 33.34% in Egypt [47], 40.7%–64.8% in Belgium [34], 41% in Hungary [33] and more than 90% in Germany and Brazil [35,46]. The average value of zearalenone in the Romanian sample was 3.44 µg/kg. Other studies have identified lower average values, such as 2.39 µg/kg or higher in Egypt and 175.5 µg/kg in Belgium [34,47]. The highest zearalenone content in a tested sample was 18.565 µg/kg, while other studies indicate maximum values of 3613 µg/kg [35].

For fumonisins, the incidence in this study was 70% for maize and 25% for products derived from maize. In previous studies, fumonisin incidences ranged from 2.5% in Belgium to 100% in Serbia, Italy, Spain and Canada [34,45,58–60]. The average fumonisin value in the tested samples of maize was 154.55 µg/kg. Other studies have identified an average for fumonisins from 1.5 µg/kg to 10,900 µg/kg [34,58]. For the analyzed products derived from maize, the average value of fumonisins was 37.68 µg/kg. Other studies have identified an average of fumonisins between 14.2 µg/kg and 104.1 µg/kg in corn flakes [59,61].

5. Conclusions

In the present research, regarding contamination with deoxynivalenol, zearalenone and fumonisins in maize and local maize-derived foods, the levels of the three mycotoxins investigated have always been within legal limits. From the three mycotoxins investigated, fumonisins were those with the highest presence, contaminating 70% of the corn samples analyzed. Deoxynivalenol contaminated 25% of the corn samples investigated and zearalenone had the lowest percentage (8%), with the risk of contamination being minimal. Regarding the samples of maize-based foods investigated, fumonisins were detected in 25% of the samples, deoxynivalenol was detected in 20% of the samples and no sample was contaminated with zearalenone.

The investigations carried out in this study, and the comparisons with previously published studies, provide information on the contamination of maize-based foods in Romania by three of the most important mycotoxins, as well as information on their overall distribution. The low percentage of foodstuffs derived from maize contaminated by mycotoxins identified in this study shows that good manufacturing practices are essential for safe food production.

Author Contributions: Conceptualization, A.M. and S.A., methodology, A.M. and S.A., software, S.A., validation, A.M. and S.A., formal analysis A.M. and S.A., investigation A.M., resources A.M. and S.A., data curation, A.M. and S.A., writing—original draft preparation A.M., writing—review and editing, A.M. and S.A., visualization A.M., supervision, S.A., project administration S.A., funding acquisition S.A. All authors have read and agreed to the published version of the manuscript.

Funding: This research was funded by "Stefan cel Mare", University of Suceava.

Institutional Review Board Statement: Not applicable.

Informed Consent Statement: Not applicable.

Data Availability Statement: Not applicable.

Acknowledgments: This research was supported by "Stefan cel Mare", University of Suceava.

Conflicts of Interest: The authors declare no conflict of interest.

References

1. Food and Agriculture Organization of the United Nations. Statistic Division. Available online: https://www.fao.org/worldfoodsituation/csdb/en/ (accessed on 28 October 2021).
2. Smith, L.E.; Stoltzfus, R.J.; Prendergast, A. Food chain mycotoxin exposure, gut health, and impaired growth: A conceptual framework. *Adv. Nutr.* **2012**, *3*, 526–531. [CrossRef]
3. Tola, M.; Kebede, B. Occurrence, importance and control of mycotoxins: A review. *Cogent Food Agric.* **2016**, *2*, 1191103. [CrossRef]
4. Misihairabgwi, J.M.; Ezekiel, C.N.; Sulyok, M.; Shephard, G.S.; Krska, R. Mycotoxin contamination of foods in Southern Africa: A 10-year review (2007–2016). *Crit. Rev. Food Sci. Nutr.* **2019**, *59*, 43–58. [CrossRef] [PubMed]
5. Alshannaq, A.; Yu, J.H. Occurrence, toxicity, and analysis of major mycotoxins in food. *Int. J. Environ. Res. Public Health* **2017**, *14*, 632. [CrossRef] [PubMed]
6. Mousavi Khaneghah, A.; Fakhri, Y.; Gahruie, H.H.; Niakousari, M.; Sant'Ana, A.S. Mycotoxins in cereal-based products during 24 years (1983–2017): A global systematic review. *Trends Food Sci. Technol.* **2019**, *91*, 95–105. [CrossRef]
7. Nyangi, C.; Mugula, J.K.; Beed, F.; Boni, S.; Koyano, E.; Sulyok, M. Aflatoxins and fumonisin contamination of marketed maize, maize bran and maize used as animal feed in northern tanzania. *Afr. J. Food Sci.* **2016**, *16*, 11054–11065. [CrossRef]
8. Mannaa, M.; Kim, K.D. Influence of temperature and water activity on deleterious fungi and mycotoxin production during grain storage. *Mycobiology* **2007**, *45*, 240–254. [CrossRef]
9. Chulze, S. Strategies to reduce mycotoxin levels in maize during storage: A review. *Food Addit. Contam.* **2010**, *27*, 651–657. [CrossRef]
10. Bennett, J.W. Mycotoxins, mycotoxicoses, mycotoxicology and Mycopathologia. *Mycopathologia* **1987**, *100*, 3–5. [CrossRef]
11. Winter, G.; Pereg, L. A review on the relation between soil and mycotoxins: Effect of aflatoxin on field, food and finance. *Eur. J. Soil Sci.* **2019**, *70*, 882–897. [CrossRef]
12. International Agency for Research on Cancer. *Monograph on the Evaluation of Carcinogenic Risk to Humans, World Health Organization, Some Traditional Herbal Medicines, Some Mycotoxins, Naphthalene and StyreneIn Summary of data Reported and Evaluation*; IARC: Lion, France, 2002; Volume 82, pp. 171–175.
13. Ostry, V.; Malir, F.; Toman, J.; Grosse, Y. Mycotoxins as human carcinogens—The IARC Monographs classification. *Mycotoxin Res.* **2017**, *33*, 65–73. [CrossRef] [PubMed]
14. European Commission. Regulation No 1881/2006 setting maximum levels for certain contaminants foodstus. *Off. J. Eur. Union* **2006**, *50*, 8–12.
15. European Commission. Regulation No 1126/2007 amending Regulation (EC) No 1881/2006 setting maximum levels for certain contaminants in foodstuffs as regards Fusarium toxins in maize and maize products. *Off. J. Eur. Union* **2007**, *255*, 14.
16. European Commission. Regulation No 165/2010 amending Regulation (EC) No 1881/2006 setting maximum levels for certain contaminants in foodstuffs as regards aflatoxins. *Off. J. Eur. Union* **2010**, *50*, 8–12.
17. Agriopoulou, S.; Stamatelopoulou, E.; Varzakas, T. Advances in Occurrence, Importance, and Mycotoxin Control Strategies: Prevention and Detoxification in Foods. *Foods* **2020**, *9*, 137. [CrossRef]
18. Amariei, S.; Mihalcea, A. Study on Ochratoxin A and Zearalenone content in corn grains from different areas of Bacau county. *Food Environ. Saf.* **2020**, *19*, 116–121.
19. Pereira, V.L.; Fernandes, J.O.; Cunha, S.C. Mycotoxins in cereals and related foodstus: A review on occurrence and recent methods of analysis. *Trends Food Sci. Technol.* **2014**, *36*, 96–136. [CrossRef]
20. Xie, L.; Chen, M.; Ying, Y. Development of Methods for Determination of Aflatoxins. *Crit. Rev. Food Sci. Nutr.* **2016**, *56*, 2642–2664. [CrossRef]
21. Dzuman, Z.; Zachariasova, M.; Lacina, O.; Veprikova, Z.; Slavikova, P.; Hajslova, J. A rugged high-throughput analytical approach for the determination and quantification of multiple mycotoxins in complex feed matrices. *Talanta* **2014**, *121*, 263–272. [CrossRef]
22. Santos Pereira, C.; C Cunha, S.; Fernandes, J.O. Prevalent Mycotoxins in Animal Feed: Occurrence and Analytical Methods. *Toxins* **2019**, *11*, 290. [CrossRef]
23. European Food Safety Authority. Deoxynivalenol in food and feed: Occurrence and exposure. *EFSA J.* **2013**, *11*, 3379.
24. Pascari, X.; Marín, S.; Ramos, A.J.; Molino, F.; Sanchis, V. Deoxynivalenol in cereal-based baby food production process. A review. *Food Control* **2019**, *99*, 11–20. [CrossRef]

25. Udovicki, B.; Audenaert, K.; De Saeger, S.; Rajkovic, A. Overview on the mycotoxins incidence in Serbia in the period 2004–2016. *Toxins* **2018**, *10*, 279. [CrossRef] [PubMed]

26. Ji, F.; Xu, J.; Liu, X.; Yin, X.; Shi, J. Natural occurrence of deoxynivalenol and zearalenone in wheat from Jiangsu province, China. *Food Chem.* **2014**, *157*, 393–397. [CrossRef] [PubMed]

27. Joint Food and Agriculture Organization; World Health Organization Expert Committee on Food Additives (JECFA). *Evaluation of Certain Food Additives and Contaminants (Seventy-third Report of the Joint FAO/WHO Expert Committee on Food Current Additives)*; WHO: Geneva, Switzerland, 2010; p. 227.

28. European Food Safety Authority. Risks to human and animal health related to the presence of deoxynivalenol and its acetylated and modified forms in food and feed. *EFSA J.* **2017**, *15*, 4718.

29. Lee, H.J.; Ryu, D. Worldwide Occurrence of Mycotoxins in Cereals and Cereal-Derived Food Products: Public Health Perspectives of Their Co-occurrence. *J. Agric. Food Chem.* **2017**, *65*, 7034–7051. [CrossRef]

30. Bianchini, A.; Horsley, R.; Jack, M.M.; Kobielush, B.; Ryu, D.; Tittlemier, S.; Wilson, W.W.; Abbas, H.K.; Abel, S.; Harrison, G.; et al. DON Occurrence in Grains: A North American Perspective. *Cereal Foods World* **2015**, *60*, 32–56. [CrossRef]

31. International Agency for Research on Cancer (IARC). Some naturally occurring substances: Food items and constituents, heterocyclic aromatic amines and mycotoxins. In *IARC Monographs on the Evaluation of Carcinogenic Rirsks to Humans*; World Health Organization: Lyon, France, 1993; Volume 56, pp. 1–609.

32. Torovi'c, L. Fusarium toxins in corn food products: A survey of the Serbian retail market. *Food Addit. Contam. Part A Chem. Anal. Control Exp. Risk Assess.* **2018**, *35*, 1596–1609. [CrossRef]

33. Tima, H.; Brückner, A.; Mohácsi-Farkas, C.; Kiskó, G. Fusarium mycotoxins in cereals harvested from Hungarian fields. *Food Addit. Contam. Part B Surveill.* **2016**, *9*, 127–131. [CrossRef]

34. Vandicke, J.; De Visschere, K.; Croubels, S.; De Saeger, S.; Audenaert, K.; Haesaert, G. Mycotoxins in Flanders' Fields: Occurrence and Correlations with Fusarium Species in Whole-Plant Harvested Maize. *Microorganisms* **2019**, *7*, 571. [CrossRef]

35. Birr, T.; Jensen, T.; Preußke, N.; Sönnichsen, F.D.; De Boevre, M.; De Saeger, S.; Hasler, M.; Verreet, J.-A.; Klink, H. Occurrence of Fusarium Mycotoxins and Their Modified Forms in Forage Maize Cultivars. *Toxins* **2021**, *13*, 110. [CrossRef] [PubMed]

36. Eckard, S.; Wettstein, F.E.; Forrer, H.-R.; Vogelgsang, S. Incidence of Fusarium Species and Mycotoxins in Silage Maize. *Toxins* **2011**, *3*, 949–967. [CrossRef] [PubMed]

37. Xing, F.; Liu, X.; Wang, L.; Selvaraj, J.N.; Jin, N.; Wang, Y.; Zhao, Y.; Liu, Y. Distribution and variation of fungi and major mycotoxins in pre-and post-nature drying maize in North China Plain. *Food Control* **2017**, *80*, 244–251. [CrossRef]

38. De Almeida, A.P.; Lamardo, L.C.A.; Shundo, L.; da Silva, S.A.; Navas, S.A.; Alaburda, J.; Ruvieri, V.; Sabino, M. Occurrence of deoxynivalenol in wheat flour, instant noodle and biscuits commercialized in Brazil. *Food Addit. Contam. Part B Surveill.* **2016**, *9*, 251–255. [CrossRef]

39. Rai, A.; Das, M.; Tripathi, A. Occurrence and toxicity of a fusarium mycotoxin, zearalenone. *Crit. Rev. Food Sci. Nutr.* **2020**, *60*, 2710–2729. [CrossRef]

40. Joint Food and Agriculture Organization; World Health Organization Expert Committee on Food Additives (JECFA). *Evaluation of Certain Food Additives and Contaminants: Fifty-Fifth Report of the JOINT/FAO/WHO Expert Committee on Food Additives*; World Health Organization: Geneva, Switzerland, 2001; p. 701.

41. Zinedine, A.; Soriano, J.M.; Moltó, J.C.; Mañes, J. Review on the toxicity, occurrence, metabolism, detoxification, regulations and intake of zearalenone: An oestrogenic mycotoxin. *Food Chem. Toxicol.* **2007**, *45*, 1–18. [CrossRef]

42. Ünüsan, N. Systematic review of mycotoxins in food and feeds in Turkey. *Food Control* **2019**, *97*, 1–14. [CrossRef]

43. European Food Safety Authority. Scientific Opinion on the risks for human and animal health related to the presence of modeified forms of certain mycotoxins in food and feed. *EFSA J.* **2011**, *9*, 2197. [CrossRef]

44. Calori-Domingues, M.A.; Bernardi, C.M.G.; Nardin, M.S.; de Souza, G.V.; dos Santos, F.G.R.; de Abreu Stein, M.; da Gloria, E.M.; dos Santos Dias, C.T.; de Camargo, A.C. Co-occurrence and distribution of deoxynivalenol, nivalenol and zearalenone in wheat from Brazil. *Food Addit. Contam. Part B Surveill.* **2016**, *9*, 142–151. [CrossRef]

45. Kos, J.; Hajnal, E.J.; Škrinjar, M.; Mišan, A.; Mandić, A.; Jovanov, P.; Milovanović, I. Presence of Fusarium toxins in maize from Autonomous Province of Vojvodina, Serbia. *Food Control* **2014**, *46*, 98–101. [CrossRef]

46. Queiroz, V.A.V.; de Oliveira Alves, G.L.; da Conceição, R.R.P.; Guimarães, L.J.M.; Mendes, S.M.; de Aquino Ribeiro, P.E.; da Costa, R.V. Occurrence of fumonisins and zearalenone in maize stored in family farm in Minas Gerais, Brazil. *Food Control* **2012**, *28*, 83–86. [CrossRef]

47. El-Desouky, T.A.; Naguib, K. Occurrence of zearalenone contamination in some cereals in Egypt. *J. Agroaliment. Process. Technol.* **2013**, *19*, 445–450.

48. Han, Z.; Jiang, K.; Fan, Z.; Di Mavungu, J.D.; Dong, M.; Guo, W.; Fan, K.; Campbell, K.; Zhao, Z.; Wu, Y. Multi-walled carbon nanotubes-based magnetic solid-phase extraction for the determination of zearalenone and its derivatives in maize by ultra-high performance liquid chromatography-tandem mass spectrometry. *Food Control* **2017**, *79*, 177–184. [CrossRef]

49. Rheeder, J.P.; Marasas, W.F.O.; Vismer, H.F. Production of Fumonisin Analogs by Fusarium Species. *Appl. Environ. Microbiol.* **2002**, *68*, 2101–2105. [CrossRef] [PubMed]

50. Alberts, J.F.; van Zyl, W.H.; Gelderblom, W.C.A. Biologically based methods for control of fumonisin-producing Fusarium species and reduction of the fumonisins. *Front. Microbiol.* **2016**, *7*, 201600548. [CrossRef]

51. Li, L.; Chen, W.; Li, H.; Iqbal, J.; Zhu, Y.; Wu, T.; Du, Y. Rapid determination of fumonisin (FB1) by syringe SPE coupled with solid-phase fluorescence spectrometry. *Spectrochim. Acta Part A Mol. Biomol. Spectrosc.* **2020**, *226*, 117549. [CrossRef]
52. Cendoya, E.; Nichea, M.J.; Monge, M.P.; Sulyok, M.; Chiacchiera, S.M.; Ramirez, M.L. Fumonisin occurrence in wheat-based products from Argentina. *Food Addit. Contam. Part B Surveill.* **2019**, *12*, 31–37. [CrossRef]
53. Cendoya, E.; Chiotta, M.L.; Zachetti, V.; Chulze, S.N.; Ramirez, M.L. Fumonisins and fumonisin-producing Fusarium occurrence in wheat and wheat by products: A review. *J. Cereal Sci.* **2018**, *80*, 158–166. [CrossRef]
54. Stępień, Ł.; Waśkiewicz, A.; Urbaniak, M. Wildly Growing Asparagus (*Asparagus ocinalis* L.) Hosts Pathogenic Fusarium Species and Accumulates Their Mycotoxins. *Microb. Ecol.* **2016**, *71*, 927–937. [CrossRef]
55. Mogensen, J.M.; Frisvad, J.C.; Thrane, U.; Nielsen, K.F. Production of fumonisin B2 and B4 by aspergillus niger on grapes and raisins. *J. Agric. Food Chem.* **2010**, *58*, 954–958. [CrossRef]
56. World Health Organization (WHO). Safety evaluation of certain mycotoxins in food (WHO food additives series 47). In *International Programme on Chemical Safety*; World Health Organization: Geneva, Switzerland, 2001; pp. 103–279.
57. Anfossi, L.; Giovannoli, C.; Baggiani, C. Mycotoxin detection. *Curr. Opin. Biotechnol.* **2016**, *37*, 120–126. [CrossRef] [PubMed]
58. Berardo, N.; Lanzanova, C.; Locatelli, S.; Laganà, P.; Verderio, A.; Motto, M. Levels of total fumonisins in maize samples from Italy during 2006–2008. *Food Addit. Contam. Part B* **2011**, *4*, 116–124. [CrossRef] [PubMed]
59. Castells, M.; Marín, S.; Sanchis, V.; Ramos, A.J. Distribution of fumonisins and aflatoxins in corn fractions during industrial cornflake processing. *Int. J. Food Microbiol.* **2008**, *123*, 81–87. [CrossRef]
60. Tran, S.T.; Smith, T.K. Determination of optimal conditions for hydrolysis of conjugated deoxynivalenol in corn and wheat with trifluoromethanesulfonic acid. *Anim. Feed. Sci. Technol.* **2011**, *163*, 84–92. [CrossRef]
61. Li, F.; Jiang, D.; Zheng, F.; Chen, J.; Li, W. Fumonisins B1, B2 and B3 in corn products, wheat flour and corn oil marketed in Shandong province of China. *Food Addit. Contam. Part B Surveill.* **2015**, *8*, 169–174. [CrossRef]
62. Luo, Y.; Liu, X.; Li, J. Updating techniques on controlling mycotoxins—A review. *Food Control* **2018**, *89*, 123–132. [CrossRef]
63. Adebiyi, J.A.; Kayitesi, E.; Adebo, O.A.; Changwa, R.; Njobeh, P.B. Food fermentation and mycotoxin detoxification: An African perspective. *Food Control* **2019**, *106*, 106731. [CrossRef]
64. European Commission. *The Rapid Alert System for Food and Feed (RASFF) Annual Report 2020*; Publications Office of the European Union: Luxembourg, 2021; Available online: https://ec.europa.eu/food/system/files/2021-08/rasff_pub_annual-report_2020.pdf (accessed on 28 December 2021).

Article

Assessment of Potentially Toxic Elements and Associated Health Risk in Bottled Drinking Water for Babies

Elena L. Ungureanu [1,2], **Gabriel Mustatea** [1,*] **and Mona Elena Popa** [2,*]

[1] National Research & Development Institute for Food Bioresources, 020323 Bucharest, Romania; elena_ungureanu93@yahoo.com

[2] Faculty of Biotechnology, University of Agronomic Science and Veterinary Medicine, 011464 Bucharest, Romania

* Correspondence: gabi.mustatea@bioresurse.ro (G.M.); monapopa@agral.usamv.ro (M.E.P.)

Abstract: Potentially toxic elements are chemical pollutants which are dangerous to human health, especially for babies and children. Because their presence has been detected in baby food and baby drinking water, exposure to these elements is mainly due to ingestion. For this reason, the main objective of this study was quantification of 12 potentially toxic elements, including Ba, Co, Cu, Zn, Mn, Ni, Li, Fe, Pb, Cd, Cr, Sb, by ICP–MS, from 19 samples of bottled baby water. Based on the levels obtained, a health risk assessment was performed of the risk caused by their consumption, as well as an analysis of the quality of the samples. Excep iron, the values obtained for all other metals were below the limits imposed by the legislation in force. The risk analysis shows that Hazard index values were included in Risk Class 1, with a very low hazard level. The order of Cancer Risk values is as follows, Cd < Cr < Ni < Pb. As a general conclusion, we can say that the samples can be intended for consumption by children and infants.

Keywords: baby water; potentially toxic elements; health risk assessment; water quality

Citation: Ungureanu, E.L.; Mustatea, G.; Popa, M.E. Assessment of Potentially Toxic Elements and Associated Health Risk in Bottled Drinking Water for Babies. *Appl. Sci.* **2022**, *12*, 1914. https://doi.org/10.3390/app12041914

Academic Editors: Mircea Oroian and Georgiana Gabriela Codină

Received: 19 January 2022
Accepted: 10 February 2022
Published: 11 February 2022

Publisher's Note: MDPI stays neutral with regard to jurisdictional claims in published maps and institutional affiliations.

1. Introduction

Heavy metals represent a group of potentially toxic elements, metals or metalloids with a density higher than 5 g/cm^3, which are able to induce toxicity even at low level of exposure [1,2]. Potentially toxic elements contamination represents a high concern due to their inherent toxicity, non-degradability, bio-accumulation and persistence in the environment and frequent detection in different matrices [1,3].

Environmental pollution with these chemical contaminants is due to both natural processes and anthropogenic activities. Natural sources include volcanic eruptions, forest fires, weathering of minerals, soil erosion, biogenic sources, airborne dust, while anthropogenic activities which are the major cause of pollution with these elements are burning of fossil fuels, mining operations, smelting, metallurgical processes, industrial discharges, agriculture, vehicular emissions [4–6].

Potentially toxic elements can be stored and bioaccumulated in the soil and can have negative consequences on human health or the environment, as they may be mobilized from the soil, or not, depending on the characteristics of the soil [4].

Moreover, more than 99% of these elements entering in aquatic systems and can be accumulated and stored in sediments [6], leading to various degrees of contamination of groundwater.

Chemical contaminants from sediments can affect water quality and aquatic ecosystem, when their concentrations exceed imposed limits, directly or indirectly having a negative impact for agricultural production and people's lives [1,7].

Soil contamination could lead to drinking water pollution, which can be transferred to human food chain which brings bioaccumulation and bio-magnification [6].

Humans are exposed to potentially toxic elements through ingestion of contaminated food or water, by inhalation of atmospheric particles enriched with metals, or absorption through the skin [8,9]. Even if potentially toxic elements reach the body at trace levels, they can accumulate and causes disorders like, neurological disease, hormone imbalance, respiratory problems, endocrine disorders, infertility, cardiovascular problems, kidney diseases, digestive problems [8].

The quality of drinking water is very important, being one of the major factors which can affect human health and, for this reason, is strongly inter-related with human heath [10,11].

The risk due to exposure to polluted water can be calculated through health risk assessment, and the results obtained may require measures to reduce the risk caused by contaminants. The most commonly used health risk assessment model is the one recommended by United State Environmental Protection Agency (USEPA) [10]. This model includes a carcinogenic and non-carcinogenic analysis, based on parameters like estimated daily intake (EDI), target hazard quotient (THQ), hazard index (HI) and cancer risk (CR) [12].

The quality of water can be evaluated by parameters like water quality index (WQI), entropy water quality index (EQWI) [10], water pollution index (WPI), comprehensive pollution index (CPI), organic pollution index (OPI), trace metal pollution index (TPI) [13], contamination factor (C_f), degree of contamination (mC_d), pollution load index (PLI) [8].

The health risk assessment of the water quality helps quantify the risk of exposure to polluted water, and, in this mode, by quantitative description of the hazards, decisions related to water source protection and management can be implemented [10].

In a study conducted by Beal, 2020 [14] on a series of baby food products and baby bottled water purchased from 15 retail chains in major cities in the United States, it was observed that 95% of products contain at least one toxic element (arsenic, lead, cadmium, mercury).

Moreover, in the study conducted by Decharat and Pan-in, 2020 [15] on a number of 48 samples of bottled water intended for consumption in schools, it was observed that lead and cadmium levels were between 2.5–18 μg/L, respectively 0.16–1.3 μg/L. In the case of lead, only 4 samples exceeded the 10 μg/L limit imposed by the WHO guide [16], while cadmium did not exceed the 3 μg/L limit. As a concerning Hazard index for lead and cadmium, the values were less than 1, but if other elements had been analyzed, the HI levels would probably have been higher.

In the study conducted by Peleka et al., 2021 [17], by estimating the hazard quotients based on the values of potentially toxic elements analyzed, for the infants and children population categories, HQ values were higher than 1 for both lead and cadmium, which show a relatively high hazard due to the consumption of the samples by these age groups.

The presence of potentially toxic elements in bottled water may also be due to contamination of raw water, as well as due to the packaging material, because some elements may remain in plastic packaging as catalyst or reaction residues, such as Pb-, Cd-, Ba-, Ca-Zn dicarboxylates and other, which can migrate into the packaged product under certain conditions [18]. Thus, in a study conducted by Allafi, 2020 [19], it was observed that the increase in temperature positively influences the migration of antimony from PET. The same was observed in other studies [20–24]. In addition to temperature, storage time can influence the migration of some elements [20,22–25], but also pH, as observed in the study by Chapa-Martinez et al., 2016 [20].

Due to these mentioned aspects, the main aim of this research is evaluation of the chemical quality of 19 samples of bottled drinking water for baby through quantification of 12 potentially toxic elements (Ba, Co, Cu, Zn, Mn, Ni, Li, Fe, Pb, Cd, Cr, Sb), as well as carcinogenic and non-carcinogenic health risk assessment for age group infants and children, based on the mentioned elements levels.

2. Materials and Methods

2.1. Sample Collection

The bottled baby water samples were purchased from local supermarkets and hypermarkets in Romania and included 10 samples of mineral water, 8 samples of spring water and 1 sample of table water. Also, 5 baby water are of Romanian origin and 14 samples are imported from France (3), Italy (1), Croatia (1), Greece (4), Hungary (1), Austria (1), Germany (1), Poland (1) and Bulgaria (1). The samples were bottled in polyethylene terephthalate (PET), with high density polyethylene screw cap and maintained at room temperature prior analysis.

2.2. Sample Preparation

The analysis of potentially toxic elements content in bottled water were performed according EPA3010A standard [26], and include that 300 mL of well mixed sample, initial acidified with a few drops of HNO_3, were evaporated to dryness on a water bath. After evaporation, the samples were treated with 2.5 mL of concentrated HNO_3, evaporated to dryness, and then, another 2.5 mL of concentrated HNO_3 were added. The extract with HNO_3 is covered with ribbed watch glass and left on the water bath for 15 min. The solution is then transferred to a 50 mL volumetric flask with ultrapure water and made up to the mark. Each of water samples was analyzed in triplicate to ensure data quality.

2.3. Equipment

Potentially toxic elements detection was performed using an Inductively Coupled Plasma–Mass Spectrometer NexION 300Q (Perkin Elmer Inc., Waltham, MA, USA). Also, for evaporation of water samples, a water bath GFL 1032 (Gesellschaft für Labortechnik, Germany), was used. ICP–MS operating conditions are listed in Table 1.

Table 1. ICP–MS operating conditions.

Parameter Description	Current Value
Sample introduction	Spray chamber
Radio Frequency Generator	Free running type, 40 MHz
Radio Frequency power	1000 W
Nebulizer Gas Flow	0.89 L/min
Auxiliary Gas Flow	1.20 L/min
Plasma Gas Flow	16 L/min
Sample Uptake Flow	1.0 mL/min
Dual Detector Mode	Pulse
Torch Z position	0.00 mm
Analytical masses	Standard mode
Ba, Co, Cu, Zn, Mn, Ni, Li,	138, 59, 63, 66, 55, 27, 60, 7, 57,
Fe, Pb, Cd, Cr, Sb mass	208, 111, 52, 121
Scanning mode	Peak hopping
Number of replicates	3
Wash	Time (45), speed (±rpm)–24
Sample flush	Time (35), speed (±rpm)–24
Read delay	Time (15), speed (±rpm)–20

2.4. Analytical Method

For quantification of analytes, calibration curves for all the tested elements were performed, using a Multielement Standard Solution 6, with a concentration of 100 mg/L, from which standard solution of 10 μg/L, 20 μg/L, 30 μg/L, 40 μg/L and 50 μg/L were obtained. The validation of the method was performed according to Magnusson and Ornemark, 2014-Eurachem Guide [27] and DIN EN ISO/IEC 17025:2018 [28]. The correlation coefficients of calibration curves obtained for a working range between 10 μg/L and 50 μg/L were higher than 0.996. Detection limits, estimated as LOD = 3 SD/b, where SD is standard deviation obtained after 10 replicates measurements of blank samples and b in the slope of the calibration curve, were 0.07 µg/L for Sb and Pb, 0.09 µg/L for Ba, Cu, Mn, Ni and Cd, 0.10 µg/L for Co and Cr, 0.11 µg/L for Li, 0.13 µg/L for Zn and 0.15 µg/L for Fe. To check the measurement precision, standards' reference solutions spiked at 10 µg/L were used, the repeatability obtained being between 2.18 and 3.86. To assess the quality and accuracy of the results, standard reference solutions and a certified reference material (CRM) SPS-SW2 Surface water were used. Recovery rates for each element were between 91.50% and 108.40%, being considered as acceptable. The mean extended compound uncertainty (U) of the tested elements was of 10.31%.

2.5. Estimation of Potentially Toxic Elements Content

Estimation of potentially toxic elements content was performed using Equation (1).

$$C = [(Reading \times F) - M] \times V]/m, \tag{1}$$

where C is analyte concentration, in µg/L, Reading represent concentration obtained from the calibration curve, F is dilution factor, M is blank sample, V is sample volume, in mL (50 mL) and m is sample weight, in mg.

2.6. Health Risk Assessment

Risk assessment represent a method of estimation of risk of exposure to a pollutant and its probability to produce harmful effects on human health. This includes carcinogenic and non-carcinogenic analysis [29].

2.6.1. Non-Carcinogenic Analysis

For non-carcinogenic estimation, the model proposed by PHA Guidance, 2005 [30] was used, considering as the main source of exposure, ingestion. In this case, the evaluation was performed for 2-year children, with a weigh of 10 kg and an ingestion rate of 1 L/day. This analysis involves calculations of parameters, like exposure dose (D), hazard quotient (HQ) and hazard index (HI). The calculation of these indices was done using Equations (2)–(4).

$$D = (C \times IR \times EF)/BW, \tag{2}$$

where D is exposure dose, in mg/kg/day, C is contaminant concentration, in mg/L, IR is intake rate of water, in L/day, EF is exposure factor (unitless) and BW is body weight, in kg.

$$HQ = D/RfD, \tag{3}$$

where HQ is hazard quotient (unitless), D is exposure dose, in mg/kg/day and RfD is reference dose, in mg/kg/day. The reference dose represents the tolerable daily intake of the metal via ingestion. The values of this parameters are presented in Table 2. Hazard quotient represent the ratio between contaminant exposure and the respected reference values, specify for each exposure way (ingestion, inhalation or dermal contact) [16].

$$HI = HQ_{Pb} + HQ_{Cd} + HQ_{Cr} + HQ_{Cu} + HQ_{Zn} + HQ_{Mn} + HQ_{Ni} + HQ_{Ba} + HQ_{Co} + HQ_{Li} + HQ_{Fe} + HQ_{Sb}, \tag{4}$$

where HI is hazard index (unitless), and HQ is hazard quotient (unitless) of all heavy metals tested. Hazard index represent the total potential risk produced by chemicals in a mixture [31].

2.6.2. Carcinogenic Analysis

This analysis represents the probability that a chemical could produce a form of cancer, after exposure of a specified dose [29]. The model used for estimation of carcinogenic analysis, was the one proposed by PHA Guidance Manual, 2005 [30].

The carcinogenic analysis evaluation involves estimating the parameter cancer risk (CR), which has been calculated using Equation (5). This parameter, has been estimated for Pb, Cd, Cr and Ni, because only these have the set CSF values.

$$CR = D/CSF, \tag{5}$$

where CR represent cancer risk (unitless), D is exposure dose in mg/kg/day and CSF is Cancer Slope Factor, in mg/kg/day. The values of CSF for oral exposure are presented in Table 2.

For the 4 elements studied, total cancer risk (TCR) was calculated, which means the result of exposure to multiple heavy metals [32], using Equation (6).

$$TCR = Cr_{Pb} + Cr_{Cd} + Cr_{Cr} + Cr_{Ni},\tag{6}$$

where TCR is total cancer risk (unitless), Cr represent cancer risk (unitless) for the four elements tested.

Table 2. RfD and CSF values for oral ingestion.

Element	RfD_{oral} (mg/kg)	CSF_{oral} (mg/kg/day)
Ba	7.0×10^{-2} [a]	-
Co	2.0×10^{-2} [b]	-
Cu	3.7×10^{-2} [b]	-
Zn	3.0×10^{-1} [b]	-
Mn	4.7×10^{-2} [c]	-
Ni	2.0×10^{-2} [b]	8.4×10^{-1} [a]
Li	2.8×10^{-2} [d]	-
Fe	7.0×10^{-2} [e]	-
Pb	3.6×10^{-3} [b]	8.50 [a]
Cd	5.0×10^{-4} [b]	6.10 [a]
Cr	3.0×10^{-3} [b]	4.1×10^{1} [a]
Sb	3.5×10^{-4} [f]	-

[a] [29]; [b] [33]; [c] [34]; [d] [35]; [e] [36]; [f] [37].

3. Results

3.1. Estimation of Potentially Toxic Elements Content

Toxic elements (metals) concentrations of the tested water samples are presented in Table 3. All the samples were analyzed in triplicate, the values being reported as mean value and standard deviation.

With the exception of Fe, all the tested elements are below the maximum imposed limits by national and international legislation in force. In case of Fe, only 4 samples were lower than imposed limit (200 µg/L).

3.2. Health Risk Assessment

3.2.1. Non-Carcinogenic Analysis

The values of exposure dose (D) and hazard quotient (HQ) are presented in Table 4. The ascendent trend of exposure dose, based on potentially toxic elements concentrations, is as following Cd > Cr > Sb > Co > Mn > Pb > Cu > Ni > Li > Zn > Ba > Fe. In case of HQ, the values decrease in order Cd > Co > Cr > Sb > Zn > Mn > Ni > Ba > Li > Pb > Cu > Fe. The interpretation and description of HQ and HI is presented in Table 5.

Except Fe, the estimated levels of hazard quotient were lower than 1. This means that repeated exposure of children does not cause side effects. However, in case of Fe, only one sample had the value lower than 1. This implies the appearance of some non-carcinogenic side effects, after repeated exposure to tested samples.

Regarding the hazard index values, obtained for hazard quotient of all analyzed elements, ranged between 9.10×10^{-1} and 2.42×10^{1}, with an average value of 7.10. Of the total of 19 samples, the sample that had the value HQ < 1 also had the value HI < 1. The rest of samples recorded HI values greater than 1. This means that after repeated exposure, some non-carcinogenic side effects can appear.

Table 3. Potentially toxic elements levels in baby drinking water samples.

Element	Heavy Metals Levels (µg/L)			Detection Rate (%)	Law No. 311/2004 (µg/L) [38]	Directive EU 2020/2184 (µg/L) [39]
	Min	Max	Mean			
Ba	<0.09	$1.68 \times 10^1 \pm 0.32$	3.87 ± 0.19	68.42	-	-
Co	<0.10	$2.50 \times 10^{-1} \pm 0.01$	$7.0 \times 10^{-2} \pm 0.01$	47.37	-	-
Cu	<0.09	$1{,}73 \pm 0.03$	$9.60 \times 10^{-1} \pm 0.02$	100	100	2000
Zn	$9.60 \times 10^{-1} \pm 0.07$	4.47 ± 0.22	2.04 ± 0.07	100	5000	-
Mn	<0.09	4.17 ± 0.20	$3.50 \times 10^{-1} \pm 0.04$	42.11	50	50
Ni	$3.10 \times 10^{-1} \pm 0.01$	2.25 ± 0.06	1.02 ± 0.04	100	20	20
Li	<0.11	7.09 ± 0.43	1.69 ± 0.08	84.21	-	-
Fe	$6.24 \times 10^1 \pm 6.13$	$1.69 \times 10^3 \pm 3.92 \times 10^1$	$4.95 \times 10^2 \pm 1.82 \times 10^1$	100	200	200
Pb	$1.10 \times 10^{-1} \pm 0.01$	1.79 ± 0.02	$4.30 \times 10^{-1} \pm 1.0 \times 10^{-2}$	100	10	5
Cd	<0.09	< 0.09	-	0	5	5
Cr	<0.10	$1.60 \times 10^{-1} \pm 0.01$	$1.0 \times 10^{-2} \, 0.00$	10.53	50	25
Sb	<0.07	$1.30 \times 10^{-1} \pm 0.00$	$3.0 \times 10^{-2} \pm 0.00$	31.58	5	10

Table 4. Exposure dose (D), Hazard quotient (HQ) and Cancer risk (CR) from baby drinking water samples.

Element	Exposure Dose (µg·kg^{-1}·day^{-1})			Hazard Quotient			Cancer Risk		
	Min	Max	Mean	Min	Max	Mean	Min	Max	Mean
Ba	0.00	1.68×10^{-3}	3.87×10^{-4}	0.00	2.39×10^{-2}	5.52×10^{-3}	-	-	-
Co	0.00	2.50×10^{-5}	7.95×10^{-6}	0.00	1.25×10^{-3}	4.19×10^{-4}	-	-	-
Cu	3.80×10^{-5}	1.75×10^{-4}	9.6×10^{-5}	4.25×10^{-3}	4.73×10^{-2}	2.54×10^{-2}	-	-	-
Zn	9.60×10^{-4}	4.47×10^{-4}	2.04×10^{-4}	3.20×10^{-4}	1.49×10^{-3}	6.18×10^{-4}	-	-	-
Mn	0.00	4.17×10^{-4}	3.51×10^{-5}	0.00	9.07×10^{-3}	7.62×10^{-4}	-	-	-
Ni	3.10×10^{-5}	2.25×10^{-4}	1.06×10^{-4}	1.55×10^{-3}	1.13×10^{-2}	5.12×10^{-3}	2.60×10^{-5}	1.89×10^{-4}	8.61×10^{-5}
Li	0.00	7.28×10^{-4}	1.72×10^{-4}	0.00	2.60×10^{-2}	6.13×10^{-3}	-	-	-
Fe	6.20×10^{-3}	1.69×10^{-1}	4.93×10^{-2}	8.86×10^{-1}	2.41×10^1	7.05	-	-	-
Pb	1.10×10^{-5}	1.79×10^{-4}	4.36×10^{-5}	3.06×10^{-3}	4.97×10^{-2}	1.21×10^{-2}	9.35×10^{-5}	1.52×10^{-3}	3.63×10^{-4}
Cd	0.00	0.00	0.00	0.00	0.00	0.00	0.00	0.00	0.00
Cr	0.00	1.60×10^{-5}	1.32×10^{-6}	0.00	5.33×10^{-3}	4.39×10^{-4}	0.00	6.56×10^{-4}	5.39×10^{-5}
Sb	0.00	1.30×10^{-5}	1.63×10^{-6}	0.00	4.33×10^{-3}	5.44×10^{-4}	-	-	-

Table 5. Hazard quotient, hazard index, cancer risk and total cancer risk classification and description (adapted from [14]).

Hazard Quotient	Description	Hazard Index	Description	Cancer Risk	Description	Total Cancer Risk	Description
≤ 1	repeated exposure not cause side effects	≤ 1	repeated exposure not cause side effects	$<1 \times 10^{-6}$	tolerable	$<1 \times 10^{-6}$	insignificant
>1	repeated exposure causes a potential risk	>1	repeated exposure can produce side effects	1×10^{-6}–1×10^{-4}	tolerable range	1×10^{-5}	acceptable level
-	-	-	-	$>1 \times 10^{-4}$	intolerable	$>1 \times 10^{-4}$	harmful

3.2.2. Carcinogenic Analysis

The values of cancer risk are presented in Table 4. The interpretation and description of this parameter is shows in Table 5. The results obtained in case of Pb were in a tolerable range, while Cd values are tolerable, due to low detection in the samples. Regarding Cr, 11% of samples were in a tolerable range and the rest of the samples were tolerable. Concerning Ni, 89% of results were in tolerable range and 11% were classed as intolerable. This means that, after repeated exposure to intolerable samples, Ni can produce some carcinogenic side effects.

The ascendent trend of cancer risk, estimated for the 4 studied elements, is as following Cd > Cr > Ni > Pb.

The levels of CR_{total} were between 1.69×10^{-4} and 1.58×10^{-3}, and a mean value of 5.03×10^{-4}. As can be seen, all the results were in tolerable range. This means that after repeated exposure to tested samples, heavy metals analyzed can't produce carcinogenic side effects in babies and children.

4. Conclusions

In this study, an Inductively Coupled Plasma Mass Spectrometry (ICP–MS) analytical method was used for detection and quantification of 12 potentially toxic elements from 19 samples of baby drinking water. The obtained results were below the maximum imposed limits by National and European legislation, with exception of Fe, where only 4 samples were below the limit. Cadmium was not detected in any sample. Based on tested elements levels, a carcinogenic and non-carcinogenic analysis was performed. The results of hazard quotient were lower than 1, except Fe, where only 1 sample had the value below the reference value. If in case of Fe the results demonstrated that after repeated exposure, some non-carcinogenic side effects could appear, the rest of elements can't produce adverse effects. Hazard index values were higher than 1, except only one sample. This means that after repeated exposure, can appear some adverse effects, but non-carcinogenic. The levels of carcinogenic risk of the four tested elements increase in the following order Cd > Cr > Ni > Pb, while total cancer risk results were between 1.69×10^{-4} and 1.58×10^{-3}, being framed in tolerable range. From the point of view of tested elements and health risk assessment, the samples can be intended for consumption by children and babies. In order to determine exactly whether the samples endanger the health of children, further, more detailed studies are needed, to detect as many potential contaminants as possible, for which a risk assessment can be performed.

Author Contributions: Conceptualization, E.L.U.; methodology, E.L.U.; validation, E.L.U.; formal analysis, E.L.U.; resources, G.M.; writing—original draft preparation, E.L.U.; writing—review and editing, G.M. and M.E.P.; visualization, M.E.P.; supervision, M.E.P.; project administration, G.M.; funding acquisition, G.M. All authors have read and agreed to the published version of the manuscript.

Funding: This research was funded through Core Program, with the support of the Ministry of Research, Innovation and Digitization, contract 22N/2019, project PN 19 02 03 02 and CNCS/CCCDI–UEFISCDI, project number PN-III-P3-3.6-H2020-2020-0014/Ctr. 2/2021.

Institutional Review Board Statement: Not applicable.

Informed Consent Statement: Informed consent was obtained from all subjects involved in the study.

Data Availability Statement: Not applicable.

Acknowledgments: In this section, you can acknowledge any support given which is not covered by the author contribution or funding sections. This may include administrative and technical support, or donations in kind (e.g., materials used for experiments).

Conflicts of Interest: The authors declare no conflict of interest.

References

1. Wu, J.; Lu, J.; Zhang, C.; Zhang, Y.; Lin, Y.; Xu, J. Pollution, sources and risks of heavy metals in coastal waters of China. *Hum. Ecol. Risk Assess.* **2019**, *26*, 2011–2026. [CrossRef]
2. Tchounwou, P.B.; Yedjou, C.G.; Patlolla, A.K.; Sutton, D.J. Heavy metals toxicity and the environment. *Exp. Suppl.* **2012**, *101*, 133–164. [PubMed]
3. Alidadi, H.; Sany, S.B.T.; Oftadeh, B.Z.G.; Mohamad, T.; Shamszade, H.; Fakhri, M. Health risk assessments of arsenic and toxic heavy metal exposure in drinking water in northeast Iran. *Environ. Health Prev. Med.* **2019**, *24*, 59. [CrossRef] [PubMed]
4. Likuku, A.S.; Mmolawa, K.B.; Gaboutloeloe, G.K. Assessment of heavy metal enrichment and degree of contamination around the copper-nickel mine in the Selebi Phikwe Region, Eastern Botswana. *Environ. Ecol. Res.* **2013**, *1*, 32–40. [CrossRef]
5. Masindi, V.; Muedi, K.L. Environmental Contamination by Heavy Metals. In *Heavy Metals*, 1st ed.; Aglan, R.F., Ed.; IntechOpen: London, UK, 2018; pp. 115–133.
6. Shen, F.; Mao, L.; Sun, R.; Du, J.; Tan, Z.; Ding, M. Contamination evaluation and source identification of heavy metals in the sediments from the Lishui River Watershed, Southern China. *Int. J. Environ. Res.* **2019**, *16*, 336. [CrossRef] [PubMed]

7. Zhang, Y.; Xion, Y.; Chao, Y.; Fan, R.; Ren, R.; Xu, B.; Liu, Z. Hydrogeochemistry and quality assessment of groundwater in Jinghui canal irrigation district of China. *Hum. Ecol. Risk Assess.* **2020**, *26*, 2349–2366. [CrossRef]

8. Shirani, M.; Afzali, K.N.; Jahan, S.; Strezov, V.; Soleimani-Sardo, M. Pollution and contamination assessment of heavy metals in the sediments of Jazmurian playa in southeast Iran. *Nat. Sci. Rep.* **2020**, *10*, 4775. [CrossRef]

9. Rehman, A.U.; Nazir, S.; Irshad, R.; Tahir, K.; Rehman, K.; Ul Islam, R.; Wahab, Z. Toxicity of heavy metals in plants and animals and their uptake by magnetic iron oxide nanoparticles. *J. Mol. Liq.* **2021**, *321*, 114455. [CrossRef]

10. Yin, Z.; Duan, R.; Li, P.; Li, W. Water quality characteristics and health risk assessment of main water supply reservoirs in Taizhou City, East China. *Hum. Ecol. Risk Assess.* **2021**, *27*, 2142–2160. [CrossRef]

11. Rishi, M.S.; Kaur, L.; Sharma, S. Groundwater quality appraisal for non-carcinogenic human health risks and irrigation purposes in a part of Yamuna sub-basin, India. *Hum. Ecol. Risk Assess.* **2019**, *26*, 2716–2736. [CrossRef]

12. Alsafran, M.; Usman, K.; Rizwan, M.; Ahmed, T.; Al Jabri, H. The Carcinogenic and Non-Carcinogenic Health Risks of Metal(oid)s Bioaccumulation in Leafy Vegetables: A Consumption Advisory. *Front. Environ. Sci.* **2021**, *9*, 742269. [CrossRef]

13. Son, C.T.; Giang, N.T.H.; Thao, T.P.; Nui, N.H.; Lam, N.T.; Cong, V.H. Assessment of Cau River water quality assessment using a combination of water quality and pollution indices. *J. Water Supply Res. Technol.-Aqua* **2020**, *69*, 160–172. [CrossRef]

14. Beal, J.A. Heavy metals in baby food: What providers and parents need to know. *MCN* **2020**, *45*, 125. [CrossRef] [PubMed]

15. Decharat, D.; Pan-in, P. Risk assessment of lead and cadmium in drinking water for school use in Nakhon Si Thammarat Province, Thailand. *Environ. Anal. Health Toxicol.* **2020**, *35*, e2020002. [CrossRef]

16. World Health Organization (WHO). *Guidelines for Drinking-Water Quality*, 4th ed.; World Health Organization: Geneva, Switzerland, 2011.

17. Peleka, J.C.M.; Diop, C.; Foko, R.F.; Saffe, M.L.; Fall, M. Health risk assessment of trace metals in drinking water consumed in Dakar, Senegal. *J. Water Resour. Prot.* **2021**, *13*, 915–930. [CrossRef]

18. Turner, A.; Filella, M. Hazardous metal additives in plastics and their environmental impacts. *Environ. Int.* **2021**, *156*, 106622. [CrossRef]

19. Allafi, A.R. The effect of temperature and storage time on the migration of antimony from polyethylene terephthalate (PET) into commercial bottled water in Kuwait. *Acta Biomed.* **2020**, *91*, e2020105.

20. Chapa-Martínez, C.A.; Hinojosa-Reyes, L.; Hernández-Ramírez, A.; Ruiz-Ruiz, E.; Maya-Treviño, L.; Guzmán-Mar, J.L. An evaluation of the migration of antimony from polyethylene terephthalate (PET) plastic used for bottled drinking water. *Sci. Total Environ.* **2016**, *565*, 511–518. [CrossRef]

21. Qiao, F.; Lei, K.; Li, Z.; Liu, Q.; Wei, Z.; An, L.; Cui, S. Effects of storage temperature and time of antimony release from PET bottles into drinking water in China. *Environ. Sci. Pollut. Res.* **2018**, *25*, 1388–1393. [CrossRef]

22. Filella, M. Antimony and PET: Checking facts. *Chemosphere* **2020**, *261*, 127732. [CrossRef]

23. Fan, Y.Y.; Zheng, J.L.; Ren, J.H.; Luo, J.; Cui, X.Y.; Ma, L.Q. Effects of storage temperature and duration on release of antimony and bisphenol A from polyethylene terephthalate drinking water bottles of China. *Environ. Pollut.* **2014**, *192*, 113–120. [CrossRef] [PubMed]

24. Faraji, M.; Taherkhani, A.; Nemati, S.; Mohammadi, A. Challenges in the use of polyethylene terephthalate bottles for packaging drinking water. *J. Biomed. Health* **2017**, *2*, 224–229.

25. Payan, L.; Poyatos, M.T.; Munoz, L.; La Rubia, M.D.; Pacheco, R.; Ramos, N. Study of the influence of storage conditions on the quality and migration levels of antimony in polyethylene terephthalate-bottled water. *Food Sci. Technol. Int.* **2017**, *23*, 318–327. [CrossRef] [PubMed]

26. Environmental Protection Agency. *SW-846 Test Method 3010A: Acid Digestion of Aqueous Samples and Extracts for Total Metals for Analysis by Flame Atomic Absorption Spectroscopy (FLAA) or Inductively Coupled Plasma Spectroscopy (ICP)*; United States Environmental Protection Agency: Washington, DC, USA, 1992.

27. Magnusson, B.; Örnemark, U. (Eds.) *Eurachem Guide: The Fitness for Purpose of Analytical Methods—A Laboratory Guide to Method Validation and Related Topics*, 2nd ed.; Eurachem: Teddington, UK, 2014.

28. *EN ISO/IEC 17025*; General Requirements for the Competence of Testing and Calibration Laboratories. International Organization for Standardization: Geneva, Switzerland, 2018.

29. Mohammad, A.A.; Zarei, A.; Majidi, S.; Ghaderpoury, A.; Hashempour, Y.; Saghi, M.Y.; Hosseingholizadeh, N.; Ghaderpoori, M. Carcinogenic and non-carcinogenic health risk assessment of heavy metals in drinking water of Khorramabad, Iran. *MethodsX* **2019**, *6*, 1642–1651. [CrossRef] [PubMed]

30. *Public Health Assessment Guidance Manual*; Agency for Toxic Substances and Disease Registry (ATSDR): Atlanta, GA, USA, 2005.

31. Goumenou, M.; Tsatsakis, A. Proposing new approaches for the risk characterisation of single chemicals and chemical mixtures: The source related Hazard Quotient (HQS) and Hazard Index (HIS) and the adversity specific Hazard Index (HIA). *Toxicol. Rep.* **2019**, *6*, 632–636. [CrossRef]

32. Sultana, M.; Rana, S.; Yamazaki, S.; Aono, T.; Yoshida, S. Health risk assessment for carcinogenic and non-carcinogenic heavy metal exposures from vegetables and fruits of Bangladesh. *Cogent. Environ. Sci.* **2017**, *3*, 1291107. [CrossRef]

33. Kamunda, C.; Mathuthu, M.; Madhuku, M. Health risk assessment of heavy metals in soils from Witwatersrand Gold Mining Basin, South Africa. *Int. J. Environ. Res.* **2016**, *13*, 663. [CrossRef]

34. Manganese Chemical Update Worksheet. Available online: https://www.michigan.gov/documents/deq/deq-rrd-chem-ManganeseDatasheet_527860_7.pdf (accessed on 3 January 2022).

35. Lithium Chemical Update Worksheet. Available online: https://www.michigan.gov/documents/deq/deq-rrd-chem-LithiumDatasheet_527862_7.pdf (accessed on 3 January 2022).
36. Anyawu, A.D.; Onyele, O.G. Human health risk assessment of some heavy metals in a rural spring, Southeastern Nigeria. *Afr. J. Environ. Nat. Sci. Res.* **2018**, *1*, 15–23.
37. Antimony Chemical Update Worksheet. Available online: https://www.michigan.gov/documents/deq/deq-rrd-chem-AntimonyDatasheet_527730_7.pdf (accessed on 3 January 2022).
38. LAW No. 311 of June 28, 2004 for the Amendment and Completion of Law No. 458/2002 on Water Quality, Issuer: Parliament, Published in: Official Monitor NO. 582 of June 30, 2004. Available online: http://www.namr.ro/wp-content/uploads/2013/02/legea311.pdf (accessed on 4 January 2022).
39. Directive (EU) 2020/2184 of the European Parliament and of the Council of 16 December 2020 on the Quality of Water Intended for Human Consumption. Available online: http://data.europa.eu/eli/dir/2020/2184/oj (accessed on 4 January 2022).

MDPI
St. Alban-Anlage 66
4052 Basel
Switzerland
Tel. +41 61 683 77 34
Fax +41 61 302 89 18
www.mdpi.com

Applied Sciences Editorial Office
E-mail: applsci@mdpi.com
www.mdpi.com/journal/applsci